Introduction to High-Speed Electronics and Optoelectronics

WILEY SERIES IN MICROWAVE AND OPTICAL ENGINEERING

KAI CHANG, Editor
Texas A&M University

FIBER-OPTIC COMMUNICATION SYSTEMS • *Govind P. Agrawal*

COHERENT OPTICAL COMMUNICATIONS SYSTEMS • *Silvello Betti, Giancarlo De Marchis, and Eugenio Iannone*

HIGH-FREQUENCY ELECTROMAGNETIC TECHNIQUES: RECENT ADVANCES AND APPLICATIONS • *Asoke K. Bhattacharyya*

COMPUTATIONAL METHODS FOR ELECTROMAGNETICS AND MICROWAVES • *Richard C. Booton, Jr.*

MICROWAVE SOLID-STATE CIRCUITS AND APPLICATIONS • *Kai Chang*

MULTICONDUCTOR TRANSMISSION LINE STRUCTURES • *J. A. Brandão Faria*

MICROSTRIP CIRCUITS • *Fred Gardiol*

HIGH-SPEED VLSI INTERCONNECTIONS: MODELING, ANALYSIS AND SIMULATION • *A. K. Goel*

HIGH-FREQUENCY ANALOG INTEGRATED-CIRCUIT DESIGN • *Ravender Goyal, Editor*

OPTICAL COMPUTING: AN INTRODUCTION • *Mohammad A. Karim and Abdul Abad S. Awwal*

MICROWAVE DEVICES, CIRCUITS, AND THEIR INTERACTION • *Charles A. Lee and G. Conrad Dalman*

ANTENNAS FOR RADAR AND COMMUNICATIONS: A POLARIMETRIC APPROACH • *Harold Mott*

SOLAR CELLS AND THEIR APPLICATIONS • *Larry D. Partain*

ANALYSIS OF MULTICONDUCTOR TRANSMISSION LINES • *Clayton R. Paul*

INTRODUCTION TO ELECTROMAGNETIC COMPATIBILITY • *Clayton R. Paul*

INTRODUCTION TO HIGH-SPEED ELECTRONICS AND OPTOELECTRONICS • *M. Leonard Riaziat*

NEW FRONTIERS IN MEDICAL DEVICE TECHNOLOGY • *Arye Rosen and Harel Rosen, Editors*

OPTICAL SIGNAL PROCESSING, COMPUTING, AND NEURAL NETWORKS • *Francis T. S. Yu and Suganda Jutamulia*

Introduction to High-Speed Electronics and Optoelectronics

M. L. RIAZIAT

Senior Scientist
Varian Research Center
Palo Alto, California

A WILEY-INTERSCIENCE PUBLICATION
JOHN WILEY & SONS, INC.
NEW YORK / CHICHESTER / BRISBANE / TORONTO / SINGAPORE

This text is printed on acid-free paper.

Copyright © 1996 by John Wiley & Sons, Inc.

All rights reserved. Published simultaneously in Canada.

Reproduction or translation of any part of this work beyond that permitted by Section 107 or 108 of the 1976 United States Copyright Act without the permission of the copyright owner is unlawful. Requests for permission or further information should be addressed to the Permissions Department, John Wiley & Sons, Inc., 605 Third Avenue, New York, NY 10158-0012.

Library of Congress Cataloging in Publication Data:
Riaziat, M. L. (Majid Leonard)
    Introduction to high speed electronics and optoelectronics / M.L. Riaziat.
       p.   cm.
    Includes bibiographical references and index.
    ISBN 0-471-01582-2 (acid-free paper)
    1. Optoelectronic devices. 2. Microwave communication systems--Equipment and supplies. 3. Optical communications--Equipment and supplies. I. Title.
TA1750.R53 1996
621.381'045--dc20   95-10083

Printed in the United States of America

10 9 8 7 6 5 4 3 2

Contents

Preface		xi
1	**Introduction**	**1**
	1.1 Applications	1
	1.1.1 Radar	2
	1.1.2 Radio Astronomy	2
	1.1.3 Spectroscopy	4
	1.1.4 Imaging	5
	Problems	8
	References	8
2	**Solid State Electronic Active Devices**	**10**
	2.1 Semiconductor Junctions	10
	2.2 Metal–Semiconductor Junctions	12
	2.3 Avalanche Transit Time Diodes	17
	2.4 Field Effect Transistors	21
	2.4.1 Small-Signal Equivalent Circuit Model	25
	2.4.2 Noise Performance	30
	2.4.3 Short-Gate Effects	31
	2.4.4 Power–Speed Trade-offs	32
	2.5 High Electron Mobility Transistors	32
	2.5.1 GaAs/AlGaAs HEMTs	33
	2.5.2 Other HEMT Structures	36
	2.6 Heterojunction Bipolar Transistors	38
	2.7 Bulk Negative Differential Resistivity Devices	40
	2.8 Tunnel Diodes	44
	Problems	46
	References	46

3 Optical Active Devices 48

- 3.1 Carrier Recombination in Semiconductors 48
- 3.2 Light Emitting Diodes 49
  - 3.2.1 Various Types of LEDs 50
  - 3.2.2 Speed Limitations 52
- 3.3 Semiconductor Lasers 53
  - 3.3.1 Basic Structure 53
  - 3.3.2 Laser Materials 55
  - 3.3.3 Active Region 56
  - 3.3.4 Optical Field Confinement 57
  - 3.3.5 Longitudinal Modes 61
  - 3.3.6 Laser Emission 62
  - 3.3.7 Distributed Feedback and Bragg Reflector Lasers 64
  - 3.3.8 Surface Emitting Laser Diodes 67
  - 3.3.9 Direct Modulation of Laser Diodes 68
- 3.4 Photodetectors 72
  - 3.4.1 Photodiodes 73
  - 3.4.2 Phototransistors 76
  - 3.4.3 Photoconductors 76
  - 3.4.4 MSM Detectors 77
- 3.5 Electro-Optic Modulators and Switches 82
  - 3.5.1 Phase Modulators 84
  - 3.5.2 Amplitude Modulators 89
  - 3.5.3 Waveguide Switches 91
  - 3.5.4 Electro-Optic Polymers 94
- Problems 95
- References 96

4 Microwave Circuits and Integrated Transmission Lines 99

- 4.1 Microwave Integrated Circuits 99
- 4.2 MESFET Characteristics 100
- 4.3 Reactively Matched Single-Stage Amplifiers 102
  - 4.3.1 Lumped-Element Matching 103
  - 4.3.2 Distributed-Element Matching 109
  - 4.3.3 Generalized Scattering Parameters 116
  - 4.3.4 Conditional Stability 117
  - 4.3.5 Gain-Bandwidth Limitations 119
  - 4.3.6 Noise Matching 119
- 4.4 Balanced Configuration 120
- 4.5 Feedback Amplifiers 121
- 4.6 Distributed Amplifiers 123
- 4.7 FET Oscillators 127

	4.7.1 Oscillation Condition	128
	4.7.2 Stabilization of FET Oscillators	129
4.8	Integrated Transmission Lines	131
	4.8.1 Microstrip	132
	4.8.2 Coplanar Waveguide	135
	4.8.3 Other Transmission Lines	147
4.9	Passive Components	147
	4.9.1 Microwave Junctions	148
	4.9.2 Directional Couplers	150
	Problems	156
	References	157

5 Optical Waveguides and Passive Components 161

5.1	Optical Fibers	162
	5.1.1 Step-Index Fibers	163
	5.1.2 Graded-Index Fibers	164
	5.1.3 Single-Mode Fibers	167
5.2	Integrated Optical Waveguides	169
	5.2.1 Propagation Modes of Planar and Channel Waveguides	170
5.3	Integrated Passive Components	174
	5.3.1 Bends and Corners	174
	5.3.2 Branching Waveguides	175
	5.3.3 Directional Couplers	177
	5.3.4 Wavelength and Polarization Separators	178
	Problems	179
	References	180

6 Short-Pulse Generation 181

6.1	Optical Pulse Forming	181
	6.1.1 Mode Locking	185
	6.1.2 Active Mode Locking	186
	6.1.3 Passive Mode Locking	187
	6.1.4 Colliding-Pulse Mode Locking	189
	6.1.5 Passive Mode Locking by Kerr-Lens Modulation	190
	6.1.6 Dispersion Compensation	191
	6.1.7 Pulse Compression	195
6.2	Electrical Pulse Forming	199
	6.2.1 Pulse Forming Networks	199
	6.2.2 Microwave Radiation Bursts	202
	6.2.3 Swept-Beam Generators	210
	6.2.4 Shock Wave Generators	213

Problems	217
References	218

7 High Speed and Long-Distance Communications — 220

7.1 Communications Link	221
7.1.1 Noise and Distortion	222
7.1.2 Modulation and Demodulation	224
7.1.3 Multiplexing and Transmission Hierarchies	229
7.2 Satellite Communications	231
7.2.1 Satellite Orbits	232
7.2.2 Current Trends in Satellite Communications	233
7.3 Fiber-Optic Communications	234
7.3.1 Transmission Rate and Distance	234
7.3.2 Soliton Transmission	238
7.3.3 Trends in Fiber-Optic Communications	240
Problems	241
References	241

8 Measurement Techniques — 243

8.1 Test Fixtures and Surface Contacting Probes	243
8.2 Noncontact Probing	245
8.2.1 Electro-Optic Sampling	245
8.2.2 Other Optical Probing Techniques	251
8.3 Network Analyzers	252
8.3.1 Calibration	253
8.3.2 Fixture Deimbedding	256
8.4 Six-Port Network Analyzer	257
8.4.1 Calibration	261
8.4.2 Dual Six-Port Network Analyzer	261
8.5 Optical Pulsewidth Measurement	262
8.6 Optical Pulse-Shape Characterization	265
Problems	266
References	266

Appendix A Waveforms and Spectra — 269

A.1 Time-Bandwidth Product	270

Appendix B Noise — 272

B.1 Types of Noise	272
B.2 Noise Figure	274
B.3 Effective Input Noise Temperature	274

B.4 Cascaded Noisy Networks	275
References	276

Appendix C Scattering Parameters and Smith Chart 277

C.1 Impedance Matrix	277
C.2 Scattering Matrix	278
C.3 Smith Chart	280
Reference	

Appendix D Five-Port Symmetrical Junction 282

D.1 S-Parameter Sensitivity to Matching of Star Junction	283
References	285

Appendix E Effect of Feedback on FET Gain 286

E.1 A More General Approach	287
References	288

Appendix F Carrier Transport in Semiconductors 289

Index 293

Preface

This book is intended to serve as a physics/electrical engineering text at the senior to first-year graduate level. It is also designed to be used as a reference for scientists and engineers working in the area of high speed optoelectronics. This intensive introduction enables the reader to develop and identify analytical and experimental tools for addressing the issues associated with high speed phenomena in electronics and optics. The complementary nature of the two fields and the expanding field of high speed optoelectronics served as motivations in the preparation of this text.

Topics covered in this book range from devices and circuits used at microwave and millimeter-wave frequencies to optical components and optoelectronic integrated circuits. In every chapter the emphasis is on the issues encountered at high speeds. Owing to the broad range of topics, detailed descriptions had to be avoided where possible. It is expected that the references to more detailed essays will fill this unavoidable gap.

The introduction to the book attempts to build a perspective and an appreciation for the broad field of analog high speed technology. The following chapters sequentially introduce electronic components, optical components, circuits, and subsystems. One application area—communications—has been chosen for more detailed analysis owing to the high level of interest and applicability of its concepts to other fields. Finally, an important area that is often neglected in similar publications is the measurement and characterization of high speed phenomena. This topic is covered in the last chapter.

I would like to thank Dr. S. Bandy and Dr. A. Emami-Naeini for their reviews and valuable suggestions. My thanks are also due to Dr. M. Rodwell for providing useful reference material.

M.L. RIAZIAT

Palo Alto, CA
August 1995

Introduction to High-Speed Electronics and Optoelectronics

CHAPTER ONE

Introduction

High speed in electronics and optoelectronics is referred to either high capacity for information transmission or ultrafast transient behavior. In this book we refer to an analog circuit as high speed if it operates above 1.0 GHz (10^9 Hz). This is the area of microwave circuits. (In the present nomenclature, microwaves cover the frequency range of 1–30 GHz and millimeter waves cover 30–300 GHz.) It should be pointed out that an electronic oscillator generating microwave or millimeter-wave frequencies is considered a high speed circuit, while an oscillator operating at much higher frequencies (1 THz or higher) falls into the category of lasers and is not considered a high speed device. In optoelectronics, a high speed laser oscillator is either one that can be modulated at microwave rates or one that can generate very short pulses, that is, pulses with appreciable Fourier components in the microwave and millimeter-wave region or above. As can be seen, the notion of high speed most often relates to microwave and millimeter-wave frequencies in either narrow-band or broadband sense, broadband being the alternate description of fast transient phenomena.

1.1 APPLICATIONS

Applications of high speed analog devices range from communications and radar to chemistry and imaging. In communications, higher frequency signal sources, modulators, and receivers lead to increased transmission capacity. In satellite communications, high frequencies can be transmitted and received using smaller size antennas. This makes direct broadcast satellite (DBS) and other direct satellite-to-customer links possible. Also, as lower frequency bands become more crowded, there is a pressure to utilize higher frequencies for emerging applications. In digital fiber-optic communications, high speed analog techniques are used in subcarrier multiplexing, which is presently the technique of choice for achieving the highest channel capacity.

2 INTRODUCTION

Telecommunication applications are explained in more detail in Chapter 7. Other applications of high speed optoelectronics such as radar, radio astronomy, spectroscopy, and imaging are briefly discussed under separate headings below.

1.1.1 Radar

In radar, there is interest in a broad range of microwave and millimeter-wave frequencies for various applications. One of the most obvious advantages of shorter wavelengths for radar is the possibility of obtaining narrow, high directivity beams with physically small antennas.[1] Narrow beams increase spatial resolution of imaging radar, and small size antennas are important in airborne radar and missile guidance systems where antenna size is constrained [1].

Scattering of radar signal from the target depends strongly on the ratio of wavelength (λ) to the characteristic physical dimension of the object (a). In the regime where $a \ll \lambda$ (Rayleigh scattering), the radar cross section of the target is very small. In this regime the information that can be extracted from the reflected signal is minimal. The scattering cross section in the Rayleigh regime is proportional to $(a/\lambda)^4$. Therefore, by operating at shorter wavelengths, scattering from small targets can be increased significantly. For example, scatterers such as cloud particles that cannot be studied by microwave radar present a significant cross section to millimeter-wave radar, making it useful for cloud-physics studies.

Another important consideration in the operation of millimeter-wave radar is atmospheric absorption. Figure 1.1 shows a plot of absorption as a function of frequency in atmosphere. The absorption peaks at 60 and 118 GHz are due to molecular oxygen; the other peaks are due to water. The plot shows the well-known atmospheric windows for radar operation at 94 and 140 GHz. Atmospheric attenuation is not necessarily a disadvantage for radar. It helps to reduce ground clutter as well as multipath interference in airborne radar. Automobile collision avoidance radar benefits from reduced interference and normally operates at a high atmospheric absorption frequency.

1.1.2 Radio Astronomy

Radio astronomy has always been a driving force behind ultra low noise detection system development in the microwave and millimeter-wave range. In the study of spectral lines, a large number of interstellar organic molecules (with emission lines ranging from microwave to submillimeter-wave frequencies) are found with blackbody emission temperatures of a few hundred de-

---

[1] The solid angle of the antenna beam is proportional to λ^2/A_e, where λ is the radiation wavelength and A_e is a parameter known as effective aperture, related to the radiating area of the antenna.

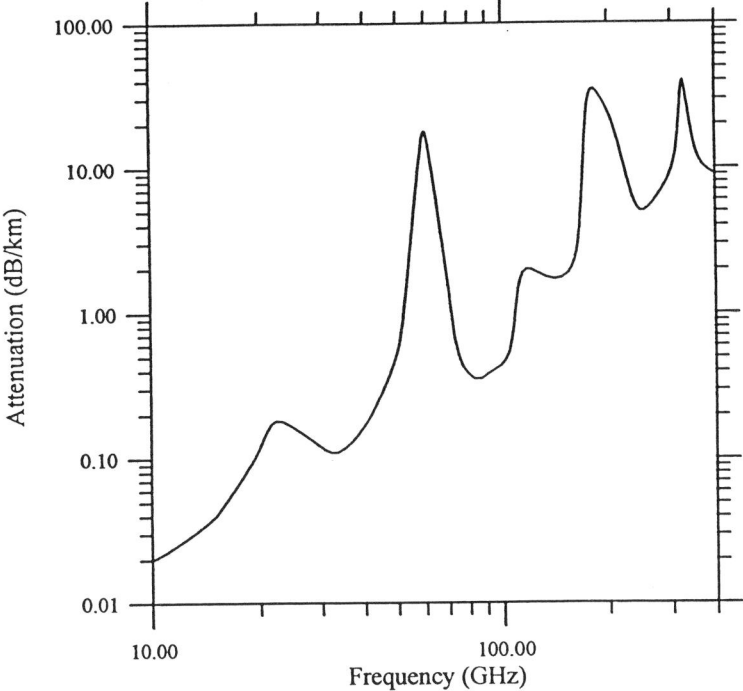

FIGURE 1.1 Atmospheric absorption as a function of frequency.

grees Kelvin. In dark clouds that are distant from radiation sources, emission temperature can drop even below 10° Kelvin [2]. This has to compete with a receiver noise temperature that is normally one to two orders of magnitude higher. Very narrow band detection with large averaging and data taking times are often employed, requiring ultrastable tunable local oscillators for heterodyne detection and low noise amplifiers before the detection stage. Masers and cooled parametric amplifiers were heavily used in radio astronomy for a number of years. These devices are being gradually replaced by solid state and in particular high electron mobility (HEMT) amplifiers that surpass the noise performance of all other types of high speed amplifiers.

The study of spectral lines in radio astronomy started with the detection of the 21-cm emission line of ionized hydrogen in 1951 [2]. The first molecular line to be detected was the 1.667-GHz OH absorption line in 1963. Thermal emission of interstellar ammonia followed in 1968. Presently, radio astronomical study of interstellar chemistry with a large number of identified molecular species provides a valuable probe into the evolution of the universe. Table 1.1 lists examples of interstellar emission and the associated frequencies [3]. Some of the same molecular absorption lines are also used for the monitoring of the atmosphere and Earth's surface from satellites.

4 INTRODUCTION

TABLE 1.1 Examples of Interstellar Emission Sources

Molecule	Frequency (GHz)
OH	13.441
NH_3	23.694
CH_3C_2H	85.457
CO	115.271
H_2O	22.235
CH_3OH	36.169
HCNO	109.906
H_2S	168.763

Note: Many of these sources have multiple emission lines in the microwave and millimeter-wave range.

1.1.3 Spectroscopy

High speed spectroscopy is subdivided here into time domain microwave spectroscopy, time domain optical spectroscopy, and chemical reaction monitoring techniques.

Time domain microwave spectroscopy is a technique to replace swept wavelength measurements with an approximate impulse response measurement that, through the Fourier transform, yields the same information [4]. Potential advantages of this technique are fast data acquisition and reduced cost. The input pulse is formed by an optically triggered switch that converts its electrical bias to a sharp electrical pulse (see Chapter 6). The electrical pulse is further sharpened by a nonlinear transmission line or a shock wave generator. A broadband antenna radiates the pulse that propagates through the sample. Another broadband antenna receives the signal. The process is repetitive, and the output waveform is detected by electro-optic sampling. The accuracy and usefulness of this technique increases with the sharpness of the pulse that impinges on the test material and the fine sampling of the outcoming waveform.

Time domain optical spectroscopy uses short pulses of light to populate electronic excited states and monitors absorption or fluorescence spectra in time, thereby giving information not only about the presence of atomic transitions but also about excited-state lifetimes. Interest in high speed spectroscopy accelerated soon after short laser pulses became available in the 1960s. Its first application to the study of nonradiative transitions of dye molecules was reported in 1967 [5].

The same technique is also used to study carrier dynamics in semiconductors. In these so called "pump-probe" experiments carriers are generated near the surface of the test sample by an above-bandgap "pump" pulse. A "probe" pulse of below-bandgap wavelength detects any changes in the refractive index of the semiconductor due to the presence of the carriers (Figure 1.2). Depth profiles and free carrier lifetimes can be determined by this method [6].

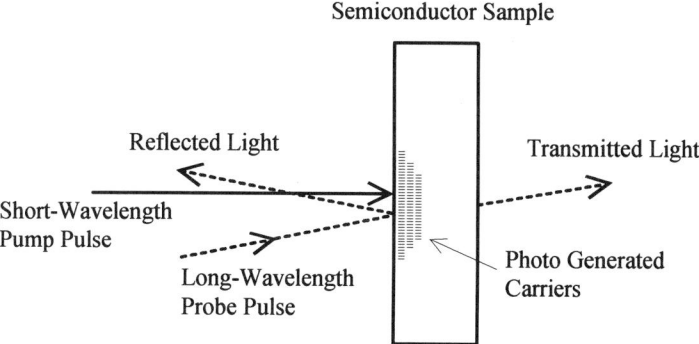

FIGURE 1.2 Pump-probe experiment for studying carrier dynamics in semiconductors.

A very specialized application of high speed spectroscopy is in chemistry, where fast optical pulses allow the monitoring of chemical reactions in progress. This yields information about the time scales of various chemical interactions as well as the detail of chemical processes including the formation of intermediate short-lifetime compounds known as transition states. Transition state lifetimes are in the range of 10–100 fs. Therefore, resolutions in the same time scale are needed for their monitoring in chemical reactions.

Controlling chemical reactions is an integral part of research in chemistry. The controlling effects of heat, light, and pressure on chemical reactions were studied extensively since the early days. Over time, as measurements became more quantitative, experiments were refined. Molecular beam technologies (shown in Figure 1.3) developed in the 1960s allowed chemists to confine chemical interactions to a small location in space and vary the interaction parameters precisely [7, 8]. The chemical reaction of interest is normally triggered by a short optical pulse that also sets the reference time. Absorption and emission characteristics of the interacting molecules, from that point on, reveal detailed information about the dynamics of the chemical interaction.

1.1.4 Imaging

Time-gated reflectometry is a technique used in radar to resolve targets in highly scattering media such as fog. With picosecond and femtosecond optical pulses, this idea was extended to time-resolved imaging with major biomedical applications. A beam of light propagating through a scattering medium is attenuated due to both absorption and scattering. Attenuation coefficients μ_a and μ_s characterize the two processes. Radiation amplitude in the beam as a function of position, given by

$$A(x) = A(0)e^{-(\mu_a+\mu_s)x}, \qquad (1.1)$$

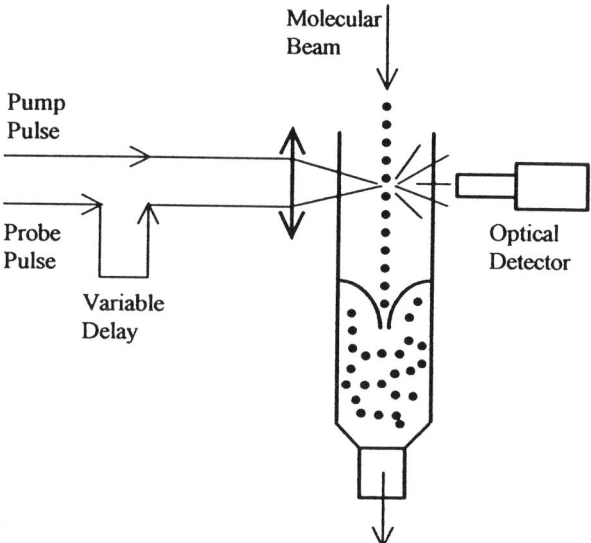

FIGURE 1.3 Molecular beam chemistry. Pump pulse initiates chemical reaction whose progress is monitored by probe beam.

is sometimes referred to as the Beer-Lambert law. Propagation of long-wavelength visible and near-infrared (600–1300 nm wavelength) radiation through human tissue is limited by light scattering rather than absorption (i.e., $\mu_s \gg \mu_a$). Scattering reduction techniques make it possible to form an image of the interior of the body. A conceptually simple scattering reduction approach is to use only the photons that take the shortest path through the tissue. This technique is referred to as time-of-flight imaging and is an effective method for sharpening an image that is blurred by scattering (Figure 1.4).

Efforts to use visible light for imaging are not new. An imaging technique known as optical transillumination, or diaphanography, has been available for biomedical applications for a few decades. This technique was developed mainly for mammography, but it never gained widespread clinical acceptance. The main motivation behind diaphanography was the fact that it used visible light and, therefore, could yield information about blood concentration in various areas of the image. Abnormally high local blood supply is often associated with malignant lesions, which appear dark in a transillumination image. On the other hand, a cyst that causes the displacement of blood-containing tissue by its fluid-filled cavity appears bright. This advantage of diaphanography was completely marred by its lack of sufficient resolution, which made "early" detection of tumors impossible.

The low resolution of the transillumination image is caused by light scattering in the tissue. Figure 1.5 shows how the extra path length taken by scattered light gives rise to this lack of resolution. "Time-resolved imaging" recovers the image from the scattered background by using pulsed radiation

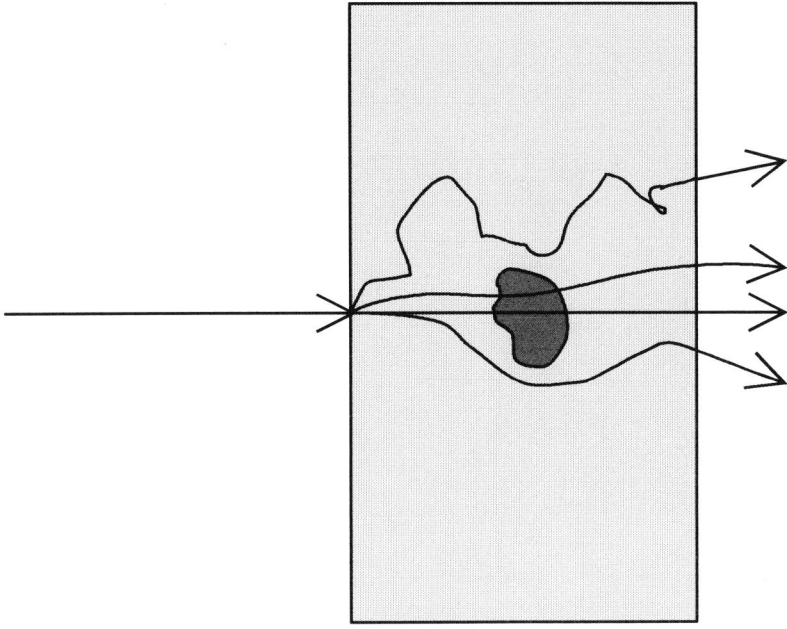

FIGURE 1.4 In time-of-flight imaging, photons that arrive late are discarded.

and selectively receiving the information that arrives in a particular time gate [9]. The time gate restricts the total deviation in path length of the signal from a direct line-of-sight path. In this way the time gate determines the size of the smallest resolvable object. An approximate expression relating the two quantities for human tissue is given elsewhere [10]. A modified version of that expression is

$$3.4 D_{\min} \approx R_a + L_a \sqrt{1 - \frac{L^2}{L_a^2 + R_a^2}}, \qquad (1.2)$$

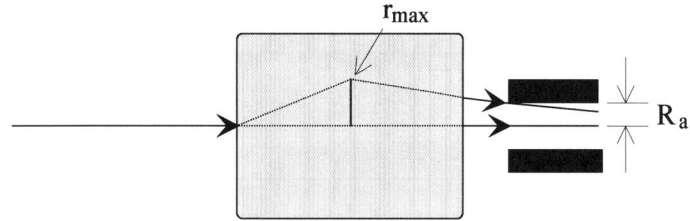

FIGURE 1.5 Resolution of time-gated image is related to maximum photon path length allowed.

8 INTRODUCTION

where, D_{min} is the diameter of the smallest resolvable object, R_a is the diameter of the detection aperture, L_a is the longest path length accepted by the time gate, and L is the shortest path length through the scattering medium. The time gate Δt is given by

$$\Delta t = \frac{(L_a - L)\eta}{c}, \qquad (1.3)$$

where η is the refractive index of the scattering medium (1.3–1.5 for human soft tissue) and c is the speed of light. For a 5-cm-thick scattering tissue, and a time gate of 20 ps, the diameter of the smallest resolvable object is approximately 5 mm. Equation (1.2) was derived from a very simple model and includes a heuristic multiplication factor.

Equations (1.2) and (1.3) establish that higher resolution is possible with a shorter gating time. Here is where the trade-off with signal-to-noise ratio has to be made. As the gating time is reduced, signal strength goes down. The noise level of the detector ultimately limits the gating time and thereby the size of the smallest resolvable object. For a given detection system, signal-to-noise ratio can be improved by either increased signal level or reduced noise by averaging. Signal power level has an absolute maximum determined by safe exposure level of human skin to light, and averaging is limited by the amount of time that human body can be kept motionless. Examples of technological advances in this field are given in references [11] and [12].

PROBLEMS

1.1 Referring to Appendix A, calculate the time–bandwidth product for the exponential waveform $V(t) = e^{-|t|}$. How does it compare with the time–bandwidth product of a Gaussian?

1.2 Repeat problem 1.1, this time using 2σ as the width of the waveform, where σ is the standard deviation defined as $\sigma^2 = \int_{-\infty}^{\infty}(x - \bar{x})^2 V(x)\, dx$ and $\bar{x}$ is the mean value of the waveform.

REFERENCES

[1] M. I. Skolnik, *Introduction to Radar Systems*, 2nd ed., McGraw-Hill, New York, 1980.

[2] D. R. W. Williams, "Fundamentals of Spectral-Line Measurements," in M. L. Meeks (Ed.), *Methods of Experimental Physics*, Vol. 12: Astrophysics, Part C: Radio Observations, Academic, New York, 1976.

[3] P. F. Goldsmith (Ed.), *Instrumentation and Techniques for Radio Astronomy*, IEEE, New York, 1988.

[4] Y. Pastol, G. Arjavalingam, J. M. Halbout, and G. V. Kopcsay, "Coherent Broadband Microwave Spectroscopy Using Picosecond Optoelectronic Antennas," *Appl. Phys. Lett.,* Vol. 54, No. 4, pp. 307–309, Jan. 1989.

[5] A. Laubereau, "An Introduction to Picosecond Spectroscopy," in F. T. Arecchi, F. Strumia, and H. Walther, (Eds.), *Advances in Laser Spectroscopy*, Plenum, New York, 1983.

[6] R. S. Miranda et al., "Use of Time-Resolved IR Reflection and Transmission as a Probe of Carrier Dynamics in Semiconductors," *Opt. Lett.,* Vol. 16, No. 23, pp. 1859–1861, Dec. 1991.

[7] A. H. Zewail, "The Birth of Molecules," *Sci. Am.,* pp. 76–82, Dec. 1990.

[8] I. W. M. Smith, "Exposing Molecular Motions," *Science,* Vol. 343, No. 6260, pp. 691–692, Feb. 1990.

[9] S. Anderson-Engles, R. Berg, S. Svanberg, and O. Jarlman, "Time-Resolved Transillumination for Medical Diagnosis," *Optics Lett.,* Vol. 15, No. 21, pp. 1179–1181, Nov. 1990.

[10] J. C. Hebden and R. A. Kruger, "Transillumination Imaging Performance: A Time-of-Flight Imaging System," *Med. Phys.,* Vol. 17, No. 3, pp. 351–356, May/June 1990.

[11] M. Hee, J. Izatt, E. Swanson, and J. Fujimoto, "Femtosecond Transillumination Tomography in Thick Tissues," *Optics Lett.,* Vol. 18, No. 13, pp. 1107–1109, July 1993.

[12] B. Das, K. Yoo, and R. Alfano, "Ultrafast Time-Gated Imaging in Thick Tissues: A Step Toward Optical Mammography," *Optics Lett.,* Vol. 18, No. 13, pp. 1092–1094, July 1993.

CHAPTER TWO

Solid State Electronic Active Devices

Limitations on the speed of solid state devices arise from either device dimensions and material characteristics (intrinsic) or parasitics associated with the structure of the device and electrical connections made to the device (extrinsic). High speed operation of most active devices are limited by one or the other of these factors. Sometimes it is possible to quantify the effect of parasitic elements through proper modeling. This allows one to distinguish between intrinsic and extrinsic speeds of the device. This distinction, however, is not always possible.

In this chapter, a few high speed device structures are described, namely, diodes, MESFETs, HEMTs, and HBTs. In each case, both the intrinsic and the extrinsic limitations on device speed are discussed, and some common trade-offs are established. For a brief overview of the electronic characteristics of semiconductors refer to Appendix F.

2.1 SEMICONDUCTOR JUNCTIONS

In the absence of any thermal excitation, the electrons in a solid occupy all available energy levels up to the *Fermi* level. In an intrinsic semiconductor, the Fermi level is close to the middle of the bandgap. In an n-doped material this level gets closer to the conduction band. In a p-doped material it gets closer to the valence band edge. When a semiconductor sample has an n-doped region and a p-doped region sharing a common boundary, a p–n junction is formed. At this junction, carriers will diffuse across the boundary, and a reduction of the number of carriers occurs through electron–hole recombination. The region of reduced carrier density is called the *depletion region*. The diffusion of charged carriers gives rise to a built-in electric field that counteracts the diffusion process, and an equilibrium is reached. The presence of the built-in electric field bends the conduction and the valence bands in the vicinity of the junction. In equilibrium, the band bending causes the Fermi levels on both sides of the junction to match (Figure 2.1)

The built-in potential V_{bi} at the junction is found by letting the diffusion current and the drift current due to the presence of the potential be equal. The result is given here without derivation [1, 2]:

$$V_{bi} = V_T \ln \frac{N_d N_a}{n_i^2}, \tag{2.1}$$

where $V_T = kT/q$ is equal to 25.6 mV at 300° Kelvin. Since both the p-type and the n-type materials have both majority and minority carriers, N_a and N_d are used here to denote the carrier densities due to the donors and the acceptors, respectively. The width of the depletion region is an important parameter that, as will be seen later, determines the junction capacitance. This width is a function of the doping levels and the built-in potential. A simple algebraic expression exists for the width of the depletion region at the junction if one side is doped much higher than the other. If $N_d \gg N_a$, the width W is given by

$$W = \sqrt{\frac{2\epsilon_0 \epsilon_r (V_{bi} \pm |V|)}{qN_d}}. \tag{2.2}$$

This expression also gives the width of the depletion layer under external

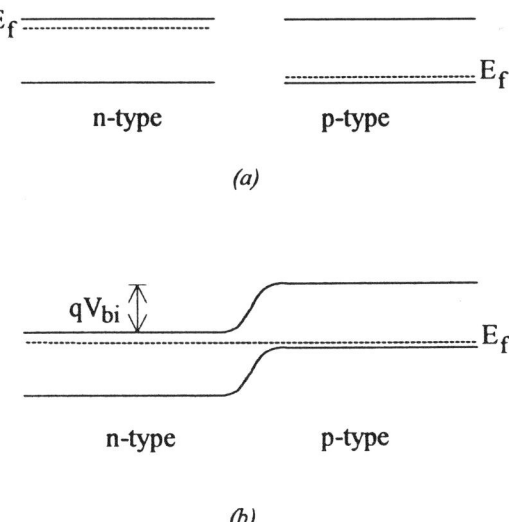

FIGURE 2.1 Energy band diagram of semiconductor p–n junction: (a) before contact; (b) after contact.

12 SOLID STATE ELECTRONIC ACTIVE DEVICES

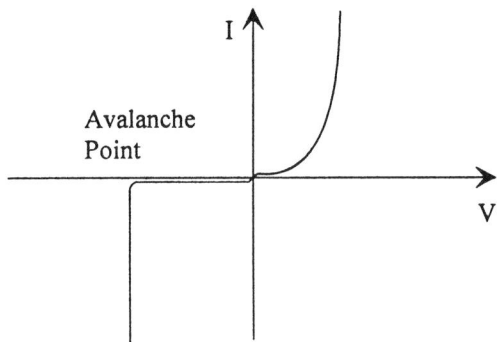

FIGURE 2.2 Current–voltage relationship of an ideal *p–n* junction.

biasing. Any small external potential V applied to the junction is either added to or subtracted from V_{bi} to represent forward or reverse biasing.

When an external electric field is applied to a semiconductor *p–n* junction, it can either add to the built-in potential or oppose it. If it is applied in the same direction as the built-in potential, the junction is said to be *reverse biased*. The applied field widens the depletion region and no current flows. If the applied field opposes the built-in potential, the junction is *forward biased*. The depletion region shrinks as the external power supply replenishes the charge that is recombined at the junction. As the forward-bias voltage increases, the current exceeds the recombination current, and positive and negative charge carriers are injected in both directions across the junction. The injected charge in the form of minority carriers is swept across the doped semiconductor by the electric field. The current increases exponentially with the applied field. The form of this exponential behavior is often written as

$$I = I_0(e^{V/V_T} - 1). \qquad (2.3)$$

The constant I_0 is the current that would ideally flow under reverse-biased conditions. Figure 2.2 shows the current–voltage relationship of a *p–n* junction. This type of a junction is often used in *p–n* diodes for current rectification. Another feature of the *p–n* junction shown in Figure 2.2 is that the reverse voltage cannot be increased indefinitely. A process known as avalanche breakdown sharply limits this voltage. Refer to Section 2.3 for applications of this process.

2.2 METAL–SEMICONDUCTOR JUNCTIONS

A diffusion of carriers similar to that of a *p–n* junction occurs at a semiconductor–metal junction. If a junction is formed between a semiconductor and a metal whose Fermi level is lower than that of the semiconductor, the electrons diffuse into the metal and a depletion region forms. The relevant energy band diagram is shown in Figure 2.3. The reference level for energy in this arrangement is usually taken to be that of a free electron in vacuum. The energy level of an electron in the conduction band of a metal can be deduced from the *work function*, $(q\phi_m)$ of the metal. This quantity is the energy needed to release an electron from the Fermi level to free space. The built-in potential V_{bi} at the metal–semiconductor junction is related to the difference between the work functions of the two materials through

$$V_{bi} = \phi_m - \phi_s, \qquad (2.4)$$

where $q\phi_s$ is the work function of the semiconductor. Since in a doped semiconductor the Fermi level lies below the conduction band, another quantity χ_s is used to characterize the metal–semiconductor $(m-s)$ barrier height for conduction band electrons to pass through. Here, χ_s is called the *electron affinity*. The term $q\chi_s$ is the energy required to release an electron from the bottom of the conduction band to the vacuum level. Using this quantity the $m-s$ barrier height ϕ_b is defined as

$$\phi_b = \phi_m - \chi_s. \qquad (2.5)$$

This is the barrier that an electron going from the metal to the semiconductor encounters.

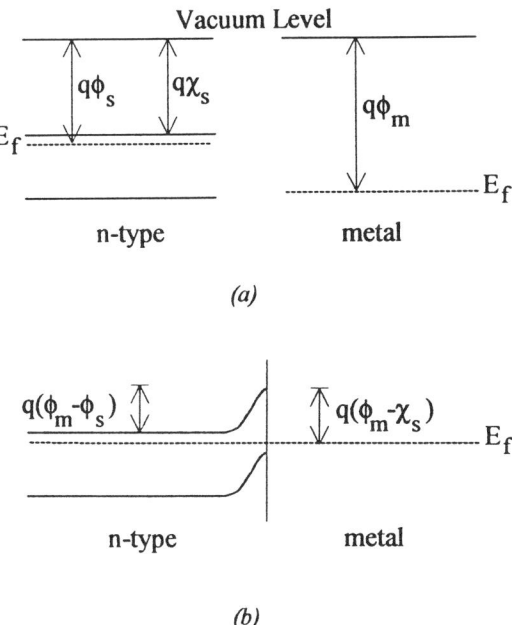

FIGURE 2.3 Metal–semiconductor junction energy band diagram: (*a*) before contact; (*b*) after contact.

Example For aluminum, $q\phi_m = 4.2$ eV. We will now calculate the barrier height of an aluminum Schottky contact on n-type GaAs. From Table 2.1 we find the electron affinity of GaAs to be 4.07 V. The Schottky barrier height is expected to be $4.2 - 4.07 = 0.13$ V. This value is not in agreement with the experimentally observed value of 0.8 V for aluminum on GaAs. ∎

Table 2.1 lists observed Schottky barrier heights for Si and GaAs. The observed barrier heights are not in agreement with Eq. (2.5), the reason being the presence of what are known as *interface states*. Unlike a *p–n* junction that is formed without any disruption of the crystal lattice, a metal–semiconductor junction creates a discontinuity in the periodic lattice of the semiconductor. This discontinuity creates a large number of local energy states, known as *interface states*, in the energy range that would normally be the forbidden gap. The distribution of the interface states can either raise or lower the Fermi level. If the density of these states is large, the Fermi level is said to be *pinned down* by the interface states. The Schottky barrier height becomes independent of ϕ_m or χ_s. Experimentally, even though barrier heights do depend on the particular metal and the semiconductor used, they are dominated by interface states (Table 2.1).

The doping level of the semiconductor does not affect the barrier height. It does, however, determine the width of the depletion region formed next to the metal contact. The width of the depletion layer can be approximated by Eq. (2.2), which was derived for a one-sided *p–n* junction. It can be seen that this width is reduced as the doping of the semiconductor is increased. At very high doping levels the width of this region becomes thin enough for charge carriers to tunnel through. In this regime, the current through the junction becomes more symmetrical with respect to the direction of the applied field. The junction is then said to approximate an *ohmic contact*. High doping levels used for ohmic contact formation are denoted as n^+ or p^+.

Current–voltage characteristics of the Schottky barrier diode are similar to those of the *p–n* junction shown in Figure 2.2 and are governed by Eq. (2.3). This equation is normally rewritten as

$$I = I_0(e^{V/nV_T} - 1) \qquad (2.6)$$

in order to account for the nonideal behavior of the *m–s* junction. The factor n is known as the *ideality factor*, which is close to unity when semiconductor

TABLE 2.1 Observed Schottky Barrier Heights for Si and GaAs

Metal	ϕ_m (eV)	Si	GaAs
Au	4.7	0.81	0.9
Al	4.2	0.67	0.8
Cu	4.4	0.71	0.82

doping is low. The ideality factor deviates from unity at high doping levels and at low temperatures. Typical values of n for Schottky diodes fall in the range of 1.02–1.05. Despite the similarities between the I–V characteristics of m–s and p–n junctions, the following differences are noteworthy:

- As the potential barrier is lowered by forward biasing the junction, electrons are emitted from the n-type semiconductor to the metal. Forward-bias current is carried by majority carriers. This is in sharp contrast to the p–n junction where current is carried by minority carriers. The presence of minority carriers is a charge storage mechanism. This is due to the fact that when electric field polarity reverses, a finite time is required for minority carriers to be swept out. The p–n junction diode is therefore intrinsically slower in responding to high speed signals than the Schottky diode.
- A typical knee voltage in the I–V curve is 0.3 V for an m–s junction as opposed to approximately 0.6 V for a p–n junction. For the same forward voltage, the m–s diode carries a higher current. Also, for the same junction area, the m–s diode has a higher saturation current because it is carried by majority charge carriers.
- The reverse-bias current of the Schottky diode is characterized by a soft breakdown. This is partly due to the high electric field regions formed near the edges of the Schottky metal. There are fabrication methods that keep the edges away from the depletion region, thereby reducing the reverse-bias current and sharpening the breakdown knee [2].

The nonlinear relationship that exists between the voltage and the current at the terminals of both p–n and Schottky diodes makes them usable as detectors, mixers, varactors, and so on. Since the Schottky diode is a majority-carrier diode, it is particularly useful in high frequency operation. Another high frequency advantage of the Schottky diode is that its fabrication process is simple and compatible with microwave integrated circuits. The diode's equivalent circuit model is shown in Figure 2.4. Here, $R_j(V)$ is the voltage-dependent resistance of the junction which is found from

$$R_j(V) = \left(\frac{dI}{dV}\right)^{-1}. \tag{2.7}$$

The term $C_j(V)$ is the variable junction capacitance, which is a function of depletion layer thickness, and subsequently a function of the applied voltage [Eq. (2.2)]: $W(V) = W(0)\sqrt{1 - V/V_{bi}}$. This gives the junction capacitance the same functional dependence on the applied voltage:

$$C_j(V) = \frac{C_j(0)}{\sqrt{1 - V/V_{bi}}}. \tag{2.8}$$

SOLID STATE ELECTRONIC ACTIVE DEVICES

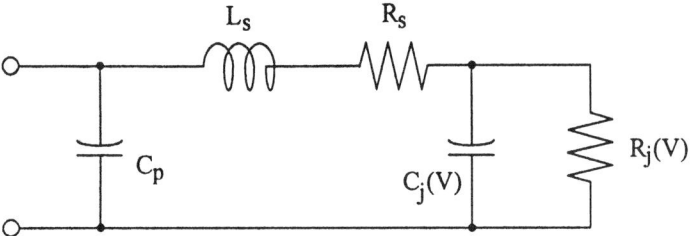

FIGURE 2.4 Equivalent circuit model of a Schottky diode.

Here, R_s and C_j are the intrinsic parameters of the diode. The other three elements in the equivalent circuit model of the diode are the extrinsic parameters, which are functions of the geometry, contacting leads, and the packaging of the device. In most practical applications, other external reactive elements are added for matching in order to maximize the voltage across the intrinsic capacitance and resistance. A useful figure of merit that allows the user to estimate the high frequency performance of the diode is the parameter known as the cutoff frequency f_c, defined as

$$f_c = \frac{1}{2\pi R_s C_j(V_b)}, \tag{2.9}$$

where V_b is the bias voltage.

Example A $5 \times 100\ \mu\text{m}$ Schottky contact to n-type GaAs has a built-in voltage of 0.4 V, and a series resistance of 6 Ω. If the doping concentration is 3.5×10^{17} cm^{-1}, find the intrinsic cutoff frequency of the diode at zero bias.

Solution The junction capacitance is due to the depletion layer thickness, which is given by Eq. (2.9):

$$W = \sqrt{\frac{2(8.85 \times 10^{-12})(12.5)(0.4)}{(1.6 \times 10^{-19})(3 \times 10^{21})}} = 0.43\ \mu\text{m}.$$

Using this thickness, the capacitance of the junction can be determined:

$$C_j(V=0) = \epsilon_0 \epsilon_r \frac{\text{area}}{W} = (8.85 \times 10^{-12})(12.5)\frac{(5)(100 \times 10^{-12})}{0.43 \times 10^{-6}} = 0.13\ \text{pF}.$$

The intrinsic cutoff frequency is

$$f_c = [2\pi(6)(0.13 \times 10^{-12})]^{-1} = 204\ \text{GHz}. \qquad\blacksquare$$

Schottky diodes can operate at extremely high frequencies. Cutoff frequencies approaching 3000 GHz are achievable [3]. High frequency diodes

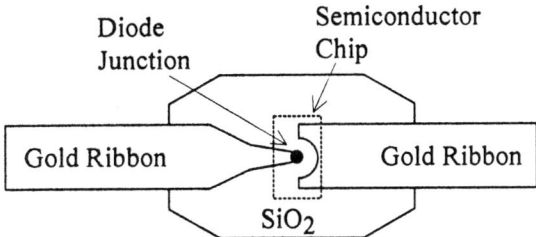

FIGURE 2.5 External connections to a high frequency, low parasitic Schottky diode.

are normally very small devices (e.g., 2.0 μm circular dot) with a small junction capacitance (10–50 fF). It is important that connecting leads to such a device have enough metal thickness to keep R_s as low as possible and at the same time be kept short and far enough from ground such that L_s is low and $C_p < C_j$. Figure 2.5 shows the overall structure of a high frequency, low parasitic diode.

A very high frequency application of Schottky diodes is in millimeter-wave signal detection with integrated antenna arrays, where Schottky diodes receive both the signal and the local oscillator power from free space. Both of these can be at very high frequencies, for which integrated circuit amplifiers are not available, but their difference frequency, *IF* (heterodyne terminology, see Chapter 7) is low enough to be amplified by circuits integrated with the detectors on the antenna surface. An example of a Schottky diode integrated antenna is described in the literature [4].

2.3 AVALANCHE TRANSIT TIME DIODES

A reverse-biased *p–n* junction presents a very high impedance to an externally applied field. The small current that flows through the junction is mainly due to thermal generation of electron–hole pairs in the depletion region. When the applied field reaches a critical level, electrons traveling through the depletion layer gain enough momentum to generate secondary electron–hole pairs in their collision with lattice cites. This effect is known as *impact ionization* and starts the process of *avalanche breakdown*. The current rises rapidly, and in devices not designed to operate in this regime, the excess heat causes the destruction of the device. Avalanche breakdown is, however, routinely used in Zener diodes for voltage regulation and in avalanche transit time diodes discussed in this section.

The avalanche process gives rise to a current buildup, that has a phase lag with respect to the applied voltage. This is known as the avalanche delay. In a device called the IMPATT diode, avalanche delay, augmented by extra lag due to carrier transit time through a drift region, is used to generate negative resistance and signal amplification. IMPATT stands for impact ionization avalanche transit time diode. Closely related to IMPATT are a number of other diodes or modes of operation with names ending in TT, such as TRAPATT and BARITT, all using transit time delay to control the negative resistance. IMPATT diodes are commonly made of either silicon or GaAs and can operate at frequencies exceeding 100 GHz. For over two decades they have kept their lead as the highest power solid state sources of millimeter waves.

An IMPATT diode consist of an abrupt p–n junction next to a drift region, which may be doped or undoped; the undoped version is known as the Reed diode (Figure 2.6). The p–n junction is biased very close to the avalanche breakdown point. An ac signal applied to the terminals drives the diode into the avalanche region periodically.

The principles of operation of the IMPATT diode are shown in Figure 2.7. In order to generate power gain, the current through the diode should be 180 degrees out of phase with the applied voltage. If the operation frequency is denoted by f, the current should lag the voltage by $\frac{1}{2}f$. This time lag is afforded by the avalanche delay time Δt_a. As shown graphically in Figure 2.7, avalanche charge generation continues to increase in the first half cycle of the ac signal and starts decreasing only after the field direction reverses. In this way, avalanche delay time introduces a π phase shift between the current and the applied voltage. Another time delay is added to this by carrier transit time through an undoped or depleted drift region of length d. The drift time delay is d/v_{sat} assuming saturated velocity for the carriers. The drift time

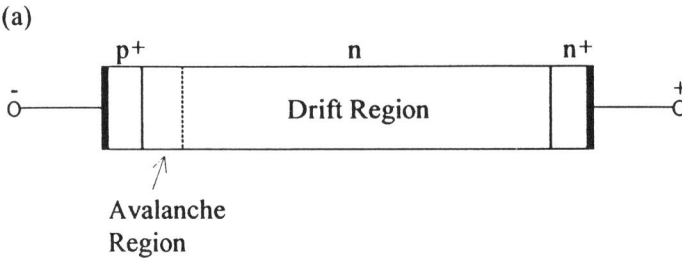

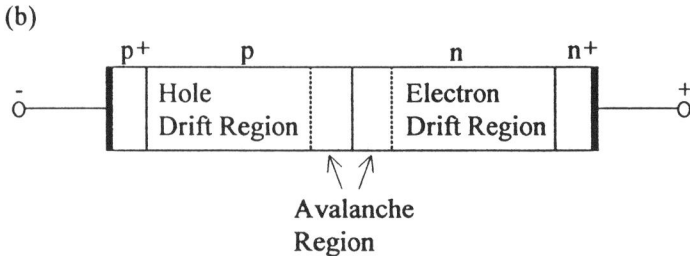

FIGURE 2.6 IMPATT (*a*) single-drift diode and (*b*) double-drift diode.

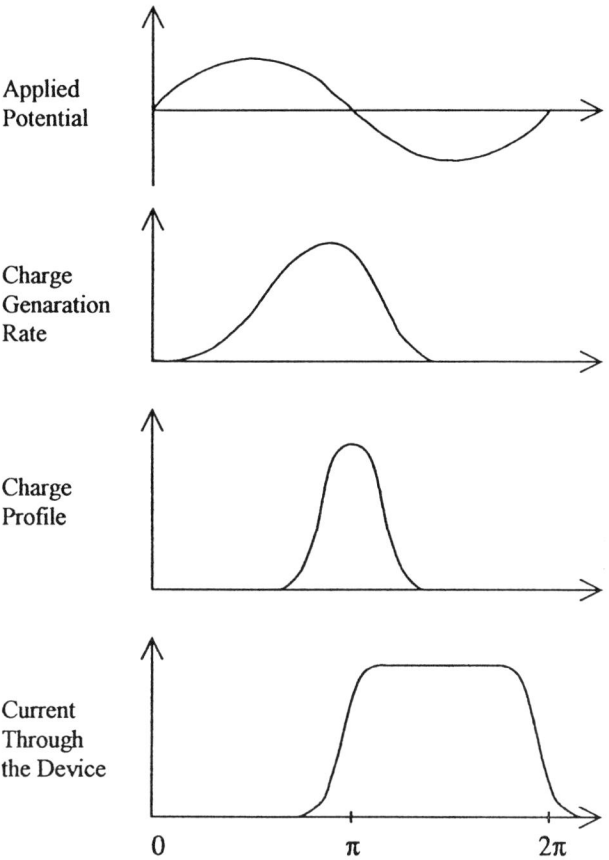

FIGURE 2.7 Steps involved in current generation in an IMPATT diode.

delay causes the current to continue to flow through the second half of the ac period. The sum of the two time delays equals one period of the ac signal:

$$\frac{1}{f} = \Delta t_a + \frac{d}{v_{sat}}. \tag{2.10}$$

For the device to manifest negative resistance, it is not necessary that the avalanche phase lag be exactly 180 degrees. However, the highest efficiency or the largest negative resistance is observed when the avalanche delay time Δt_a is equal to $\frac{1}{2}f$. Figure 2.6 shows two examples of IMPATT diode structures. In the double-drift IMPATT device, efficiency is enhanced by making use of both the electron current and the hole current. Common efficiencies of IMPATT diodes range from 20% to 40%. Efficiencies as high as 70% have been reported.

Example Determine the length of the drift region for a silicon IMPATT diode operating at 100 GHz.

Solution Using Eq. (2.10) and assuming the ideal operation condition of $\Delta t_a = 1/2f$, we need to have

$$\frac{d}{v_{\text{sat}}} = \frac{1}{2f} = \frac{1}{200 \times 10^9}.$$

For silicon $v_{\text{sat}} = 10^7$ cm/s. Therefore $d = 0.5\ \mu$m. ∎

There is a trade-off between the maximum power available from an IMPATT diode and its operation frequency. In order to establish this trade-off, note that the maximum current through the device, and thereby the maximum power generated by the device, is proportional to the area A of the junction: $P_m \propto A$. In Eq. (2.10), it was shown that the operation frequency $f \approx v_{\text{sat}}/2d$. Combining these two we get $P_m f \propto A/d$. Note that $\epsilon A/d$ is the capacitance C of the device. Therefore, $P_m f \propto C$. Maximum operation frequency is inversely proportional to device capacitance. This is due to the RC time constant of the circuit. So, device capacitance is proportional to $1/f$. Combining all these arguments we arrive at $P_m f \propto 1/f$ or

$$P_m \propto \frac{1}{f^2}. \tag{2.11}$$

IMPATT diodes can be used either as oscillators or as reflection-type amplifiers. When used as oscillators the diode is usually coupled to a resonant tank circuit or a resonant cavity that provides frequency stability. If the diode is coupled to a resonant cavity, the coupling location in the cavity can serve to partly match its impedance to that of the diode [1]. Since the negative resistance of a high frequency IMPATT diode is relatively small (3–5 Ω), the coupling design is critical.

Negative resistance can also be used to amplify signals. A device manifesting negative resistance has an impedance whose real part is negative, that is,

$$Z_s = R_s + jX_s, \qquad R_s < 0. \tag{2.12}$$

If a signal traveling along a transmission line of impedance Z_0 is incident on this device, the reflection coefficient S_{11} is given by (Appendix C)

$$S_{11} = \frac{Z_s - Z_0}{Z_s + Z_0} = \frac{-Z_0 - |R_s| + jX_s}{Z_0 - |R_s| + jX_s}. \tag{2.13}$$

The magnitude of the reflection coefficient is greater than unity. Therefore, the reflected signal is amplified. An amplifier of this type is referred to as

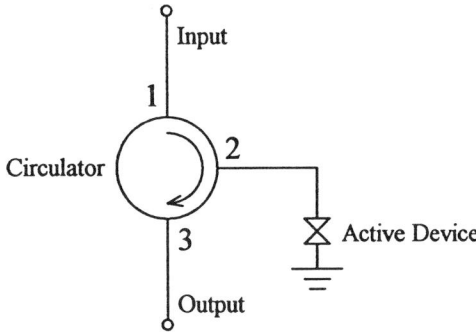

FIGURE 2.8 Operation of a reflection-type amplifier.

a *reflection-type amplifier*. A circulator[1] is normally used with this type of amplifier in order to separate the input and output signals (Figure 2.8). The highest amplification is achieved when $R_s = Z_0$. At this point the amplifier is unstable and can oscillate spontaneously depending on existing parasitic reactances.

2.4 FIELD-EFFECT TRANSISTORS

Diodes are relatively simple solid state devices to manufacture, but due to their two-terminal nature, they are not very versatile active devices for circuit design. High speed diodes have been available for many years, but it was not until the 1980s that significant advances in high speed transistors paved the way for the development of a new multitude of microwave circuits for diverse applications. Of particular importance was the success of the GaAs metal–semiconductor field effect transistor (MESFET). This device was responsible for many of the advances made in high speed solid state components, in particular in monolithic microwave integrated circuits (MMICs). The availability of MMICs led to reduced size and weight in transmit–receive modules for a variety of communications applications as well as for airborne radar, guidance systems, and electronic countermeasures. Gallium arsenide MESFETs are also finding increasing application in commercial high speed logic circuits. It appears that with new materials and fabrication techniques, the basic MESFET is evolving into a number of higher performance transistors for more specialized applications such as optoelectronics. Of these transistors, the *high electron mobility transistor* (HEMT) is of particular importance in high speed applications. This device is described in some detail in this chapter.

---

[1] A circulator is a three-port nonreciprocal device. Power incident on port 1 goes to port 2, and power incident on port 2 goes to port 3; $S_{21} = S_{32} = 0$.

The basic structure of a MESFET is shown in Figure 2.9. Current flows between source and drain, which are two ohmic contacts made to the doped semiconductor material. The gate electrode is a Schottky contact whose electric potential (bias level) controls the channel thickness and therefore the drain current. Basic I–V characteristics of an FET are shown in Figure 2.10. There are three distinct regions of operation in this figure. First, the linear region between zero drain bias and the knee voltage. In this region, drain-to-source characteristics resemble that of a variable resistor whose resistance is controlled by the gate voltage. Second is the saturation region. This is the region of operation of most transistors. Drain current is almost independent of drain voltage. It is controlled only by the gate. The third region is the breakdown region where drain current rises rapidly due to the breakdown of the gate–drain junction. Only the saturation region is of interest here.

Consider a block of semiconductor with two ohmic contacts (source and drain). The current that flows through the semiconductor doped channel is given by

$$I = \frac{q\mu N_d W_g h_c}{L} V_{ds} = g_0 V_{ds}, \tag{2.14}$$

where W_g is the width of the channel, h_c is its thickness, L is its length, and g_0 is the low field conductance of the channel. As the drain voltage V_{ds} is increased, the current saturates to

$$I_{\text{sat}} = q v_s N_d W_g h_c. \tag{2.15}$$

Next, add a Schottky gate between source and drain. The presence of the gate depletes a portion of the carriers, thereby reducing the effective channel thickness and increasing its resistance. The drain-to-source current is reduced as shown in Figure 2.11. As drain voltage is increased, carrier velocity $\mu V_d/L$ increases. At the same time the gate–drain junction is further reverse biased,

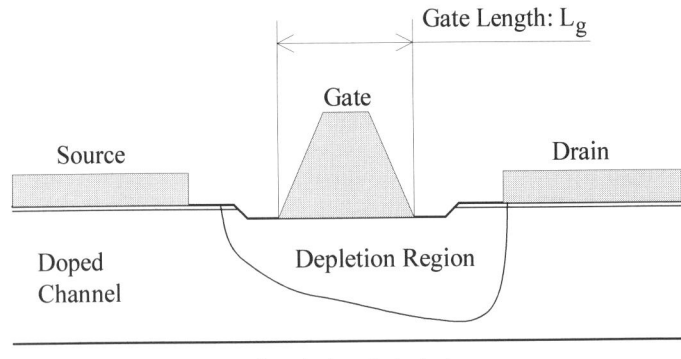

FIGURE 2.9 Basic structure of a MESFET.

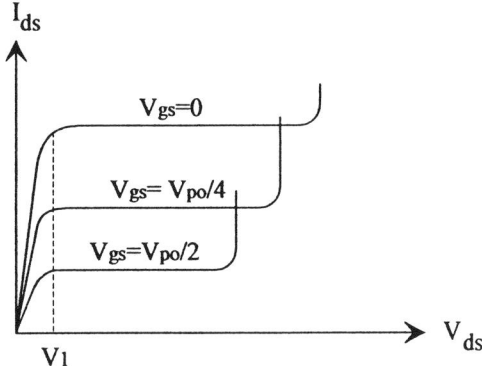

FIGURE 2.10 Idealized I–V characteristics of an FET for different gate bias voltages.

leading to a thinner channel. Current still rises but at a lower rate. An approximate expression for the current in this regime is [5]

$$I_{ds} = g_0 \left[V_{ds} - \frac{2}{3} \frac{(V_{ds} + V_{bi} - V_{gs})^{3/2} - (V_{bi} - V_{gs})^{3/2}}{V_{po}^{1/2}} \right]. \quad (2.16)$$

Here, V_{gs} is the gate-to-source voltage, V_{bi} is the built-in voltage of the Schottky gate, and V_{po} is the pinch-off voltage, which is the voltage on the gate that completely closes the channel. It is approximately given by

$$V_{po} = \frac{qN_d h_c^2}{2\epsilon}. \quad (2.17)$$

Equation (2.16) is valid before current saturation is reached. Saturation is reached by one of two mechanisms.

1. Carriers reach saturated velocity in the region under the gate where the effective channel thickness is minimum. In this region channel resistance is the highest, and so is the horizontal electric field.
2. Saturation could be dominated by gate–drain reverse-bias, causing the channel to be completely pinched off.

Of course carrier velocity does have to saturate close to pinch-off due to the high electric field. The point to be made here is that when gate and source are at the same potential, pinching off the channel by increasing drain voltage does not reduce the current. It simply saturates the current. After the channel pinch-off point is reached, any further increase in the drain potential does not significantly change the gate depletion boundary on the source side. The pinch-off point along the channel still serves as an injection point for charge carriers into the drain side of the channel, and the current remains relatively unaffected by drain voltage.

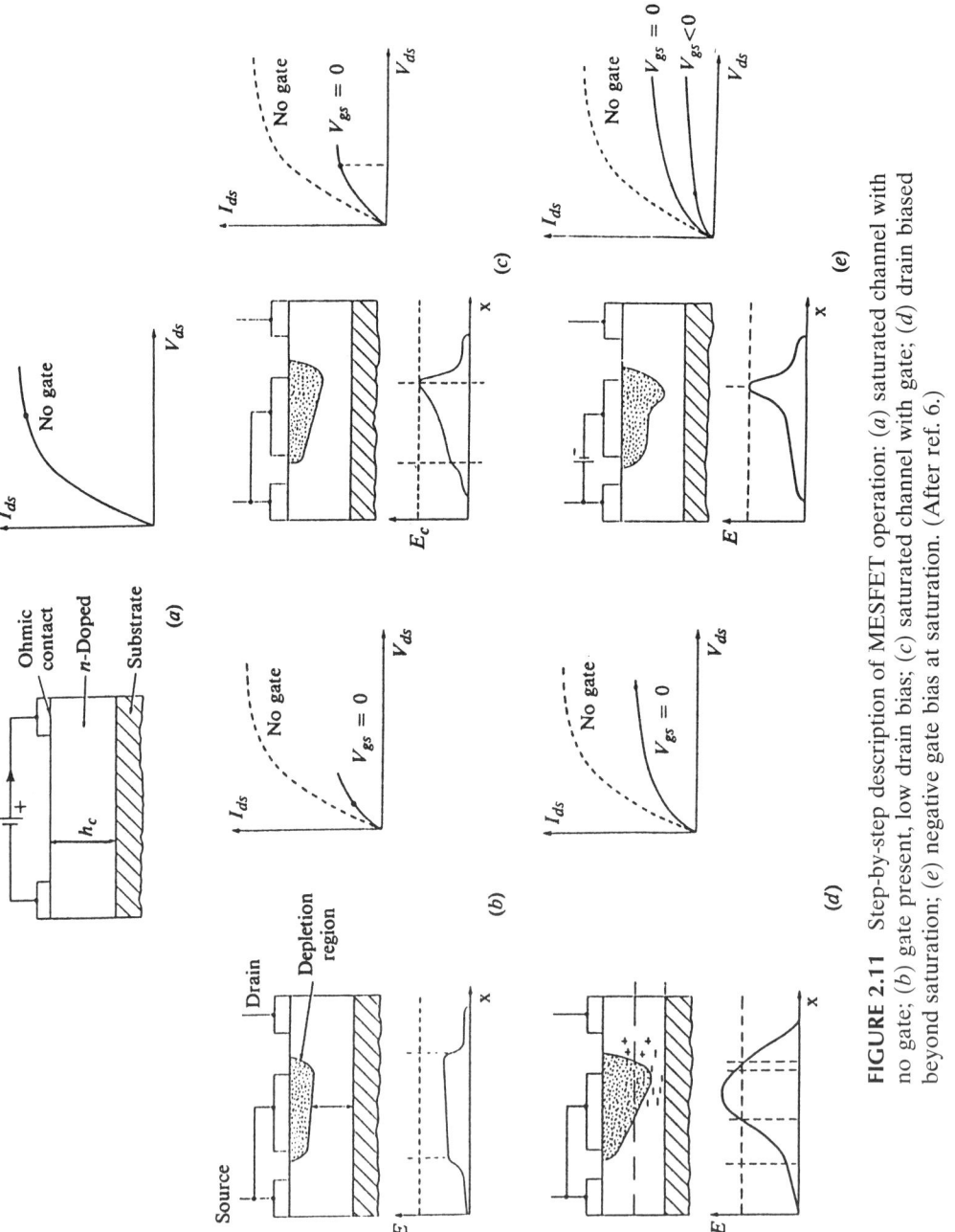

FIGURE 2.11 Step-by-step description of MESFET operation: (a) saturated channel with no gate; (b) gate present, low drain bias; (c) saturated channel with gate; (d) drain biased beyond saturation; (e) negative gate bias at saturation. (After ref. 6.)

In Eq. (2.16), when V_{ds} increases beyond the saturation voltage V_s, the drain current does not increase with drain voltage any longer. It is therefore possible to substitute V_s for V_{ds} in Eq. (2.16) and use it as an approximate expression in the saturation region:

$$I_{ds} = g_0 \left[V_s - \frac{2}{3} \frac{(V_s + V_{bi} - V_{gs})^{3/2} - (V_{bi} - V_{gs})^{3/2}}{V_{po}^{1/2}} \right]. \quad (2.18)$$

Beyond drain current saturation, the parameter that can significantly alter the current is gate potential. A negative potential on the gate broadens the gate depletion region toward the source contact, and the pinch-off region becomes wider. By this mechanism the gate potential controls the drain current, and for the particular gate potential V_{po}, it reduces the current to zero.

The rate of change of drain current with respect to gate voltage is referred to as the *transconductance* g_m, which is an important parameter in the operation of a MESFET:

$$g_m = \frac{\partial I_{ds}}{\partial V_{gs}} = g_0 \frac{(V_s + V_{bi} - V_{gs})^{1/2} - (V_{bi} - V_{gs})^{1/2}}{V_{po}^{1/2}}. \quad (2.19)$$

Since for most practical MESFET applications $V_s \ll (V_{bi} - V_g)$, the above equation may be approximated by

$$g_m \approx g_0 \frac{V_s}{2\sqrt{V_{po}(V_{bi} - V_g)}} = \left[\frac{qN_d \epsilon_0 \epsilon_r}{2(V_{bi} - V_{gs})} \right]^{1/2} v_s W_g. \quad (2.20)$$

Another important parameter that affects the performance of a MESFET, especially at high frequencies, is the gate-to-source capacitance C_{gs}. In most applications of the device this is the major fraction of the input capacitance. It was seen earlier in this section that a Schottky gate depletion layer in a transistor does not have a uniform shape. This fact complicates any rigorous calculation of the capacitance associated with it. Nevertheless, it is instructive to ignore the irregular shape of the depletion region and calculate its capacitance using the parallel-plate formula:

$$C_{gs} \approx \epsilon_0 \epsilon_r \frac{W_g L}{W} = \sqrt{\frac{qN_d \epsilon_0 \epsilon_r}{2(V_{bi} - V_{gs})}} (W_g L). \quad (2.21)$$

In the above expression W is the depletion layer thickness given by Eq. (2.9).

2.4.1 Small-Signal Equivalent Circuit Model

The main reasons for developing an equivalent circuit model for an active device are (1) device optimization by identifying problem areas and (2) scaling device size and modifying device layout in circuit design. There are also other reasons, some of which are less well known, for example, investigating high speed performance, beyond the frequencies at which the device can be characterized. This particular aspect of an equivalent circuit model is discussed in some detail in this section.

In common-source small-signal operation, the input signal places a variable voltage V_g across the gate depletion capacitor C_{gs}. This voltage controls the drain current I_{ds}. The simplest model of an FET consists of C_{gs}, a voltage-controlled current source, the input resistance R_i, and the output conductance G_{ds}. The presence of R_i is inherent in the structure of the device. The time constant $R_i C_{gs}$ represents a significant fraction of the time delay between the input and the output signals due to transit time under the gate. The portion of the delay that is not accounted for by this time constant, is explicitly assigned to the current source as a phase factor. This simple model is sometimes referred to as the intrinsic device model (Figure 2.12). Extrinsic elements need to be added to this model in order to accurately represent the operation of an FET. But before adding more detail to the model it is instructive to investigate some of the characteristics of the intrinsic device.

Speed Limitations

Current gain of the device at any given frequency is defined as

$$h_{21} = \left[\frac{\partial I_{ds}}{\partial I_{gs}}\right]_{V_{ds}=0}. \tag{2.22}$$

that is, the ratio of the short-circuited output current to the input current excluding bias currents. Current gain decreases with increasing frequency at the rate of 6 dB per octave, due to the fact that voltage across C_{gs} drops as an increasing fraction of the input voltage appears across R_i. The frequency at which the current gain becomes unity is called the *current gain cutoff frequency* f_T. For the intrinsic device it is given by

$$f_T = \frac{g_m}{2\pi C_{gs}}. \tag{2.23}$$

Under the approximation of a fully velocity saturated channel, Eqs. (2.20) and (2.21) may be used to find that $g_m/C_{gs} = v_s/L$. Therefore,

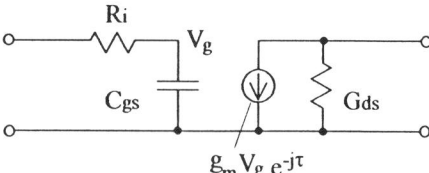

FIGURE 2.12 Intrinsic FET equivalent circuit model.

$$f_T \approx \frac{1}{2\pi} \frac{v_s}{L} = \frac{1}{2\pi\tau}. \qquad (2.24)$$

Since τ is the electron transit time under the gate, f_T is sometimes referred to as the transit time cutoff frequency.

With this simple model it can already be seen that (1) shorter gate devices operate at higher speeds and (2) the speed of the device is independent of gate width, since this parameter changes g_m and C_{gs} at the same rate. Both of these statements are valid as long as the gate length remains much longer than the channel thickness and the parasitic element associated with gate metallization resistance is negligible.

Example Consider an intrinsic FET with the following parameters:

$$C_{gs} = 0.5 \text{ pF}, \qquad R_i = 6\,\Omega, \qquad g_m = 80 \text{ mS}, \qquad G_{ds} = 0.002 \text{ S}.$$

(a) For a load resistance $R_L = 50\,\Omega$ calculate the voltage gain as a function of frequency. (b) Compare the voltage gains at dc and at $f = f_T$.

Solution The voltage across the gate capacitance is

$$V_g = \frac{V_{in}}{1 + j\omega R_i C_{gs}}.$$

Output voltage is a consequence of the current $g_m V_g$ through the load combination, $R_L/(1 + R_L G_{ds})$:

$$V_{out} = \frac{g_m V_{in}}{1 + j\omega R_i C_{gs}} \frac{R_L}{1 + R_L G_{ds}}.$$

For the given values, the voltage gain is

$$\frac{V_{out}}{V_{in}} = \frac{3.6}{1 + j(1.88)10^{-11}f}.$$

At low frequencies, the denominator is unity and the voltage gain is 3.6. The value of f_T is calculated from Eq. (2.23) to be 25.5 GHz. At this frequency, the voltage gain is

$$\frac{V_{out}}{V_{in}} = \frac{3.6}{1 + j(0.48)},$$

whose magnitude is 3.2. ∎

Another measure of high speed performance is the *maximum frequency of oscillation* f_{max}. This parameter is defined as the highest frequency at which the device is capable of providing power gain without any added external feedback. An equivalent definition of f_{max} is the frequency at which *maximum available gain* (MAG) becomes zero. Maximum available gain is the gain achieved by simultaneous conjugate matching of both input and output when the device is stable. For the intrinsic device $f_{\max}$ is given by

$$f_{\max} = \frac{f_T}{2\sqrt{R_i G_{ds}}}. \tag{2.25}$$

In microwave MESFETs, $R_i G_{ds} \ll 1$, and therefore, maximum frequency of oscillation is normally higher than current gain cutoff frequency.

MAG is a parameter that can be defined for any linear two-port device with gain regardless of its structure. A linear two-port system is completely defined by one of the 2×2 complex matrices known as impedance, admittance, scattering, hybrid, and so on. For an FET, in terms of scattering matrix, MAG is given by

$$\text{MAG} = \left|\frac{S_{21}}{S_{12}}\right| \left(k - \sqrt{k^2 - 1}\right), \tag{2.26}$$

where k is the stability factor, defined as

$$k = \frac{1 + |S_{11}S_{22} - S_{12}S_{21}|^2 - |S_{11}|^2 - |S_{22}|^2}{2|S_{12}S_{21}|}. \tag{2.27}$$

The MAG is defined only when $k > 1$. For $k < 1$, simultaneous conjugate matching may not be possible without the outbreak of oscillation. At frequencies where $k < 1$ the device is called *conditionally stable*. For a conditionally stable device, another parameter known as *maximum stable gain* (MSG) is used, defined as

$$\text{MSG} = \frac{S_{21}}{S_{12}}. \tag{2.28}$$

The main appeal of MSG is the relative ease with which it can be measured. It is, however, not as useful a parameter as MAG. Physically, MSG is the maximum gain the device is capable of providing once enough loss has been added to make it unconditionally stable. Since this is usually not the way conditionally stable devices are used in amplifiers, the definition is slightly removed from reality.

Maximum stable gain as a function of frequency drops at the rate of 3 dB/octave for the intrinsic FET. For every FET there is always a frequency known as f_k above which the device is unconditionally stable (i.e., $k > 1$). At f_k the value of the stability factor is equal to unity and MAG = MSG. Figure 2.13 shows a typical plot of available gain versus frequency for an actual MESFET as well as for the intrinsic MESFET. The difference between the two is the contribution of the parasitic elements such as contact pad capacitances, bond wire inductances, and the resistances of metal contacts.

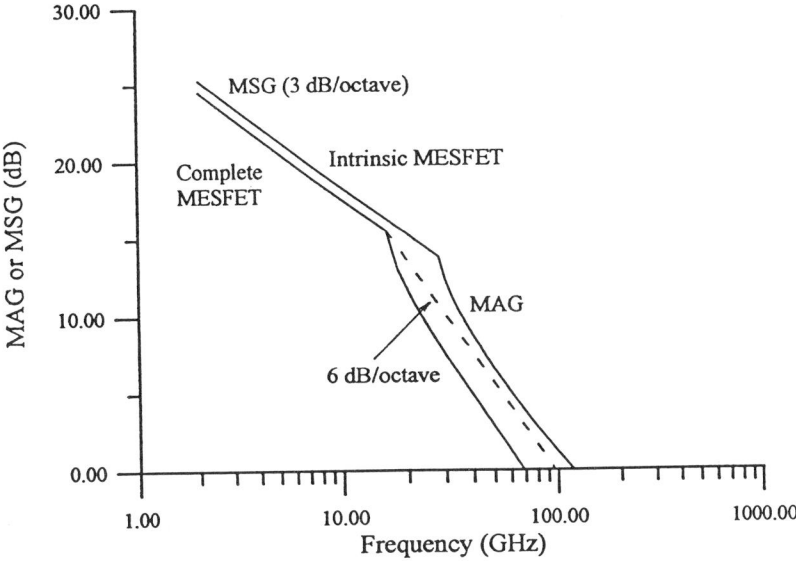

FIGURE 2.13 Available gain vs. frequency for MESFET as well as for its intrinsic segment (C_{dg} added to intrinsic model to assure finite S_{12}).

Figure 2.14 shows the equivalent circuit model of a MESFET with all the parasitic elements included.

In dealing with high speed devices, it frequently happens that f_T and f_{max} of a device fall above the highest frequency at which accurate measurements can be made. In order to estimate maximum frequency of oscillation f_{max} for these devices it is customary to use the very rough rule of extrapolating MSG at 3 dB/octave to the estimated f_k frequency and then continuing at 6 dB/octave from that point on. This, as was seen earlier, is not a dependable method. A better estimate of f_{max} at the time of measurement is obtained from the expression [7]

$$f_{max} = f \frac{\text{MSG}}{\sqrt{2(\text{MSG})k - 1}}. \qquad (2.29)$$

This is based on the assumption that MSG is inversely proportional to ω, and k is proportional to ω. This expression correctly predicts that MAG falls faster than 6 dB/octave immediately after f_k and then stabilizes to 6 dB/octave. It is possible depending on the magnitude of parasitic elements such as drain-to-source capacitance and drain resistance to cause k to grow faster than proportional to ω. In such cases MAG continues to fall faster than 6 dB/octave and f_{max} is a lot lower than predicted. The most accurate estimate of f_{max} is obtained from the complete device model.

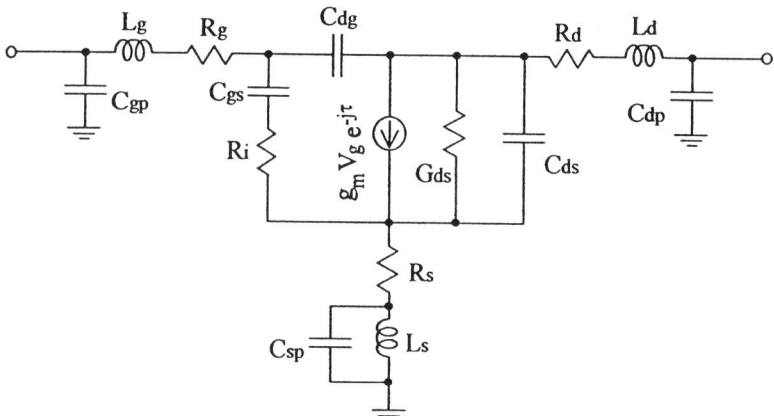

FIGURE 2.14 More complete small-signal equivalent circuit model of MESFET.

2.4.2 Noise Performance

Thermal noise dominates the noise characteristics of microwave MESFETs under most operating conditions. This section deals with this type of noise as it relates to the performance of FETs and related devices. At lower frequencies (below 100 MHz) the $1/f$ noise contribution also becomes significant for FETs. This effect is not discussed here [8]. (Appendix B gives the definitions and derivations of some basic noise parameters.)

The noise figure of a linear two-port gain element is a function of the admittance presented to the input of the device, or the source admittance Y_s. For a particular value of source admittance known as the optimum admittance, Y_{opt}, the *minimum noise figure* F_{min} of the device is achieved. In low noise circuits the matching network that is needed to equalize the gain and to reduce the reflections should also perform *noise matching*. In other words, a device is said to be noise matched if the admittance presented to its input is close to Y_{opt}.

Noise figure as a function of input admittance has a parabolic dependence on deviation from Y_{opt}:

$$F = F_{min} + \frac{R_n}{Re(Y_s)} |Y_s - Y_{opt}|^2 . \qquad (2.30)$$

In this equation, R_n is a characteristic of the device known as the *noise resistance*. As can be seen, the noise performance of the device is completely determined by F_{min}, R_n, and Y_{opt}. Sine Y_{opt} is a complex number, these are called the four noise parameters of the device. For the analysis of noise, we take the intrinsic MESFET model of Figure 2.12 and add two more elements that are significant contributors to device noise, namely gate metallization resistance R_g and source resistance R_s. Using this small-signal model for the FET, the noise parameters can be expressed as follows [9]:

$$F_{\min} = 1 + k_1 \frac{f}{f_T} g_m \sqrt{R_g + R_s}, \tag{2.31}$$

$$R_n = \frac{k_2}{g_m^2}, \tag{2.32}$$

$$R_{\text{opt}} = k_3 \left(\frac{1}{4g_m} + R_g + R_s \right), \tag{2.33}$$

$$X_{\text{opt}} = \frac{k_4}{g_m} \frac{f_T}{f}, \tag{2.34}$$

where $f_T = g_m/2\pi C_{gs}$ and $R_{\text{opt}} + jX_{\text{opt}} = Z_{\text{opt}} = 1/Y_{\text{opt}}$. The parameters $k_1, \ldots, k_4$ are numbers that are determined empirically, and vary with device type, but remain relatively invariant with changes in physical dimensions of the device. Typical k numbers for FETs are

$$k_1 = 2.5 \times 10^{-3}, \quad k_2 = 0.03, \quad k_3 = 2.2, \quad k_4 = 1000. \tag{2.35}$$

The above equations give guidelines for low noise device design and present a rough estimate of the frequency dependence of the noise parameters.

Besides a small $F_{\min}$ it is desirable for a device to have a small value of R_n. Small R_n makes accurate noise matching less critical when designing a low noise circuit. From Eqs. (2.31) and (2.32) it can be seen that reducing C_{gs} and the resistances of the gate and the source will reduce the noise figure and R_n of the device. One way of reducing C_{gs} is by shortening the gate length L_g. But a shorter gate has a higher gate metal resistance R_g. Some manufacturers use a mushroom metallization profile in order to keep R_g from rising. This is one approach to designing a low noise MESFET.

2.4.3 Short-Gate Effects

Equation (2.20) indicates that the transconductance of an FET is independent of its gate length. The model based on which Eq. (2.20) was derived is known as the gradual-channel model. One key assumption in that model is that the gate length is much larger than the channel thickness. As the gate length is reduced to achieve higher speed, if the channel thickness is kept constant, a regime is reached where the gate length and the channel thickness have similar dimensions. It has been observed that under these circumstances the efficiency of the gate potential in modulating the source-to-drain current is reduced [10]. Transconductance is degraded and drain current fails to saturate. Other adverse short-gate effects include the strong dependence of threshold voltage on channel length and biasing conditions. The analysis of carrier dynamics in the channel is more involved in this regime since the potential distribution is truly two dimensional.

It is generally accepted in device design that short-gate effects are to be avoided. Therefore it is necessary to accompany any gate length reduction with an appropriate channel thickness reduction such that channel thickness is always small compared with the gate length.

2.4.4 Power–Speed Trade-offs

Current gain cutoff frequency f_T serves as a simple measure of the speed of an FET. Although by itself, f_T does not correlate with the usability of the device at high frequencies, it is nevertheless a useful number for designing high speed devices. By definition, an increase in f_T means an increase in transconductance per unit gate capacitance. Gate length reduction increases f_T linearly as long as the gate length remains much larger than channel thickness, avoiding short-gate effects. A thinner channel means sacrificing the power handling capability of the FET, because the thinner channel carries less saturated current. This problem can be circumvented by increased semiconductor doping, which raises the current level. Higher doping increases both g_m and C_{gs} proportional to $\sqrt{N_d}$, leaving f_T unaffected. The negative side effect of high doping is that both mobility and saturated velocity of carriers are reduced due to increased scattering. Remembering that f_T is proportional to carrier transit time under the gate, this leads to a reduced f_T. High doping also reduces the breakdown voltage. If the breakdown voltage is too low, the device cannot be pinched off, and its operation is very inefficient.

Another way by which the total drain current can be increased is by widening the FET or increasing the gate periphery. To first order, f_T remains unchanged. There are however other serious side effects to this approach. Gate metal resistance increases with the width of the gate. A mushroom gate profile or an interdigitated gate layout is needed to keep gate resistance under control. But eventually, very wide gates result in higher values of other parasitics such as pad capacitance and reduced gate-to-drain isolation.

From the foregoing arguments it may be concluded that varying design parameters of a MESFET in order to increase its f_T will lead to a lower power handling capability either directly or indirectly. The reverse of this argument also holds: Lowered f_T not only makes the device less suitable for high frequency operation, but it also results in lower gain, because available power gain is a function of f_T/f. At any operation frequency, increasing the power handling capability of the device tends to reduce the available gain through the reduction of f_T.

2.5 HIGH ELECTRON MOBILITY TRANSISTORS

Based on the trade-offs presented in the previous section, some authors in the early 1970s predicted the maximum operation frequency of a "useful" GaAs FET to be about 30 GHz. Since then, GaAs MESFETs have been fabricated with respectable performance beyond 60 GHz. Nevertheless the trade-offs between power, speed, gain, and noise remain the same, and these high speed devices have simply pushed the barriers further out by advanced fabrication techniques.

The high electron mobility transistor HEMT, also known as MODFET (modulation-doped FET), avoids some of the trade-offs mentioned above through a fundamentally different approach. The basic idea is to separate the charge carriers from the dopants. The immediate consequence of this scheme is that high doping will not lead to increased scattering and degraded electron transport. The charge carriers move in an undoped medium where mobility and saturated velocity are high. The dopant is placed in a different, higher bandgap material that can form a good Schottky barrier with the gate metal. Since two materials are involved, an extra degree of freedom is introduced in the design of the transistor, thus changing the trade-offs. The first successful HEMTs were the GaAs/AlGaAs devices that demonstrated higher gains and lower noise figures than were achievable by conventional GaAs FETs. Other HEMT structures employing InGaAs channels soon followed offering incremental advantages.

Presently, HEMTs fabricated on InP are creating new possibilities both in high frequency, low noise operation and in the promise of easier integration of high speed circuits with optical components operating at fiber-optic communication wavelengths.

2.5.1 GaAs/AlGaAs HEMTs

The atomic radius of aluminum is very close to that of gallium, and since they are in the same column of the periodic table, they can be interchanged in some compound crystals such as GaAs. Fractional replacement of Ga with Al in GaAs yields the semiconductor $Al_xGa_{1-x}As$, which has a lattice constant very close to that of GaAs for every value of x, and therefore can be grown epitaxially on GaAs without any appreciable induced strain. Aluminum arsenide has a bandgap of 2.16 eV, compared to 1.42 for gallium arsenide. The bandgap of AlGaAs lies between these two values depending on the fraction x. A commonly used composition of AlGaAs is $Al_{0.3}Ga_{0.7}As$.[2]

The material structure of a GaAs/AlGaAs HEMT is shown in Figure 2.15. It consists of a doped layer of AlGaAs over an undoped GaAs channel. In order to analyze the operation of this device, it is necessary to investigate the behavior of charge carriers at the GaAs/AlGaAs interface. First, consider the simpler case where the doped and the undoped regions are made of the same semiconductor with an abrupt transition. This situation was described earlier in this chapter as a one-sided p–n junction. Carriers from the doped region diffuse into the undoped region, generating a depletion layer in the doped material whose width is given by Eqs. (2.1) and (2.2). The diffusion of charge equalizes the Fermi levels, and the built-in potential bends the conduction and

---

[2]Higher fractions of aluminum in epitaxial AlGaAs tends to generate undesirable charge trapping centers. Also, AlGaAs becomes an indirect bandgap material for $x > 0.3$.

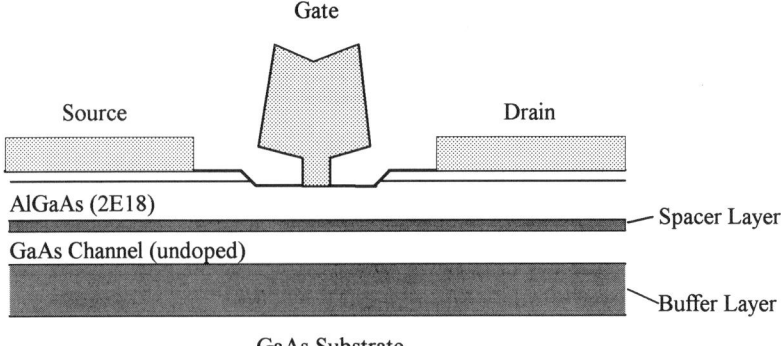

FIGURE 2.15 Material structure used in GaAs/AlGaAs HEMT.

valence bands such that they are continuous across the junction. A different effect is expected when the doped and the undoped materials have different bandgaps. The band bending caused by the built-in potential is not sufficient to assure the continuity of both bands. The situation is shown in Figure 2.16. The difference in the bandgaps shows as discontinuities in both the valence band and the conduction band. The discontinuity in the conduction band generates a potential well that can trap the electrons in a narrow region close to the junction. Electrons trapped in this potential well are referred to as the *two-dimensional electron gas*.

Electron density n_s in the two-dimensional electron gas is a function of the doping level N_d of the AlGaAs, the conduction band discontinuity ΔE_c, and the thickness of the spacer layer d_i. The following approximate expression for the electron density is often used [11]

$$n_s = \sqrt{\frac{2N_d \epsilon}{q}(\Delta E_c - E_f + (E_{cd} - E_f) + \delta) + N_d^2(d_i + \Delta d)^2} - N_d(d_i + \Delta d), \tag{2.36}$$

where Δd is approximately 80 Å in room temperature, $E_{cd} - E_f$ is the energy difference between the conduction band of AlGaAs and the Fermi level, away from the interface, and δ is a correction term on the order of 0.05 eV to bandgap discontinuity due to band bending. Equation (2.36) is plotted in Figure 2.17 and shows the interface carrier density as a function of doping level for three different spacer layer thicknesses. The presence of the spacer layer reduces the carrier density. But its incorporation is necessary in order to achieve high electron mobility in the two-dimensional electron gas. Despite the fact that electrons and donor sites are physically separated at the junction, their close proximity still gives rise to what is known as *Coulomb scattering*, thereby reducing their mobility. The spacer layer of undoped AlGaAs reduces Coulomb scattering and increases carrier mobility at the expense of

HIGH ELECTRON MOBILITY TRANSISTORS 35

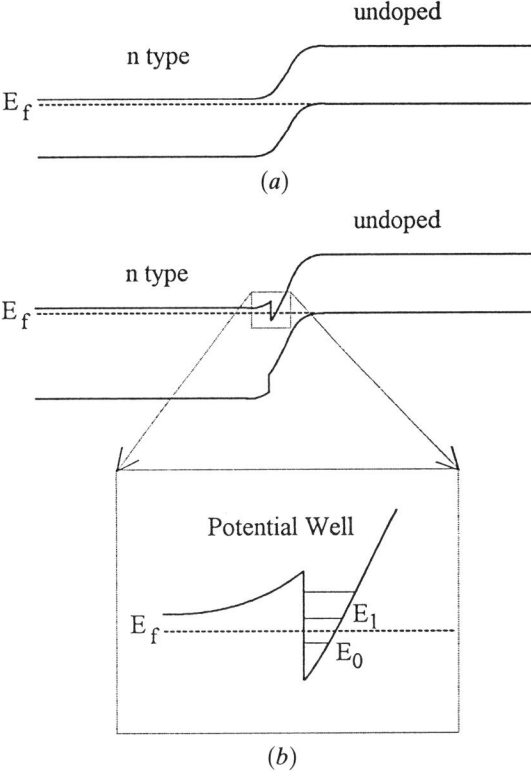

FIGURE 2.16 Potential well at a semiconductor heterojunction: (*a*) homojunction, (*b*) heterojunction.

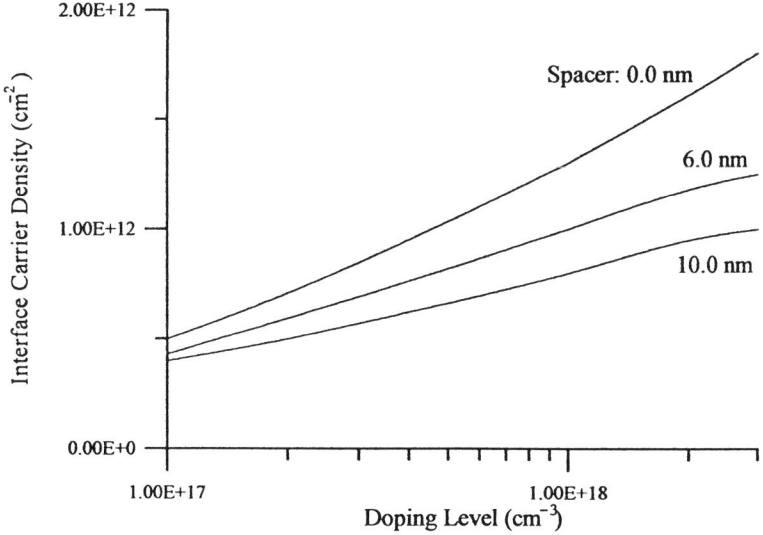

FIGURE 2.17 Interface carrier density as a function of AlGaAs doping for three different spacer layer thicknesses.

reduced electron density. The optimum spacer layer thickness depends on the application of the device, but in general it is on the order of 50 Å. Electron mobilities in AlGaAs HEMTs reach 7000 cm^2/V-s with corresponding two-dimensional carrier densities of approximately 1.2×10^{12} cm^{-2} [12].

The two-dimensional electron gas is expected to have a temperature-dependent thickness close to half of the de Broglie wavelength for electrons in the lattice. At room temperature, this dimension is approximately 150 Å in GaAs.

A typical 0.3-μm mushroom gate, low noise GaAs MESFET with an f_{min} of 2.0 dB at 20 GHz with an associated gain of 8 dB could be converted to an AlGaAs HEMT keeping its top metal structure the same. This conversion is likely to reduce f_{min} to 1.5 dB and increase the associated gain to 12 dB at the same frequency [13].

2.5.2 Other HEMT Structures

Soon after their introduction, GaAs/AlGaAs HEMTs proved to be versatile devices for high speed, low noise applications. Low noise HEMT amplifiers brought about the obsolescence of the cumbersome parametric amplifier. The concept of HEMT, however, does not stop with the GaAs/AlGaAs device. Other semiconductor materials are known to have much higher electron mobilities and saturated velocities than GaAs, making them desirable as HEMT channel materials. One such material is InGaAs. Superior electron transport properties of InGaAs were established by the early 1980s [14]. These properties are enhanced with an increased mole fraction of indium in this compound.

Incorporation of a small fraction of indium in the channel of a GaAs/AlGaAs HEMT has the twofold effect of increasing channel mobility as well as enhancing conduction band discontinuity at the heterojunction. The only adverse effect of the incorporation of indium is the resulting strain in the lattice. Unlike AlGaAs, which has a fixed lattice constant, the lattice constant of InGaAs increases with higher fractions of indium. Due to the lattice mismatch, this type of HEMT is known as *pseudomorphic*. The fraction of indium in In$_x$Ga$_{1-x}$As should not exceed approximately 0.35 in GaAs pseudomorphic HEMTs in order to avoid excessive lattice mismatch. Excessive mismatch or thick growth of mismatched layers can cause dislocations in the lattice that impede carrier transport.

Figure 2.18 shows the lattice constants and energy gaps of a selected number of semiconductor materials. Lines joining binary semiconductors represent ternary semiconductors with variable degrees of substitution of two elements. For example, the line joining GaAs and AlAs represents AlGaAs. Note that this line is almost vertical, meaning that an increased aluminum mole fraction does not change the lattice constant appreciably. On the other hand, with InGaAs (line joining GaAs and InAs) the lattice constant is a strong function of the indium mole fraction. It is important to observe that for

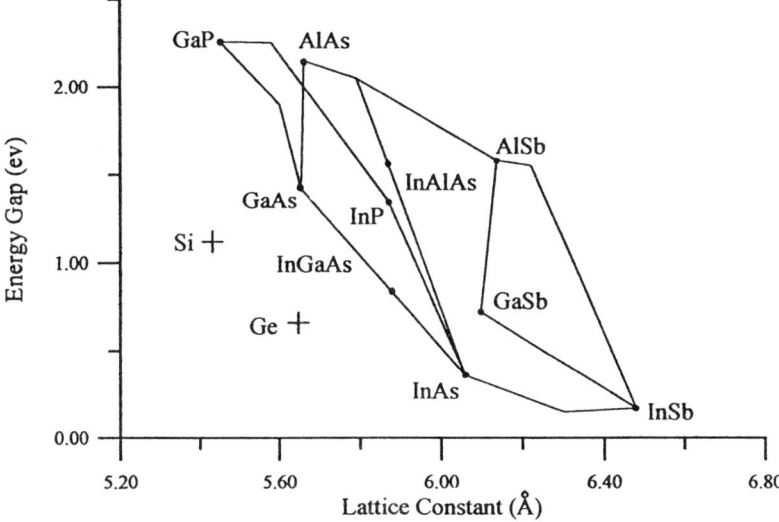

FIGURE 2.18 Lattice constants and energy gaps of semiconductors.

one particular indium fraction (53%), InGaAs can be made lattice matched to InP.

It is possible to take advantage of the improved properties of higher indium content InGaAs if the substrate material of the HEMT is not GaAs, but rather InP, which is a readily available substrate material. While electronic properties of InP are very similar to GaAs, this material is not in common use for electronic circuits. The main reason is that an acceptable Schottky contact to InP is not available; therefore the performance of InP MESFETs fall short of their GaAs counterparts. This is of no concern when the active device is a HEMT. The Schottky contact is made to a high bandgap material lattice matched to the substrate. When HEMTs are made on GaAs, the Schottky contact is made to AlGaAs. For HEMTs made on InP, $In_{0.52}Al_{0.48}As$ lattice matched to InP is available to support the Schottky barrier (on the line joining InAs and AlAs) [15]. The bandgap of this material is 1.45 eV. The $In_{0.53}Ga_{0.47}As$ has a bandgap of 0.75 eV, with an electron mobility exceeding 12,000 cm^2/V-s at room temperature [6, 8]. The material structure of an InGaAs/InAlAs HEMT on InP is shown in Figure 2.19. This device has achieved some of the highest speeds and lowest noise figures reported to date for any transistor [7, 16]. Pseudomorphic InP HEMTs have also been fabricated using still higher mole fractions of indium (up to 65%). Higher sheet carrier densities and higher mobilities have been observed [17].

The InP-based HEMT is particularly attractive when optoelectronic applications are considered. In particular (see Chapter 3), a popular type of semiconductor laser used in communications is the InGaAsP laser fabricated on InP substrate. The availability of high performance electronic circuits on

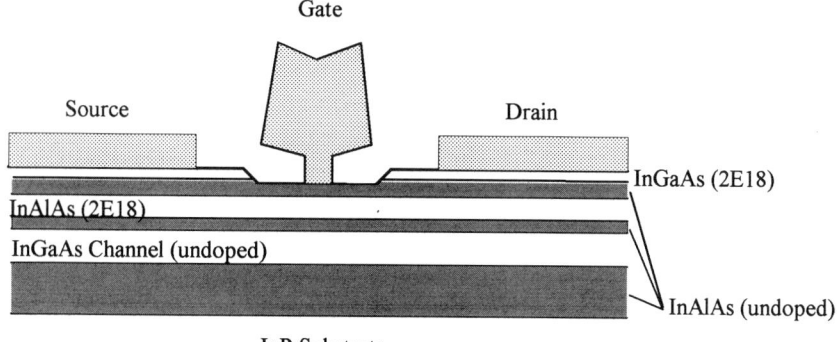

FIGURE 2.19 Material structure of InGaAs/InAlAs HEMT on InP substrate.

InP opens the way for monolithic integration of optical components and electronic circuits for use in the communications industry.

The idea of optimizing HEMT channel material, Schottky contact material, and substrate material independent of each other in lattice-matched or strained layer devices is expanding rapidly into other material groups. Some such heterostructures being studied currently are Si–SiGe [19], GaInP lattice matched to GaAs [22], and AlN lattice matched to SiC. The use of these material combinations are being explored in HEMTs, bipolar transistors, and optoelectronic devices. It is likely that a large number of such structures will lead to successful devices for diverging applications, thus eroding the domination of particular semiconductor materials over all application areas.

2.6 HETEROJUNCTION BIPOLAR TRANSISTORS

Before the advent of GaAs MESFET, the transistor used in both analog and digital high speed circuits used to be bipolar. Silicon bipolar transistors continue to be used in circuits operating in the low microwave frequencies.

A bipolar transistor consists of a thin doped layer of semiconductor known as a base sandwiched between an "emitter" and a "collector," both doped in the opposite sense to the base (Figure 2.20). The transistor is therefore either *npn* or *pnp*, depending on the doping. Let us focus attention on the *npn* transistor. In normal operation, the emitter–base junction is forward biased. Electrons flow from the emitter to the base. The very thin base region allows these electrons to permeate through, to the reverse-biased base–collector junction, where they flow as the collector current I_c. This current is controlled by the base-to-emitter potential, which raises or lowers the barrier to electron injection. (See the band diagram in Figure 2.21.) Base current I_b is a parasitic current caused by hole injection from base to emitter across the junction barrier. The ratio I_c/I_b is the current gain of the transistor. In order to achieve

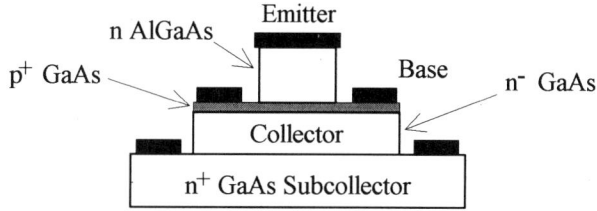

FIGURE 2.20 Material structure of a heterojunction bipolar transistor.

high current gain, base current is kept low in the conventional transistor by ensuring that emitter doping level is much higher than the base [20]. Let us call this "requirement A" for future reference.

As is the case with the FET, the current gain cutoff frequency of the bipolar transistor is limited by electron transit time through the controlling electrode region, which is the base in this case:

$$f_T \approx \frac{1}{2\pi} \frac{v_s}{t_b}, \qquad (2.37)$$

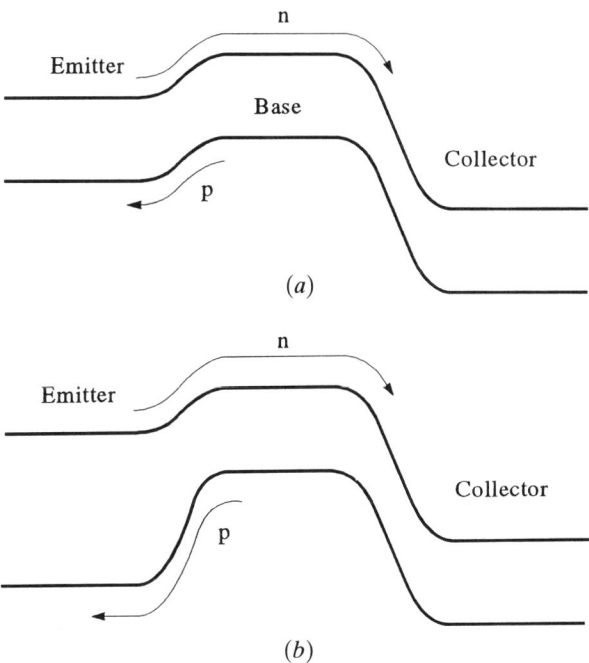

FIGURE 2.21 Energy band diagrams of bipolar transistors under external bias: (*a*) homojunction; (*b*) graded heterojunction.

where t_b is the thickness of the base layer. In high frequency transistors, electron transit time through the base is reduced by shrinking the thickness of the base. However, when base thickness is reduced in high speed devices, it is desirable to increase its doping to limit the ohmic resistance of the base, R_b. This is inconsistent with requirement A in a conventional homojunction transistor.

Another factor limiting the speed of homojuntion bipolars is high base-to-emitter capacitance due to high emitter doping needed to fulfill requirement A. These conflicting requirements limit the high speed operation capability of the silicon bipolar transistor. Recently, however, with advanced material growth techniques, the fabrication of *heterojunction bipolar transistors* (HBTs) has become feasible. As will be seen in the following discussion, HBTs are able to circumvent the conflicting requirements for high speed operation mentioned above [18]. This has rekindled the interest of high speed circuit designers in working with bipolar transistors.

Figure 2.20 shows the material structure of a typical HBT whose energy band diagram under external bias is shown in Figure 2.21. The heterojunction is graded in order to avoid the formation of a potential well. In the heterojunction transistor, due to the unequal bandgaps, the valence band barrier between base and emitter is enlarged without affecting the conduction band barrier. Consequently, electron diffusion from the emitter into the base is easier than hole injection from the base into the emitter. The device is thus capable of current gain regardless of the doping levels of the base and the emitter. Requirement A is relaxed. This adds a degree of freedom to the design of the transistor and allows the high doping of the thin base layer for low R_b as well as the low doping of the emitter for low junction capacitance. A modern GaAs/AlGaAs HBT can achieve both an f_T and an f_{max} of greater than 100 GHz [21].

2.7 BULK NEGATIVE DIFFERENTIAL RESISTIVITY DEVICES

In Section 2.4, the generation of negative resistance by avalanche delay in semiconductor junction diodes was discussed. Achieving negative resistance does not need to rely on a semiconductor junction. An effect known as negative differential resistivity (NDR), which is associated with a number of different instability mechanisms in semiconductors, can be used to fabricate a negative-resistance device. One such instability, referred to as the *Gunn effect* (or the *transferred electron effect*), is used extensively in one-port devices for oscillator and reflection-type amplifier applications. Gunn devices operate at frequencies ranging from low microwaves to beyond 100 GHz.

Negative differential resistivity is defined as $\Delta V/\Delta I < 0$ across a device. This can occur when an increase in the applied potential leads to a decrease in current, or vice versa. In a number of compound semiconductors, in particular GaAs and InP, the mobility of charge carriers may be reduced with increased applied electric field. This effect is caused by charge transfer from a high mobility, lower energy valley in the conduction band to a higher energy, low mobility valley. This so called Γ–L transition (in GaAs) gives rise to negative differential resistivity. Figure 2.22 shows electron drift velocity as a function of applied field for GaAs and InP. The electric field value called E_T corresponds to the peak electron velocity in the material. For applied electric fields greater than E_T, a further increase in the electric field leads to a reduction in the current. This can lead to ac signal generation as described below.

The transferred electron device (TED) is commonly made of a thin, n-doped semiconductor with two ohmic contact terminals. This device has two modes of oscillation. One that is mainly controlled by the geometry of the device and another that is controlled by the external circuit. First consider the device in an external resonant circuit with its equivalent circuit model shown in Figure 2.23. The TED is represented by the negative-conductance element G_T. Here, G_0 represents the power loss in the resonant circuit plus the power delivered to the external load. At stable circuit oscillation $G_0 = G_T$, and the oscillation frequency $\omega_c = 1/\sqrt{LC}$ is set mainly by the resonant circuit.

In order for the TED to demonstrate negative resistance, it is biased at a voltage V_0 higher than V_T. Figure 2.24 shows the voltage and current swings across the device. The power output of the device at the fundamental frequency is

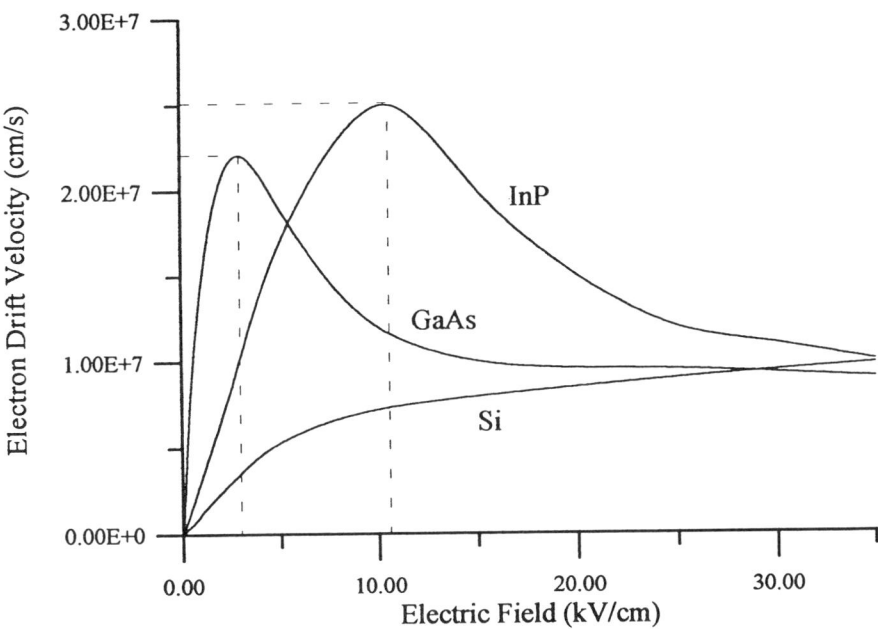

FIGURE 2.22 Electron drift velocity as function of applied electric field for GaAs and InP.

42 SOLID STATE ELECTRONIC ACTIVE DEVICES

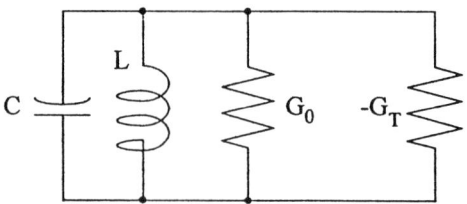

FIGURE 2.23 Equivalent model of NDR device in external resonant circuit.

$$P_0 = \frac{1}{2}\left(\frac{V_M - V_T}{2}\right)\left(\frac{I_T - I_V}{2}\right). \tag{2.38}$$

The dc power used by the device is $P_{dc} = I_0 V_0$. The efficiency of the device is therefore given by

$$\frac{P_0}{P_{dc}} = \frac{1}{8V_0 I_0}(V_M - V_T)(I_T - I_V). \tag{2.39}$$

For any given bias value, the highest efficiency is achieved when voltage and current swings are maximum.

Example Compare the efficiencies of GaAs and InP TEDs.

Solution Equation (2.39) may be rearranged as

$$\frac{P_0}{P_{dc}} = \frac{1}{8}\left(\frac{V_M - V_T}{V_0}\right)\left(\frac{I_T - I_V}{I_0}\right).$$

The factor involving voltages has a maximum value of 2 for $V_M \gg V_T$. The

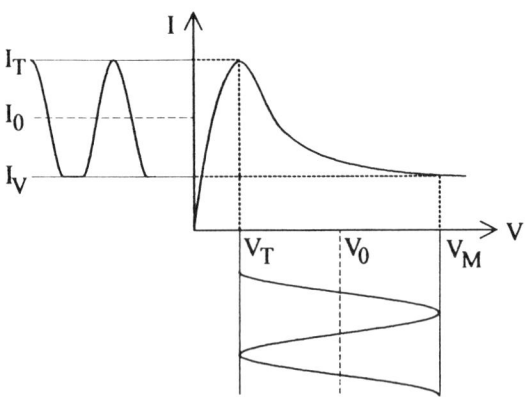

FIGURE 2.24 ac signal generation by NDR.

ratios of currents are the same as the ratios of electron drift velocities. So, the factor involving currents is simply the maximum velocity variation divided by its mean value. Maximum and minimum velocities in the NDR region are v_T values of 2.5×10^7 cm/s for InP and 2.2×10^7 cm/s for GaAs and v_V values of 0.5×10^7 cm/s for InP and 0.9×10^7 cm/s for GaAs. Inserting these values in the above equation we find approximate efficiencies of 20% for GaAs and 33% for InP.

For a more detailed analysis of the transferred electron effect in different semiconductors see reference [23]. ∎

The mode of operation just described is referred to as the *limited space charge* (LSA) mode, where the operation frequency is controlled by the external circuit. The transferred electron device can also oscillate independent of an external circuit by forming moving charge-dipole regions known as Gunn domains. If a dc voltage $V_0 > V_T$ is placed on the device, a dc current I_0 is expected to flow. However, due to the negative resistivity of the device, small current perturbations can grow, as shown in Figure 2.25. The formation of a small moving dipole generates a local high electric field region with slow moving electrons. Electrons at the leading edge of the dipole move faster due to the lower electric field. These electrons move away from the dipole and enhance its depleted leading edge. At the same time, electrons behind the dipole region catch up with its trailing edge and increase the dipole strength. The traveling-dipole region formed this way is called the Gunn domain. A fully developed Gunn domain travels at the velocity v_s across the device. The voltage drop across the Gunn domain keeps other such domains from forming. When the Gunn domain reaches the end of the device, the process starts over again. The frequency of oscillation in this mode is determined by the length L of the device and is approximated by $f = v_s/L$. In most applications of the TED, the preferred operation mode is the LSA mode due to its higher stability and efficiency.

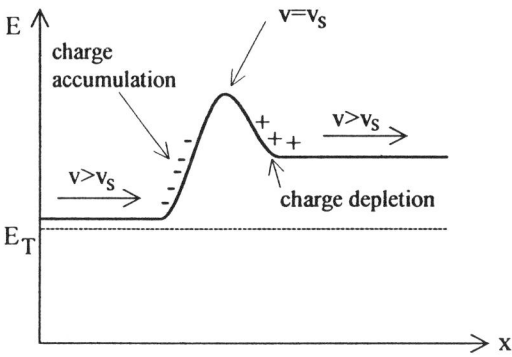

FIGURE 2.25 Gunn domain formation in a TED.

2.8 TUNNEL DIODES

Tunnel diodes achieve NDR by the mechanism of carrier tunneling through a quantum-mechanical barrier. The barrier is either a *p–n* junction, a quantum well, or a layer of dielectric.

Tunneling is the mechanism by which electrons penetrate through a "thin" potential barrier. The probability of tunneling reduces exponentially with the thickness of the barrier. The time it takes for an electron to traverse a barrier is not limited by its drift velocity in the medium; rather it is a process that is a function of electron momentum and, in general, is much faster than electron drift through the same distance. For this reason, tunnel devices are inherently capable of extremely high frequency operation (up to THz).

In its simplest form, the tunnel diode consists of a highly doped (degenerate)[3] *p–n* junction. Recall that the width of the depletion region shrinks with higher doping levels. Electron tunneling becomes significant when the thickness of the depletion region is less than or on the order of 100 Å

Energy and momentum are conserved under the tunneling process. This requires that for an electron to tunnel through a barrier there should be an unoccupied energy state available on the other side. In an unbiased *p–n* junction where the Fermi levels of both sides are equal, no tunneling current flows. As the junction is forward biased, conduction electrons on the *n* side of the junction find unoccupied energy states in the valence band of the *p* side, and tunneling current increases (Figure 2.26). With higher forward biasing, tunneling current rises to a maximum. Eventually, as the entire conduction band of the *n*-doped material rises in energy above the valence band of the *p* side, tunneling current falls to zero. Meanwhile, the normal thermal current due to the forward biasing of the junction continues to increase exponentially. A typical current–voltage relationship for the tunnel diode is shown in Figure 2.27.

The device demonstrates negative differential resistance at bias voltages between V_p and V_v. Typically, V_p ranges from 0.1 to 0.2 V, and V_v falls between 0.2 and 0.7 V. An important device parameter that is a performance gauge for the tunnel diode (as well as other NDR devices) is the peak-to-valley ratio I_p/I_v. The valley current I_v is determined by the thermal forward-bias current that competes with the tunneling current. The peak current I_p, on the other hand, is a function of semiconductor doping level. The higher doping needed for high peak current has the negative side effect of increasing junction capacitance, which limits the operation speed of the device. Commonly encountered peak-to-valley ratios in conventional tunnel diodes range between 20 and 50.

Tunnel diodes can use other barriers besides a *p–n* junction for tunneling. With new epitaxial material growth techniques, diodes with quantum well

---

[3] Degenerate *n* doping brings the Fermi level into the conduction band. In degenerate *p* doping the Fermi level is in the valence band.

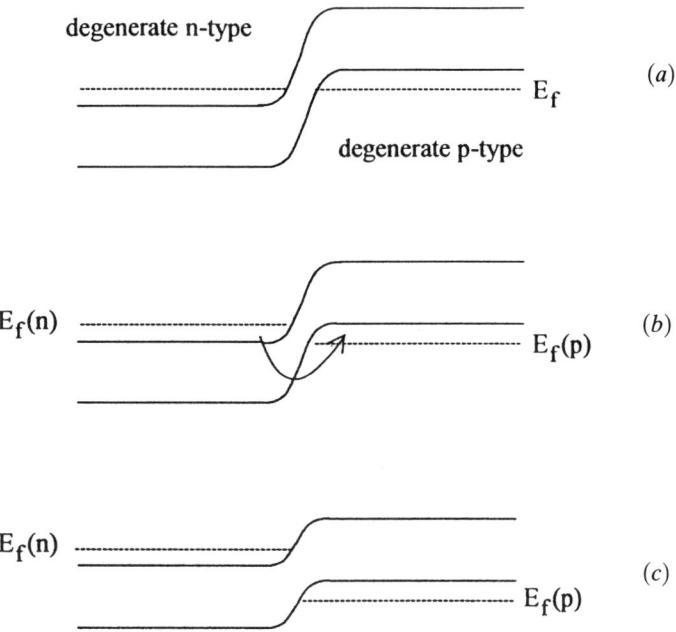

FIGURE 2.26 Band diagram of tunnel diode operation: (*a*) zero bias, (*b*) biased for maximum tunneling current, and (*c*) bias level too high for tunneling.

barriers have become practical. Particularly noteworthy is the quantum well resonant tunnel diode [24]. This device uses junctions between low bandgap and high bandgap materials as barriers for electrons to tunnel through. Furthermore, two such barriers in cascade are used to form a Fabry–Perot type resonant structure. The resonant nature of electron tunneling enhances the peak-to-valley ratio. Room temperature peak-to-valley ratios of over 100 have been demonstrated for these devices [25].

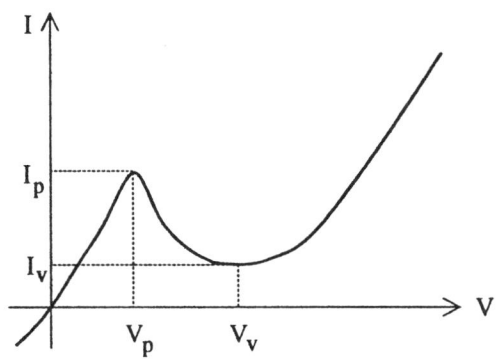

FIGURE 2.27 Current–voltage relationship of tunnel diode.

PROBLEMS

2.1 For a 10-cm-long silicon rod of 1.0 cm^2 cross section, (a) calculate the intrinsic resistance. (b) If every millionth silicon atom is substituted with phosphorus, what is the new resistance of the rod?

2.2 Compare the intrinsic room temperature resistivities of Si, GaAs, and InP. Find the resistivities of each when doped by donor impurities at the concentration of 10^{17}cm^{-3}.

2.3 Calculate the m–s junction capacitance C_j between GaAs doped at $(5)10^{17}$cm^{-3} and a rectangular aluminum pad of dimensions 2.0×100.0 μm.

2.4 Show that current gain as a function of frequency drops at the rate of 6 dB/octave for the intrinsic MESFET.

2.5 For a given MESFET the value of h_{21} is measured to be 13.5 at the frequency of 18 GHz. Estimate its f_t.

2.6 A conditionally stable FET with a k value of 0.6 at 18 GHz has $S_{21} = 2.1$ and $S_{12} = 0.02$. Estimate its value of $f_{\max}$.

REFERENCES

[1] S. M. Sze, *Physics of Semiconductor Devices*, 2nd ed., Wiley, New York, 1981.
[2] E. S. Yang, *Fundamentals of Semiconductor Devices*, McGraw-Hill, New York, 1978.
[3] S. A. Maas, *Microwave Mixers*, 2nd ed., Artech House, Boston, 1993.
[4] P. T. Parrish et al., "Printed Dipole Schottky Diode Millimeter Wave Antenna Array," *Millimeter Wave Technol.*, Vol. SPIE 337, pp. 49–52, 1982.
[5] R. S. Pengelly, *Microwave Field-Effect Transistors—Theory, Design and Applications*, Wiley, New York, 1982.
[6] C. A. Liechti, "Microwave Field-Effect Transistors—1976," *IEEE Trans. Microwave Theory Tech.*, Vol. MTT-24, No. 6, pp. 279–300, June 1976.
[7] M. Riaziat et al., "Highest Current Gain Cutoff Frequency with 0.08 micron Gate on InP," *Proceedings IEEE's Second International Conference on Indium Phosphide and Related Materials,* pp. 50–56, April 1990.
[8] A. van der Ziel, *Noise in Solid State Devices and Circuits*, Wiley, New York, 1986.
[9] H. Fukui, "Optimal Noise Figure of Microwave GaAs MESFETs," *IEEE Trans.*, Vol. ED-26, No. 7, pp. 1032–1037, May 1979.
[10] M. Reiser and P. Wolf, "Computer Study of Submicrometere FETs," *Electron. Lett.,* Vol. 8, No. 10, pp. 254–256, May 1972.
[11] M. Shur, *GaAs Devices and Circuits*, Plenum, New York, 1987.
[12] K. Lee, M. Shur, J. Klem, T. Drummond, and H. Morkoc, "Parallel Conduction Correction to Measured Mobility in AlGaAs Modulation Doped Layers," *Jpn. J. Appl. Phys.*, Vol. 23, No. 4, pp. L230–231, April 1984.

[13] C. Yuen, M. Riaziat, S. Bandy, and G. Zdasiuk, "Application of HEMT Devices to MMICs," *Microwave J.*, Vol. 31, No. 8, pp. 87–104, Aug. 1988.

[14] S. Bandy, C. Nishimoto, S. Hyder, and C. Hooper, "Saturation Velocity Determination for InGaAs Field Effect Transistors," *Appl. Phys. Lett.*, Vol. 38, No. 10, pp. 817–819, May 1981.

[15] C. M. Hanson and H. H. Wieded, "Critical Properties and Applications of InAlAs/InP," *J. Vac. Sci. Technol. B*, Vol. 5, No. 4, pp. 971–975, July 1987.

[16] M. Riaziat, I. C. Pao, C. Nishimoto, G. Zdasiuk, S. Bandy, and S. L. Weng, "HEMT Millimeter Wave Monolithic Amplifiers on InP," *Electron. Lett.*, Vol. 25, No. 20, pp. 1328–1329, Sept. 1989.

[17] G. I. Ng, D. Pavlidis, M. Tutt, J. E. Oh, and P. K. Bhattacharya, "Improved Strained HEMT Characteristics Using Double Heterojunction InGaAs/InAlAs Design," *Electron Dev. Lett.*, Vol. 10, No. 3, pp. 114–116, March 1989.

[18] H. Kroemer, "Heterojunction Bipolar Transistors and Integrated Circuits," *Proc. IEEE*, Vol. 70, No. 1, pp. 13–25, Jan. 1982.

[19] R. Taft, J. Plummer, and S. Iyer, "Demonstration of a *p*-channel BICFET in the GeSi/Si System," *IEEE Elect. Dev. Lett.*, Vol. 10, No. 1, pp. 14–16, Jan. 1989.

[20] N. Moll, "Heterojunction Bipolar Transistors—A Review," in F. Ali, I. Bahl, and A. Gupta (Eds.), *Microwave and Millimeter Wave Heterostructure Transistors and Their Applications*, Artech House, Norwood, MA, 1989.

[21] G. B. Gao, D. J. Roulston, and H. Morkoc, "Design Study of AlGaAs/GaAs HBTs," *IEEE Trans. Electron. Dev.*, Vol. 37, No. 5, pp. 1199–1208, May 1990.

[22] W. Liu et al., "Current Transport Mechanism in GaInP/GaAs Heterojunction Bipolar Transistors," *IEEE Trans. Electron Dev.*, Vol. 40, No. 8, pp. 1378–1383, Aug. 1993.

[23] B. K. Ridley, "Anatomy of the Transferred Electron Effect in III–V Semiconductors," *J. Appl. Phys.*, Vol. 48, No. 2, pp. 754–764, Feb. 1977.

[24] L. Chang, L. Esaki, and R. Tsu, "Resonant Tunneling in Semiconductor Double Barriers," *Appl. Phys. Lett.*, Vol. 24, No. 12, pp. 593–595, June 1974.

[25] D. Day, J. Lu, R. Yang, M. Sweeny, and J. Xu, "Demonstration of a DQW Resonant Interband Tunnel Diode with a 300 K Peak to Valley Ratio Over 100," Extended Abstracts, International Conf. on Solid State Devices and Materials, pp. 711–712, Tsukuba, Japan, 1992.

CHAPTER THREE

Optical Active Devices

In many high speed optical systems such as the optical communications link, the three main active devices used are lasers, detectors, and modulators. In this chapter each of these components is discussed briefly. If information is encoded on the light beam by direct modulation of the light source, issues of speed and natural resonances of the laser are significant. On the other hand, if the communication link uses modulators as distinct elements from lasers, the main requirements on the laser source are the stability, reproducibility of the wavelength, and modal purity of the output. In this chapter, the discussion of light sources includes light emitting diodes (LEDs), and semiconductor diode lasers. Under modulators, only the electro-optic types are considered due to the limited scope of the book. Finally, semiconductor detectors and their high speed issues are deliberated.

3.1 CARRIER RECOMBINATION IN SEMICONDUCTORS

In thermal equilibrium carrier concentrations in semiconductors are governed by Eq. (F.6). This equilibrium is established at any temperature between the thermal generation of electron–hole pairs and their recombination. There are two basic recombination processes: radiative and nonradiative. In *radiative recombination* energy is released through the emission of a photon. In nonradiative recombination the excess energy is delivered to lattice phonons in the form of thermal energy.

Since in the recombination process momentum is conserved, the probability of radiative recombination is maximum between electrons and holes having the same k vector. Figure 3.1 shows conduction and valence bands of two semiconductors. In one case, conduction band minimum and valence band maximum have the same k vector (momentum) associated with them. This type of semiconductor is called *direct bandgap* and lends itself to light emission applications. Gallium arsenide falls in this category. In the second

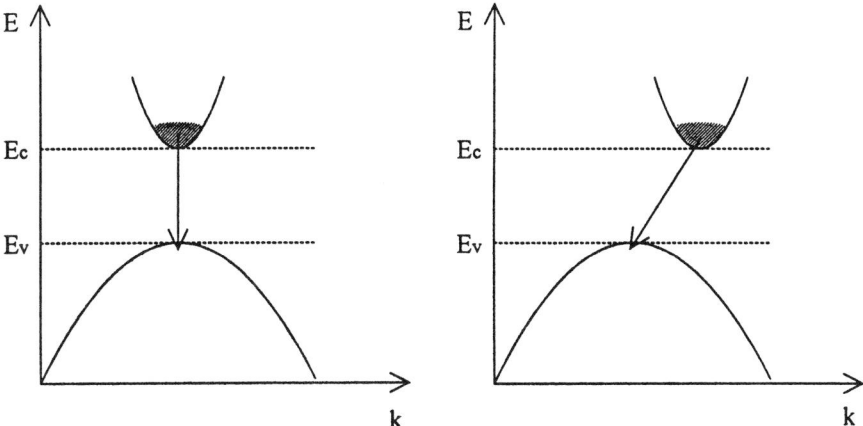

FIGURE 3.1 Energy versus momentum, k, for (a) a direct bandgap semiconductor, and (b) an indirect bandgap semiconductor.

case, the recombination of an electron from the bottom of the conduction band to valence band maximum involves a change in the k vector. This type of semiconductor is called *indirect bandgap*, as represented by silicon. An electron can recombine with a hole with a different momentum and still release a photon. In this case momentum conservation is ensured by interaction with lattice phonons. The probability of this process is very low due to the fact that it is a three-particle interaction. Silicon is therefore used for photon detection but not for its generation. In semiconductors with indirect bandgap it is more likely for electrons and holes to recombine through impurity *recombination centers*. This is called *indirect recombination* and occurs through sequential trapping of an electron and a hole by an impurity site. It is possible to increase the efficiency of radiative recombination in indirect bandgap semiconductors by artificially adding such recombination centers in the form of selected impurities. Examples of this technique are given in the next sections.

3.2 LIGHT EMITTING DIODES

In a forward-biased p–n junction, current conduction is done by minority carriers, where excess minority carriers are injected into both p and n regions. The increased minority carrier population leads to an increased recombination rate and therefore an increased photon emission. This form of emission is referred to as *spontaneous emission*. The probability of spontaneous emission is proportional to the density of excess carriers and the inverse of their lifetime τ. Carrier lifetime is the mean time for a charge carrier in the semiconductor to spontaneously recombine. It is inversely proportional to the majority carrier population (or the doping level N in most cases). The proportionality relationship is often written as

$$\tau = (B \cdot N)^{-1}, \qquad (3.1)$$

where B is the recombination constant whose value is 7.21×10^{-10} cm^3/s for GaAs and 1.79×10^{-15} cm^3/s for silicon. Note the five orders of magnitude difference in the value of B for a direct versus an indirect bandgap material. In an LED the recombination occurs in a region close to the junction within a distance given by $v_s \tau$, where v_s is the saturated velocity. More specifically, minority-carrier density as a function of distance x from the junction exponentially decays according to

$$\rho(x) = \rho(0) e^{-x/L_c}, \qquad (3.2)$$

where L_c is the characteristic diffusion length for the carriers and depends on temperature, mobility of the carriers, and the applied field. The maximum value of L_c is $v_s \tau$.

Example For a GaAs LED doped at $N = 10^{18}$ cm^{-3} carrier lifetime $\tau = 1.4$ ns according to Eq. (3.1). This corresponds to a maximum L_c of 140 μm if the saturated velocity is assumed to be 1.0×10^7 cm/s. ∎

A large fraction of photons generated in this fashion do not leave the semiconductor material. The main reason is that the wavelength of the emitted photon is close to the peak absorption wavelength of the semiconductor. In other words there is a high probability that the emitted photon is absorbed in the same material again. This probability is proportional to the path length of the photon in the semiconductor. The large mismatch between the dielectric constants of the semiconductor and that of free space further increase this path length by total internal reflection. Since the critical angle is small, a considerable fraction of photons are incident at the interface at larger than the critical angle and thereby get reflected back into the semiconductor (Figure 3.2). These two factors are the main contributors to low efficiency in LEDs. In higher efficiency diodes the active p–n junction area is kept very thin, and the diode is placed in a matching epoxy that reduces the fraction of back-reflected light at the interface (Figure 3.2). Furthermore, the epoxy is generally dome shaped such that most light rays impinge on the interface at close to normal incidence.

3.2.1 Various Types of LEDs

The most common type of near-infrared LED is made of GaAs. Gallium arsenide is a direct bandgap material with well developed p-n junction and ohmic contact technologies. In LEDs that are made of low-doped GaAs the radiation arises from band-to-band transitions with a wavelength close to 860 nm corresponding to its bandgap. Light emitting diodes using heavily doped

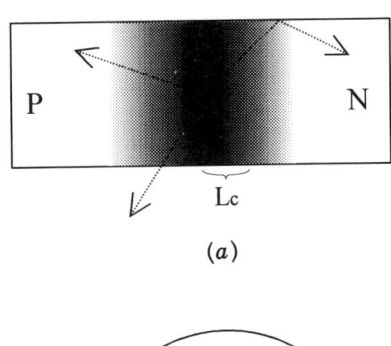

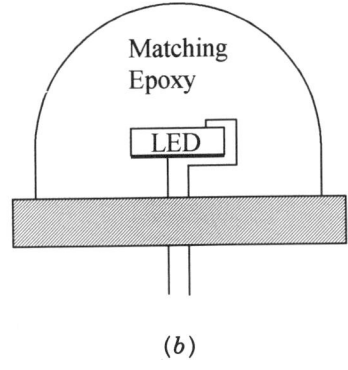

FIGURE 3.2 (*a*) Light generation in an LED. Low efficiency is caused by reabsorption of light. Total internal reflection at interface increases fraction of photons absorbed. (*b*) Encapsulating LED in a dome-shaped epoxy increases efficiency.

material rely on indirect recombination at impurity sites. These LEDs have longer emission wavelengths (900–1000 nm), with the added advantage that the radiation is not heavily absorbed by the material. Approximately 10% external efficiency is presently achievable with heavily doped GaAs LEDs.

The heterojunction LED is another type of diode where carriers are injected into a thin recombination layer through a doped higher bandgap material. A GaAs/AlGaAs heterojunction LED is shown in Figure 3.3. This type of LED offers higher efficiency due to the fact that the higher bandgap material forming the bulk of the LED is transparent to the emitted radiation.

There are a variety of LEDs with emission in the visible, primarily for application in LED displays. Visible radiation in red and yellow regions of the spectrum is normally generated by AlGaAs and GaAsP diodes. The quaternary material InGaAlP is used for green LEDs. Zinc selenide (direct bandgap) and silicon carbide (indirect bandgap) are used for blue LEDs. Short-wavelength LEDs are, in general, less efficient due to practical problems with making good ohmic contacts as well as achieving efficient *p*- and *n*-type doping in large-bandgap semiconductors.

Radiation from an LED is almost entirely due to spontaneous emission. It is however possible for some stimulated emission (see the next section) to occur and amplify the light before it leaves the active region. Diodes that are designed to take advantage of this effect are called *superluminescent* LEDs.

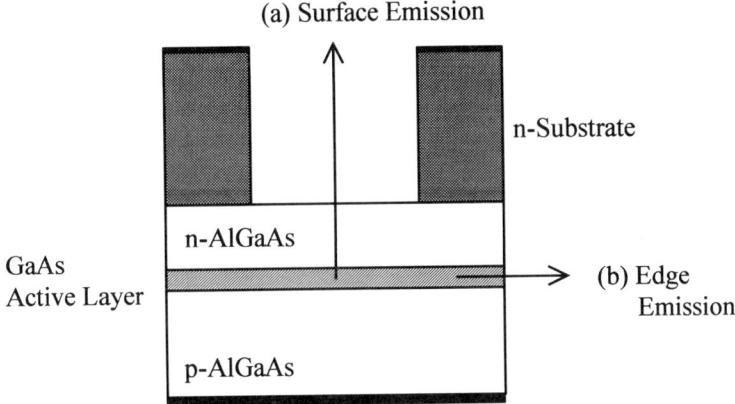

FIGURE 3.3 Material structure of GaAs/AlGaAs LED.

3.2.2 Speed Limitations

The output intensity of an LED can be modulated by its drive current for optical communication applications. It should be pointed out that the use of LEDs in communications is limited to short-distance links such as those used within a single building. For long-distance applications laser diodes are preferred due to their higher output power, spectral purity, higher efficiency, and easier coupling into optical fibers.

An important internal factor that limits the modulation frequency of the LED is the minority-carrier lifetime. Consider an operating LED that is switched off by instantly removing the applied electric field. Before the device reaches the new equilibrium, minority charge carriers injected into either side of the junction need to be removed. The reduction of this excess charge is an exponential decay with a time constant equal to minority-carrier lifetime τ. The cutoff frequency associated with this time constant is $f_\tau = 1/2\pi\tau$. If the LED is modulated at the frequency f, the modulation depth $M(f)$ of its optical output has the following frequency dependence:

$$M(f) = \frac{M(0)}{\sqrt{1 + (f/f_\tau)^2}}. \tag{3.3}$$

It is clear that in order to increase f_τ, minority-carrier lifetime τ needs to be reduced. This can be done, according to Eq. (3.1), by increasing the doping level. For GaAs doped at 10^{18} cm^{-3}, the value of f_τ is close to 110 MHz. The negative side effect of high level doping in LEDs is that it increases the probability of impurity site and nonradiative recombinations, leading to reduced efficiency.

It is possible to stretch the limitation set by Eq. (3.1), by using a moderately doped device with a thin active region under high current injection, that is, modulating the output of the device by modulating the bias current about a highly forward biased point. High level charge injection causes carrier concentrations to rise above the equilibrium level and carrier lifetimes to be reduced [1].

As is the case with most optoelectronic devices, what limits the operation speed of the device is a combination of intrinsic limitations and extrinsic or parasitic contributions from contact-pad capacitance, bias lead inductance, and so on. For a more detailed discussion, refer to Section 3.5.

3.3 SEMICONDUCTOR LASERS

A photon traveling through a semiconductor medium can get absorbed, causing the generation of an electron–hole pair (if its energy is close to or exceeds that of the bandgap). The photon can also cause the opposite effect, namely, to interact with an electron in the conduction band and induce a downward transition to the valence band. The first process attenuates the light passing through the medium, while the second gives rise to light amplification. This process of light amplification is known as *stimulated emission*. The probability of either absorption or stimulated emission depends on the density of available states for either upward or downward transition. In thermal equilibrium without any excess carriers the absorption probability far exceeds that of stimulated emission. This is due to the fact that most electronic states in the conduction band are empty and those in the valence band are full. It was seen in the previous section that in a forward-biased *p–n* junction a large number of excess carriers are injected into the active region. As the concentration of excess carriers increases, so does the probability of stimulated emission. At the same time the probability of absorption is reduced and eventually becomes smaller than that of stimulated emission. At the point where the two are equal the material is said to have reached *transparency* [2].

Gallium arsenide reaches transparency at injected carrier concentration of $\approx 1.55 \times 10^{18}$ cm^{-3}. At higher injection levels the material manifests optical gain. Figure 3.4 shows how this gain varies with optical wavelength and carrier concentration in GaAs. In general, a material provides optical gain at a particular wavelength if it has achieved a *population inversion*. This means that the material has reached a state where there is a higher probability for a photon to induce a downward transition than upward. Population inversion is achieved by *pumping* or adding energy to the medium. A semiconductor laser is pumped by an electrical current that injects excess carriers into the active region.

3.3.1 Basic Structure

The most common type of laser diode is an edge emitter with cleaved facets, as shown in Figure 3.5. Current flows vertically through the *p–n* junction. An

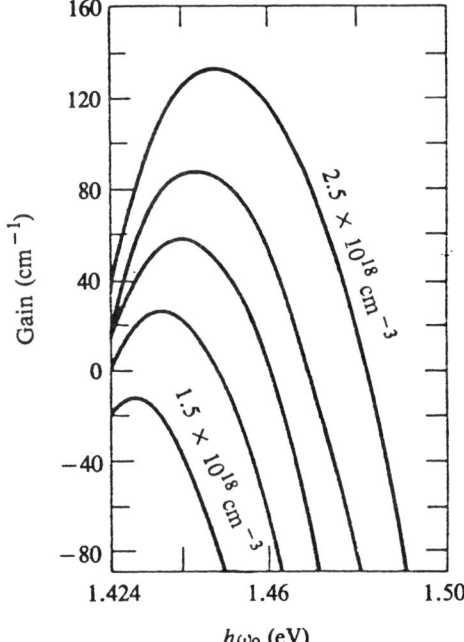

FIGURE 3.4 Optical gain in GaAs as a function of wavelength and carrier concentration. (After ref. [3].)

optical waveguide confines the light to the vicinity of the active region. The smooth cleaved facets form the high reflector and the output coupler of the Fabry–Perot optical resonator. The back facet is usually coated to increase its reflectivity. The cavity length of an edge emitting laser diode is typically between 200 and 1500 μm but can be as short as 50–100 μm for high speed diodes. The vertical thickness of the active layer is typically between 0.1 and

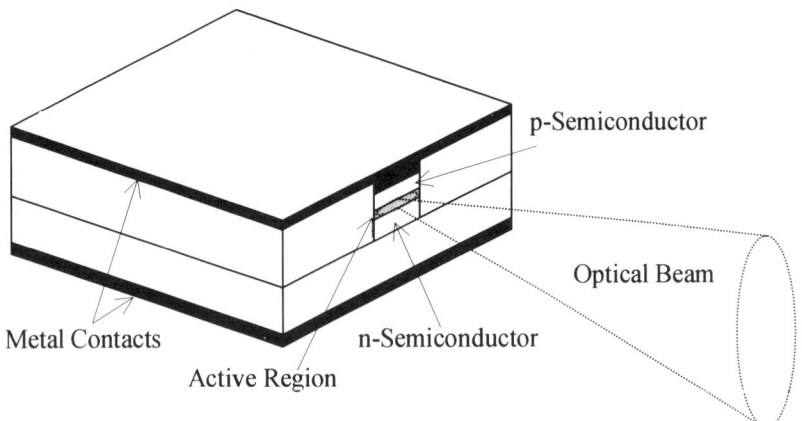

FIGURE 3.5 Schematic diagram and beam shape of an edge emitter laser diode.

1.0 μm, while the lateral width is commonly 2–10 μm or wider. The shape of the emitting area is therefore a narrow stripe, leading to an elliptical and very divergent beam (typically 10°–30°). Divergence in the plane parallel to the active layer is generally a factor of 2–10 times less than in the perpendicular plane.

The output of a laser diode is normally polarized parallel to the plane of the junction. The polarization purity varies from about 10:1 for gain-guided lasers to anywhere from 50:1 to beyond 100:1 for index-guided lasers. (These terms are defined in the following sections.) Beam ellipticity, divergence, and astigmatism are typical undesired characteristics of laser diode emission. All of these problems can be corrected by external optics.

3.3.2 Laser Materials

Table 3.1 lists semiconductor materials most commonly used in the fabrication of laser diodes and the corresponding emission wavelengths. Wavelengths of particular interest are (1) 980 and 1480 nm for pumping erbium-doped fiber amplifiers; (2) 1300 nm for optical fiber communications (zero dispersion point of silica fibers); (3) 1550 nm, also for fiber communications (wavelength of lowest attenuation in silica fibers); and (4) short-wavelength visible lasers for data storage.

Most of the materials used to fabricate laser diodes need to satisfy the following requirements:

- The light emitting material needs to have a direct bandgap for efficiency.
- The active region must have a higher refractive index than the semiconductor layers above and below it (claddings) in order to vertically confine the optical field.
- The active region must have a lower bandgap than the cladding layers for efficient charge injection and in order for light not to be absorbed in the claddings.
- It should be thermodynamically possible to grow the material layers

TABLE 3.1 Approximate Emission Wavelength Ranges Associated with Commonly Used Laser Diode Materials

Semiconductor Material	Wavelength Range (nm)
ZnSe	525
AlGaInP	630–680
GaAlAs	680–880
GaAs	780–890
InGaAs	900–1000
InGaAsP	1100–1650

of interest. Certain compound semiconductor combinations cannot be formed with present epitaxial growth techniques.
- All the materials should be lattice matched (or nearly lattice matched) to an available substrate material such as GaAs, InP, GaSb, and InAs.

3.3.3 Active Region

The active region is the portion of the device where population inversion is maintained to provide optical gain. The optical field is constrained to propagate along this region, which normally lies between two reflecting surfaces that form the optical resonator. In the early days of semiconductor lasers the active region consisted of a simple *p–n* junction made of GaAs with smooth reflecting facets. The junction was forward biased, and the thickness of the active region was loosely defined by carrier diffusion length. This type of laser (known as a homojunction laser) had a very high threshold current due to the inefficiencies cited earlier for LEDs. Methods were needed to better confine the charge carriers and the optical field. This was done by a structure known as a double heterojunction. It consists of either a doped or an undoped layer of semiconductor sandwiched between two oppositely doped materials with higher bandgaps. The initial lasers of this type were GaAs/AlGaAs lasers [4]. A forward-biased current injects carriers into the lower bandgap active region, where they remain confined due to the band discontinuity. The optical field is also confined to the lower bandgap material due to total internal reflection at the interfaces. Further enhancement of the efficiency (or reduction of threshold current) can be realized if the thickness of the active region is reduced to 20 nm or less. The active region in this case is called a *quantum well*. The thickness of the layer significantly affects carrier energy levels and the carrier distribution is essentially two dimensional. Higher density of states result from the two-dimensional nature of the distribution, and population inversion is more easily achieved. One drawback of the quantum well laser is that due to the very small thickness of the active region, the optical field cannot be closely confined to it. Most of the optical energy would be propagating outside the active region, where there is no gain. Multiple-quantum-well lasers were developed to circumvent this problem. These lasers employ a number of stacked quantum wells. The optical cavity encompasses all quantum wells, leading to an increased overlap between the optical field and the gain regions. Some of the highest efficiency lasers employ quantum well technology [5]. Increased modulation speed due to a smaller interaction volume is another advantage of quantum well lasers.

The advantages of quantum well lasers are due to carrier confinement to two dimensions. This concept is taken a step further in quantum wire lasers where carriers are confined to one dimension. Quantum wire lasers have been reported with threshold currents less than 1.0 mA [6]. The combination of high gain and small interaction volume are expected to further enhance the high frequency modulation capability of quantum wire lasers over conventional ones.

3.3.4 Optical Field Confinement

The optical cavity of an edge emitting laser is a Fabry–Perot resonator whose parallel mirrors are the cleaved facets of the laser. Between the mirrors, the optical field needs to be kept confined both vertically and laterally. Vertical confinement is achieved by the lower refractive index claddings above and below the active region. This configuration forms a dielectric slab waveguide. Light traveling along the waveguide gets totally internally reflected at the interfaces (Figure 3.6). The interference of the wavefronts produce distinct standing-wave patterns known as transverse modes. Two sets of transverse modes are allowed: transverse electric (TE), and transverse magnetic (TM). Transverse electric modes are polarized with the electric field parallel to the interfaces (y direction). Transverse magnetic modes have the magnetic field along that direction. Laser diodes generally operate in the TE mode. Therefore we will focus attention on the analysis of these modes. At the end of the discussion, some of the reasons for the preference will become apparent.

The solution for a guided mode in a slab waveguide is obtained by matching solutions in three different regions: below the slab, inside the slab, and above the slab. Starting with the solution of the wave equation in the three regions, the tangential fields are matched at the boundaries. For the TE waves this can be written as

$$\frac{d^2 E}{dx^2} + (k_0^2 \eta^2 - \beta^2) E = 0. \tag{3.4}$$

Here, $k_0 = 2\pi/\lambda_0$ is the propagation constant in free space, η has three different values in the three regions, and β is the propagation constant of the mode, which is the unknown to solve for. Matching the fields at the boundaries and assuming a symmetrical structure where $\eta_1 = \eta_3$, we find the following relationships for field variation along x:

$$E(x) = \begin{cases} e^{k_1(d/2 - |x|)}, & |x| \geq d/2, \\ \cos(k_2 x)/\cos(k_2 d/2), & |x| \leq d/2. \end{cases} \tag{3.5}$$

The electric field varies sinusoidally inside the waveguide and has an exponentially decaying evanescent tail outside, and

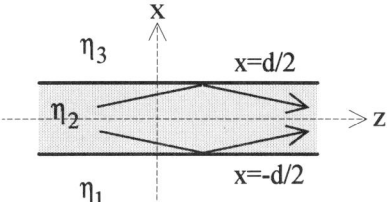

FIGURE 3.6 Optical guiding in an edge emitting laser diode and the coordinate system used for its analysis.

$$k_1 = \sqrt{\beta^2 - k_0^2 \eta_1^2}, \quad k_2 = \sqrt{k_0^2 \eta_2^2 - \beta^2}, \quad \tan(k_2 d/2) = k_1/k_2. \quad (3.6)$$

Numerous solutions may be found for β depending on the optical wavelength, the dimensions, and the refractive indices. Each of these solutions is known as a *transverse mode* of the laser cavity. Most of these modes have cutoff wavelengths and will not be excited if d/λ is small enough. However, there is a solution without cutoff that corresponds to the lowest value of β. This solution is the lowest order transverse electric mode TE_0.

Notice that the evanescent tails outside the waveguide are relatively long when the refractive index discontinuity at the boundary is small. Since this is the case in most laser diodes, it is important to quantify how much of the optical power propagates inside the waveguide. This would give the fraction of the optical field that interacts with the active region in a laser. The fraction of the optical power propagating inside the waveguide is called the *confinement factor* Γ. The value of Γ depends on the waveguide thickness d, the refractive index discontinuity, and the propagation mode. For the TE_0 mode and for values of d that are small compared with the optical wavelength in the medium, the approximate value of Γ is

$$\Gamma \approx \tfrac{1}{2} k_0^2 d^2 (\eta_2^2 - \eta_1^2). \quad (3.7)$$

The confinement factor for the TE_0 mode is the highest of all the propagation modes including TM_0. This means that it experiences the highest gain and is most likely to be the dominant mode in a laser diode. The TE_0 mode is also more desirable than TM_0 because it gives a linearly polarized output.

Up to this point we assumed that the optical waveguide coincides with the active region of the laser. In higher power lasers, in order to reduce the optical power density, the waveguide can be made larger than the active region. This type of a waveguide is known as a *large optical cavity* (LOC). The vertical refractive index profile of an LOC is shown in Figure 3.7. A more generalized version of the same idea is the *separate confinement heterostructure* (SCH), which is also shown in Figure 3.7. Finally, the SCH structure may utilize a graded refractive index profile (GRIN-SCH). In all these cases the objective is to maintain an optimum trade-off among low optical power density, high confinement factor, and low barrier to electron transport.

Lateral confinement of the optical field is achieved either by gain guiding or by index guiding. Gain guiding weakly confines the light to the stripe where current flows. Index guiding, on the other hand, generally provides a stronger confinement by surrounding the active region with lower refractive index material. Gain-guided lasers are simpler to fabricate but often do not have a high quality output beam. Also, due to the generally weaker confinement of the optical field, they tend to be less efficient. The weaker confinement is an advantage in some high power lasers, however, where a lower optical power density is desirable.

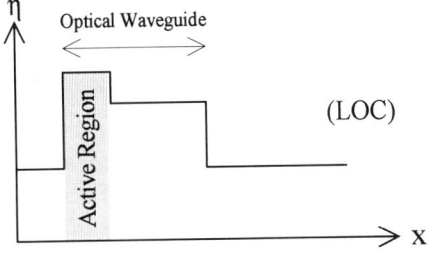

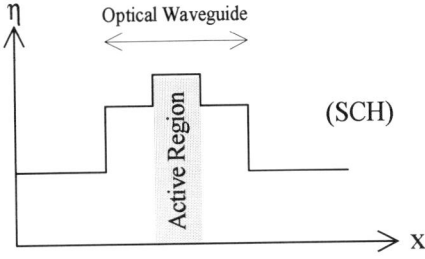

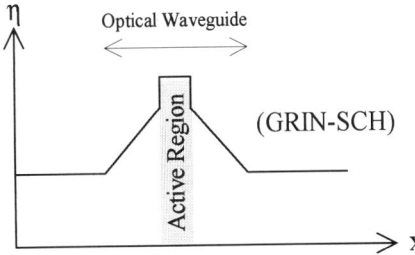

FIGURE 3.7 Methods of separate vertical confinement of carriers and optical field: (*a*) large optical cavity, (*b*) separate confinement heterostructure, and (*c*) graded-index SCH.

Gain guiding restricts the lateral spreading of the optical field to the region of high carrier density. This is done by the perturbation of the dielectric constant in the active region due to the presence of excess carriers. This perturbation is the sum of two different effects: the direct interaction of the optical field with the electrons and the modification of the spectral distribution of loss and gain in the material [7]. The presence of gain itself can be represented by the imaginary part of the dielectric constant, which is different from that of the surrounding lossy material. The main mathematical differences between gain guiding and index guiding is that in gain guiding the change in the dielectric constant is gradual and complex. Another aspect of gain guiding is that the carrier density is not controlled by the electrical current alone; rather, regions of high optical power have lower carrier density due to the higher recombination rate. This gives rise to a central dip in the

60 OPTICAL ACTIVE DEVICES

dielectric constant change. Detailed analysis of gain guiding may be found in reference [8].

Index guiding requires more advanced fabrication techniques. The optical waveguide is to be placed between lower refractive index materials at its horizontal boundaries. This cannot be accomplished by epitaxial growth alone. Some extra lithography and etching is needed to fulfill this task. Many common approaches fall into the categories of (1) ridge waveguide, (2) buried heterostructure, and (3) buried crescent (Figure 3.8). Relative merits of these techniques lie in the particular goals they are designed to achieve, such as

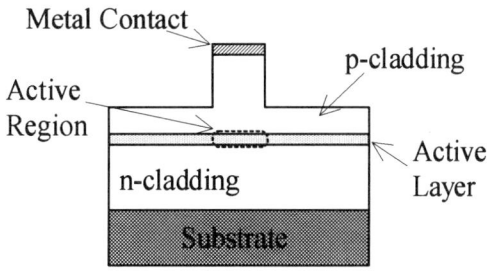

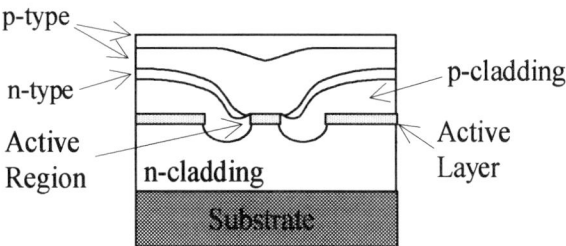

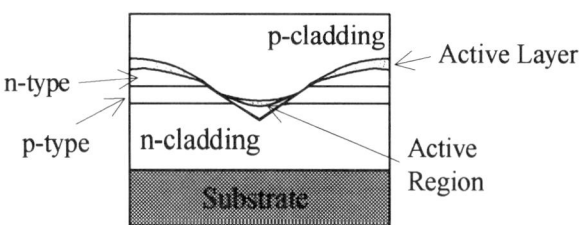

FIGURE 3.8 Three fabrication methods for index guiding: (*a*) ridge waveguide, (*b*) buried heterostructure, and (*c*) buried crescent.

a lower threshold current, a particular beam shape, higher optical power, or lower processing cost.

The vertical dimension of the optical waveguide in an edge emitting laser is normally small enough to assure single-mode propagation. The lateral dimension, on the other hand, may not be small enough, and multiple transverse modes may be excited. Transverse modes of the optical cavity determine the beam profile, its divergence, and its far-field pattern. Index-guided lasers usually operate in a single transverse mode while gain-guided lasers tend to support multiple transverse modes.

Another performance difference between gain-guided and index-guided lasers is their astigmatism. In gain-guided lasers the output beam waist perpendicular to the active layer occurs at the emission surface, while the beam waist parallel to it (in the gain-guided direction) occurs 10–50 μm behind the surface due to carrier depletion near the cleaved facet. This is the cause of astigmatism seen in many gain-guided laser diode outputs. The astigmatism, like beam divergence, is not a serious drawback and may be corrected by a weak cylindrical lens.

3.3.5 Longitudinal Modes

Resonance condition for the Fabry–Perot resonator with two parallel mirrors is that the distance L between the mirrors should be equal to an integral number of half wavelengths $(m\lambda/2)$. Noting that $\lambda = c/\eta f$, it follows that

$$\lambda = \frac{c}{\eta f} = \frac{2L}{m}, \tag{3.8}$$

where η is the refractive index of the medium and f is the oscillation frequency. Rearranging the variables,

$$f = m\frac{c}{2\eta L}. \tag{3.9}$$

Every integral value of m represents a resonance. Each one of these resonances is called a *longitudinal mode*, and the mode separation is equal to the inverse of light's round-trip time in the cavity.

Short laser cavities support longitudinal modes with relatively large frequency separations. A 250-μm-long slab of GaAs supports modes that are 168 GHz apart. This frequency separation is still small compared with the bandwidth of the semiconductor gain medium. Gallium arsenide can theoretically be a useful gain medium for photon energies from 1.38 to 1.58 eV, corresponding to wavelengths between 0.785 and 0.898 μm or frequencies between 330 and 380 THz. Almost 300 longitudinal modes can be excited in this range. Of course, in practice, only the mode (or modes) located closest to the peak of the gain curve is excited.

3.3.6 Laser Emission

Let us examine the behavior of a semiconductor laser as a function of the drive current. As current through the device is increased from zero, initially broadband radiation at a low power level is observed due to spontaneous emission. With increased current, the gain of the medium rises such that some of the radiation is amplified. At this stage, the radiation begins to be preferentially emitted at resonance frequencies of the optical cavity (longitudinal modes). Eventually, the material reaches transparency at the frequency of the longitudinal mode or modes closest to the peak of the gain curve. This is the point where laser oscillation begins.

The condition for laser oscillation to build up is for the optical gain of the cavity to compensate for all the losses including the light leaving the cavity. Denote the optical loss per unit length (attenuation coefficient) of the material in the absence of any applied current by α_l. The presence of the current adds a gain per unit length α_g. In a round trip around a cavity of length L, the amplitude of the optical field will either be attenuated or amplified by the factor $e^{2(\alpha_g - \alpha_l)L}$. Furthermore, the reflectivities R_1 and R_2 of the facets reduce the amplitude by an extra factor of $R_1 R_2$. The condition for laser oscillation is therefore $R_1 R_2 e^{2(\alpha_g - \alpha_l)L} = 1$ or

$$\alpha_g = \alpha_l + \frac{1}{2L} \ln\left(\frac{1}{R_1 R_2}\right). \tag{3.10}$$

Current through the device at the point of satisfying this condition is called the *threshold current*. Any further increase in the current does not raise the gain of the medium at the emission wavelength and leads only to higher optical output. Stimulated emission increases the recombination rate to compensate for any increase in charge injection. The gain curve is *clamped* at the threshold level.[1]

Example A 250-μm-long laser has a material loss $\alpha_l = 10$ cm^{-1}. The high reflector has a reflectivity $R_1 = 0.9$, and the reflectivity of the output coupler is $R_2 = 0.3$. Calculate the threshold gain α_g.

Solution

$$\alpha_g = 10 - \frac{1}{0.05} \ln[(0.9)(0.3)] = 36 \text{ cm}^{-1}. \qquad \blacksquare$$

Since the evanescent tails of the optical field extend beyond the gain region, the material gain is actually larger than α_g by the factor $1/\Gamma$. Material gain in semiconductor lasers is high compared with most other types of lasers. Thus the semiconductor laser cavity is generally very short and has a highly transmitting output coupler.

---

[1] This is true in narrow-stripe lasers before reaching gain saturation levels.

The effect of increased pumping beyond threshold on the longitudinal modes of the laser depends on the cavity and the contribution of spontaneous emission to laser output. In laser cavities with relatively low material gains and high reflectivity output couplers, increased pumping causes the gain to rise at other longitudinal mode wavelengths.[2] The laser begins to support multiple longitudinal modes. Figure 3.9 shows the hypothetical gain profile with respect to the longitudinal modes of such a laser. When the gain profile is above threshold, at each oscillating mode it acquires depressions. These depressions are due to the fact that, at each oscillating mode, the gain is clamped at the threshold level. The introduction of these depressions in the gain curve is called *hole burning*.[3] In cases where the width of the emission line of each mode is comparable to the mode spacing, the hole burning in the gain curve is not pronounced, and the same phenomenon leads to clamping a broad region of the gain curve at the threshold level.

In semiconductor lasers with high material gain and lower reflectivity output couplers, the contribution of spontaneous emission to the output is significant. Due to spontaneous emission, optical losses of the resonator do not have to be completely compensated by stimulated emission. In the operation of the laser, the gain of the main (zero-order) mode remains below cavity losses by a small fraction ϵ, and a number of other longitudinal modes also contribute to the output. Increased current through the device leads to a

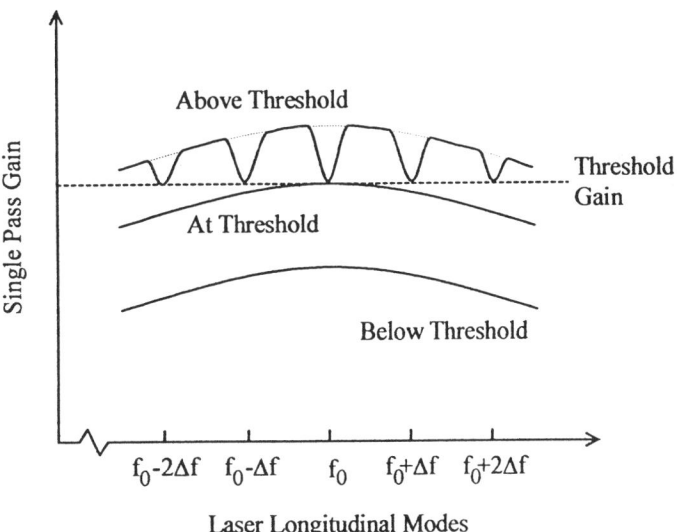

FIGURE 3.9 Semiconductor laser material gain as a function of optical frequency for different pumping levels.

---

[2]This also requires that the gain spectrum be inhomogeneously broadened, i.e., distinct groups of atoms or carriers respond to radiation at different wavelengths. This is true in case of semiconductor lasers.

[3]This is different from "spatial hole burning" encountered in wide-stripe lasers.

reduced value of ϵ, which in turn increases the fraction of optical power in the central mode. The laser operation can be "substantially single mode" for a high enough drive current. An approximate expression for the number of active longitudinal modes of the laser is [8]

$$m_{1/2} \approx 1 + 2M\sqrt{\epsilon}, \qquad (3.11)$$

where $m_{1/2}$ is the number of modes with intensity higher than half that of the central mode, and $2M$ is the total number of longitudinal modes supported by the gain spectrum.

Increasing the bias current not only changes the modal behavior of the laser, but also causes a net shift of the wavelength due to the thermal expansion of the cavity and the perturbation of the refractive index. It is due to these effects that direct modulation of the laser's drive current does not result in a simple amplitude modulation of the output.

3.3.7 Distributed Feedback and Bragg Reflector Lasers

Reflectors on the two sides of the laser cavity can be made more wavelength selective than the Fabry–Perot resonant cavity alone; that is, instead of relying on the cleaved facet of the semiconductor to reflect light back into the cavity, it is possible to use a Bragg reflector consisting of a series of equally spaced grooves. A laser utilizing such reflectors is known as a distributed Bragg reflector (DBR) laser. If the grating lies in the pumped region of the laser, then the laser is referred to as a distributed feedback (DFB) laser [9]. The grating is a wavelength-selective reflector that reduces the spectral width and temperature dependence of wavelength. With DFB and DBR lasers better control of the longitudinal modes and the polarization of the output is possible.

In order to analyze the reflection from a DBR, consider a waveguide supporting an optical field with a propagation constant $\beta = 2\pi\eta_{eq}/\lambda_0$, where λ_0 is the free-space wavelength. The evanescent tails of the optical field extend beyond the physical boundaries of the waveguide and, therefore, η_{eq} is used to represent its equivalent refractive index. If along this waveguide there are periodic perturbations to the refractive index at the separation Λ, there will be some reflected wave associated with each one. The phase difference between two consecutive reflections is of course $2\beta\Lambda$, as shown in Figure 3.10. The condition for constructive interference of the reflected waves is that the phase difference be an integral multiple of 2π, or

$$\Lambda = \frac{m\lambda_0}{2\eta_{eq}}. \qquad (3.12)$$

Here, m is a positive integer known as the order of the reflector.

Example Calculate the groove spacings for a Bragg reflector of a 1550-nm InGaAsP laser and an 800-nm GaAs/AlGaAs laser.

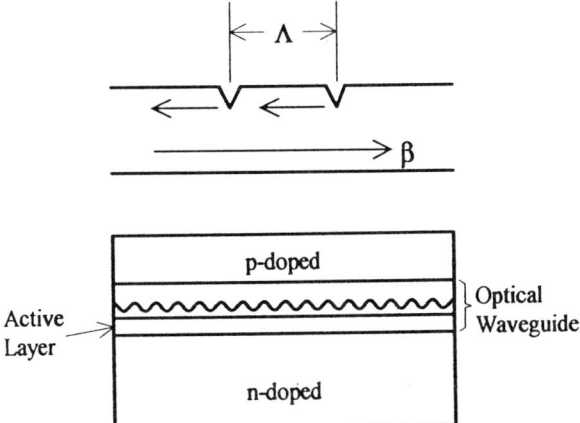

FIGURE 3.10 (*a*) Reflection of an optical wave by a grating. (*b*) Schematic diagram of a DFB laser.

Typical refractive index values are 3.4 for InGaAsP and 3.55 for GaAs. If we take $m = 1$ (first-order grating), the value of Λ is half of the optical wavelength in the medium, or 0.23 μm for InGaAsP and 0.11 μm for GaAs. ∎

Diode laser Bragg reflectors are normally made by etching grooves on the surface of the wafer between two growth steps such that the grating falls within the optical waveguide (Figure 3.10). As was seen in the above example, the groove spacing for a first-order grating is half of the optical wavelength in the material. This spacing is very small, and its fabrication is a rather challenging lithography step. The problem is more severe for shorter wavelength visible lasers.

It is possible to relax the constraint on the spacing of the grooves by increasing the order m of the reflector. A byproduct of increasing m is that the gratings then allows the optical energy to be coupled into many different angles. In general, since each groove can scatter light into all angles from 0 to 2π, radiation is coupled by the grating into all angles for which individual contributions interfere constructively. For example, a first-order grating only allows forward (0°) and back-reflected (180°) radiation, while a second-order grating also couples radiation into ±90°. The second-order grating is utilized in surface emitting lasers.

The concept of each groove being considered an isotropic scatterer of light can be taken one step further. The entire Bragg reflector may be considered as a one-dimensional array of isotropic radiators. Initially, let us assume that each groove reflects a very small fraction of the incident radiation such that it is not attenuated. Thus all the elements in the array radiate the same optical power. Referring to Figure 3.11, along any direction θ, the total optical field is the sum of the contributions from each element:

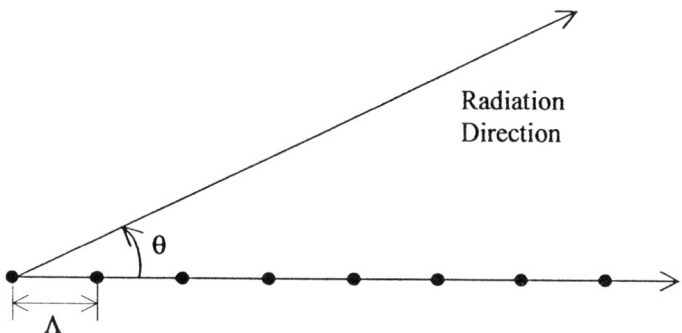

FIGURE 3.11 Linear array of isotropic point sources of equal amplitude and spacing.

$$E = E_0(1 + e^{-j\psi} + e^{-2j\psi} + \cdots + e^{-(N-1)j\psi}) = E_0 \sum_{n=1}^{N} e^{-j(n-1)\psi}, \quad (3.13)$$

where $\psi = \beta\Lambda \cos\theta + \delta$ and δ is the phase difference between two consecutive elements: $\delta = \beta\Lambda$. Therefore, $\psi = \beta\Lambda(1 + \cos\theta)$. The above summation may be recognized as a geometric progression whose value is given by

$$E = E_0 \frac{\sin(N\psi/2)}{\sin(\psi/2)} e^{-j\phi} \quad (3.14)$$

The phase angle ϕ depends on the element of the array chosen as the phase reference. If the center of the array is chosen as the reference, then $\phi = 0$, and the radiation pattern of the grating is

$$E = E_0 \frac{\sin(N\beta\Lambda(1 - \cos\theta)/2)}{\sin(\beta\Lambda(1 - \cos\theta)/2)} \quad (3.15)$$

The maximum field value is NE_0, which occurs at $\psi/2 = 0, \pi, \ldots$. For the grating spacing ($\beta\Lambda = \pi$), the maxima occur at $\theta = \pi$ and $\theta = 0$, corresponding to forward and reflected waves. If the value of $\beta\Lambda$ is selected to be 2π (second-order grating), other major lobes occur at $\theta = \pm\pi/2$.

The wavelength of maximum reflection is known as the *Bragg wavelength*. The wavelength selectivity of the grating increases with the number of grating lines, as shown in Figure 3.12. This is what gives DFB and DBR lasers their capability to select a single longitudinal mode. The discussion presented above indicates that this selectivity increases indefinitely as more grooves are added. This would indeed be the case if the forward-traveling wave did not attenuate. The attenuation cannot be ignored as higher fractions of the incident wave get reflected. This puts a limit on the wavelength selectivity of the grating.

A more general analysis of wave propagation in the presence of a grating would involve the solution of the wave equation with the constraint of a

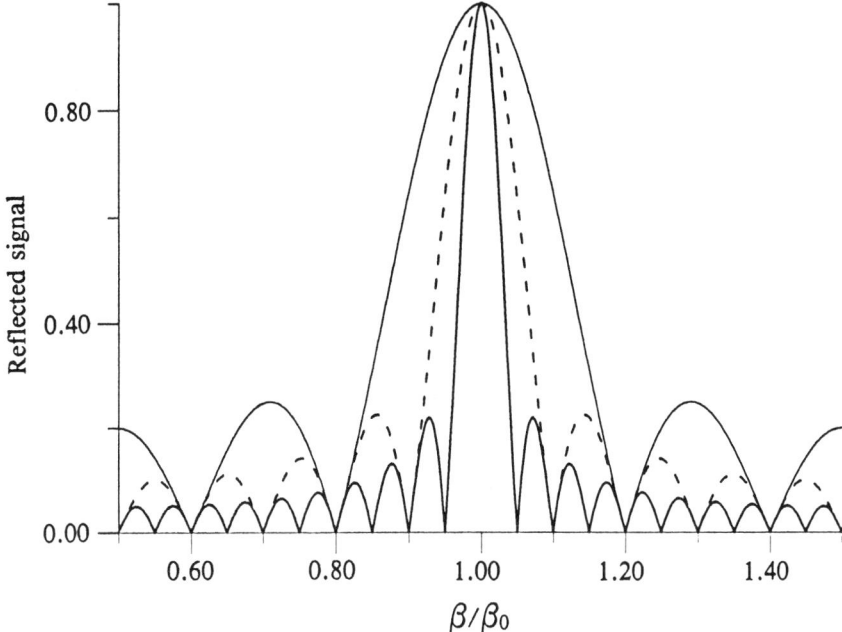

FIGURE 3.12 Reflected signal as a function of propagation constant for grating consisting of (a) 5 grooves, (b) 10 grooves, and (c) 20 grooves.

periodically varying refractive index [1, 8, 9]. This analysis will not be given here, but the general solution for a first-order grating is two counterpropagating waves with position-dependent amplitudes. In the case of a DBR, one propagation direction corresponds to the incident wave, while the other is the reflected wave. In a DFB laser where the grating covers the entire active region, the counterpropagating waves resemble the standing wave formed by multiple reflections in a Fabry–Perot cavity. A DFB laser with anti-reflection coated facets tends to oscillate in two modes on either side of the Bragg wavelength. Removing this degeneracy requires that the periodicity of the grooves be interrupted at one point. Refer to Agrawal and Dutta [9] for a more detailed analysis of the DFB laser.

3.3.8 Surface Emitting Laser Diodes

Diode lasers that emit light from the surface rather than the end facets have applications including chip-to-chip optical interconnects, high power two-dimensional arrays, and optoelectronic integration. Three methods have been commonly used to achieve surface emission: 45 degree etched mirror [10], optical coupling by a grating [11], and vertical cavity [12]. In the first approach, the output coupler of the laser, instead of being a cleaved facet, is an etched groove. One sidewall of the groove is etched at 45° and metallized in order to reflect the laser output into the vertical direction. Most of the beam characteristics of this type of surface emitting laser are the same as those of edge emitting diodes.

Grating-coupled surface emitting lasers utilize a second-order Bragg reflector as the output coupler. Second-order gratings (as was discussed in the previous section) not only reflect the light back into the cavity, but also couple the light out of the cavity perpendicular to the surface. The surface emission area is relatively broad in this case, and the divergence of the beam may be considerably reduced compared with an edge emitting laser.

Vertical cavity surface emitters offer an efficient use of semiconductor wafer area for laser emission. The optical cavity is vertical, and the reflectors are stacks of alternating materials with different refractive indices. The principle of light reflection from dielectric stacks is similar to reflection from horizontal Bragg reflectors. Another advantage of the vertical cavity lasers over the horizontal Bragg reflector variety is that the periodicity of the reflectors is controlled by precision material growth rather than by the challenging lithography. The material growth gives precise control over the length and the optical properties of each layer grown.

3.3.9 Direct Modulation of Laser Diodes

Laser diodes can in general be modulated at a higher speed than LEDs. In an LED it was seen that the modulation rate is limited by spontaneous carrier recombination lifetime. In a laser diode, recombination is dominated by stimulated recombination lifetime, which is much shorter, and leads to a wider modulation bandwidth.

If a laser diode is suddenly turned on from zero bias, a turn-on delay and an exponential rise in the optical output will be observed. The optical output initially overshoots and goes through a few cycles of damped oscillation before reaching equilibrium. This behavior is known as *relaxation oscillation*, and the oscillation frequency ω_R is an important factor in determining the frequency response of the laser diode (Figure 3.13).

Relaxation oscillation is caused by the inverse relationship between carrier density and photon density in the semiconductor. An increase in photon density, or the optical intensity, causes an increased recombination rate due to stimulated emission. The increased recombination leads to a reduction in carrier density, and thereby a reduction in optical gain, which in turn reduces the optical intensity. The reduced optical intensity reduces the recombination rate and allows carrier density to build up again, and the cycle continues.

Before studying the physics of relaxation oscillation, it is instructive to study the transient response of a laser diode, as shown in Figure 3.13. The above-threshold portion of the curve is the transient response of a standard second-order system [13]. It can be described in terms of a general second-order transfer function $Y(s)$, which is the normalized ratio of output intensity P to input current I,

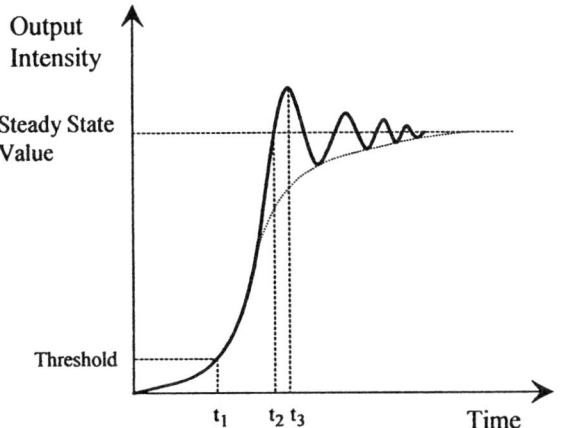

FIGURE 3.13 Response of laser diode to a step input current.

$$Y(s) = \frac{\omega_n^2}{s^2 + 2\zeta\omega_n s + \omega_n^2}, \quad (3.16)$$

where $s = (-\sigma + j\omega)$ is the complex frequency. The entire response curve (above threshold) is described in terms of the two parameters ζ and ω_n. Here, ω_n is referred to as the undamped natural frequency and ζ is called the damping coefficient. The speed or the bandwidth of the response is determined by the undamped natural frequency whereas the shape of the response is determined by ζ. The poles of the system are the roots of the characteristic polynomial [the denominator in Eq. (3.16)],

$$p_i = -\sigma \pm j\omega_d, \quad (3.17)$$

where $\sigma = \zeta\omega_n$ is the attenuation coefficient and ω_d is the damped natural frequency,

$$\omega_d = \omega_n\sqrt{1 - \zeta^2}. \quad (3.18)$$

The overshoot magnitude M_p occurs when the derivative of the response is zero. Its value is given by

$$M_p = e^{-\pi\zeta/\sqrt{1-\zeta^2}}, \quad 0 \leq \zeta < 1. \quad (3.19)$$

The transient response may be written as

$$y(t) = 1 + ce^{-\sigma t}\cos(\omega_d t + \phi), \quad (3.20)$$

where

$$c = -\frac{1}{\sqrt{1-\zeta^2}}. \qquad (3.21)$$

The transient response is confined within the envelope curves $1 \pm ce^{-\sigma t}$. Equation (3.20) characterizes the transient response of a system having no finite zeros and two complex poles with undamped natural frequency ω_n, damping ratio ζ, and real part σ.

In order to correlate the response parameters of the laser diode to its physical parameters, let us write the rate equations for the photon density P and injected carrier density N in the active region. The rate equation for N may be written as

$$\frac{dN}{dt} = \frac{I}{eV} - \frac{N}{\tau} - A(N - N_{\text{tr}})P. \qquad (3.22)$$

The first term on the right is the injected carrier rate due to the current I into the volume V. The second term is the spontaneous recombination rate, and the third term is the stimulated recombination rate. The parameter N_{tr} is the carrier density needed to induce transparency, and A is the growth constant representing the strength of the stimulated emission interaction. It has units of L^3/T. The rate equation for P is

$$\frac{dP}{dt} = \Gamma A(N - N_{\text{tr}})P - \frac{P}{\tau_p}. \qquad (3.23)$$

The first term on the right is the stimulated emission contribution. The second term represents the photons leaving the cavity or being absorbed. These effects are combined in the constant τ_p, which is photon lifetime in the cavity. Spontaneous emission is ignored in this equation due to its small contribution. The rates in the above two equations are set equal to zero in order to find the steady state solution. Moreover, the terms I, N, and P are assumed to have a harmonically variable component at the frequency ω:

$$I = I_0 + i_\omega e^{j\omega t}, \qquad (3.24)$$

$$P = P_0 + p_\omega e^{j\omega t}, \qquad (3.25)$$

$$N = N_0 + n_\omega e^{j\omega t}. \qquad (3.26)$$

Solving for modulation efficiency $M(\omega) = p_\omega/i_\omega$, the following relationship is obtained:

$$M(\omega) = \frac{-AP_0\Gamma/(eV)}{\omega^2 - j\omega(AP_0 + 1/\tau) - AP_0/\tau_p}. \qquad (3.27)$$

Equation (3.27) is recognized as the frequency response of the second-order system discussed before. It may be rewritten as

$$M(\omega) = \frac{A\omega_n^2}{-\omega^2 + j2\zeta\omega_n\omega + \omega_n^2},\qquad(3.28)$$

where

$$A = M(0) = \frac{\tau_p \Gamma}{(eV)},$$

$$\omega_n = \sqrt{\frac{AP_0}{\tau_p}},$$

$$\zeta = \frac{AP_0 + 1/\tau}{2\sqrt{AP_0/\tau_p}}.$$

Here, the response is written as a function of ω rather than the complex frequency s. The magnitude of the frequency response is

$$|M(\omega)| = \frac{|A|}{\sqrt{(1 - \omega^2/\omega_n^2)^2 + [2\zeta(\omega/\omega_n)]^2}}.\qquad(3.29)$$

The ratio of optical power modulation to input current modulation is constant at low frequencies. It rises to a maximum at $\omega = \omega_R$ and drops at a rate of 12 dB/octave thereafter (Figure 3.14). The value of ω_R is given as

$$\omega_R = \left[\frac{AP_0}{\tau_p} - \frac{1}{2}\left(\frac{1}{\tau} + AP_0\right)^2\right]^{1/2} = \omega_n\sqrt{1 - 2\zeta^2},\qquad 0 \leq \zeta \geq 0.707.\qquad(3.30)$$

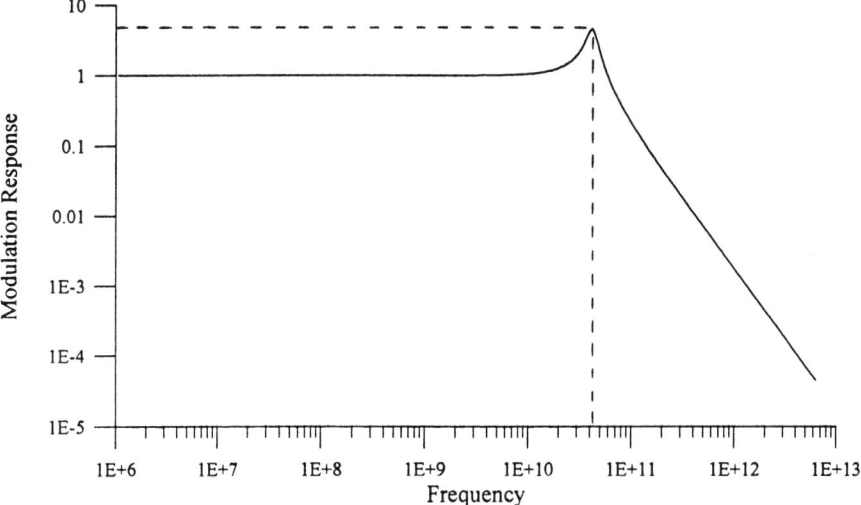

FIGURE 3.14 Direct modulation frequency response of hypothetical laser diode.

The magnitude of the resonant peak M_r is given by

$$M_r = |M(j\omega_R)| = \frac{1}{2\zeta\sqrt{1-\zeta^2}} \cong \frac{1}{2\zeta}. \tag{3.31}$$

Since in Eq. (3.30) the term AP_0/τ_p is dominant in most practical cases (small damping coefficient), it may be further simplified to

$$\omega_R \approx \omega_n = \sqrt{\frac{AP_0}{\tau_p}}. \tag{3.32}$$

For a given laser diode, increased output power improves frequency response. Direct modulation bandwidth is proportional to the square root of output power, limited by gain saturation and electrical circuit time constants.

Example A typical GaAs laser has the following parameters: growth constant times equilibrium photon density $AP_0 = 1.5 \times 10^9 \, \text{s}^{-1}$, cavity photon lifetime $\tau_p = 3.0 \, \text{ps}$, and recombination lifetime $\tau = 4 \, \text{ns}$. Calculate ω_R and the damping coefficient.

The damping coefficient is

$$\zeta = \frac{1.5 \times 10^9 + 0.25 \times 10^9}{2\sqrt{5 \times 10^{20}}} = 0.039,$$

which is much smaller than unity and can be ignored in Eq. (3.30). Therefore ω_R can be found using Eq. (3.32):

$$\omega_R = \sqrt{5 \times 10^{20}} = 22.4 \, \text{GHz}. \quad\blacksquare$$

Direct modulation of the laser diode in general does not generate a pure amplitude- or intensity-modulated output. Even in single-mode devices and in thermally stable operation, the refractive index of the active region varies as a function of charge injection or drive current. It is therefore expected that any direct amplitude modulation of the laser will have an inevitable frequency modulation associated with it. This does not pose any problem if the detection mode is direct detection. It is, however, a major issue if any form of coherent detection is used to receive the modulated signal.

3.4 PHOTODETECTORS

A wide range of devices exist that respond to optical excitation by providing an electrical output. These devices are known as photodetectors and range from the human eye to the photomultiplier tube. Here, only photodetectors that can be used at high speeds and can be integrated with electronic components will be discussed. These photodetectors fall into three general categories: (1) photodiodes, (2) photosensitive transistors, and (3) photoconductors. The merits of each category are discussed separately.

3.4.1 Photodiodes

Photodiodes consist of various types of *p–n*, *p–i–n*, and Schottky junctions that are reverse biased to form a depletion region. An above-bandgap photon absorbed in this region generates a positive charge and a negative charge (electron–hole pair) that move under the applied field and generate an external current. The current through the device is proportional to the number of optically generated electron–hole pairs. Since the photon energy needs to be close to or above that of the bandgap, there is a long-wavelength detection cutoff given approximately by

$$\lambda_c\,(\mu m) = \frac{hc}{E_g} = \frac{1.24}{E_g\,(eV)}. \tag{3.33}$$

The *quantum efficiency* of the detector is the number of electron–hole pairs generated per incident photon:

$$Q = \frac{I_p/q}{P_{opt}/h\nu}, \tag{3.34}$$

where I_p is the photocurrent, q is the elementary charge, P_{opt} is the incident optical power, and $h\nu$ is the photon energy.

Another parameter often used to characterize a photodetector is the *responsivity* R, defined as the ratio of the output photocurrent to the incident optical power:

$$R = \frac{I_p}{P_{opt}} = Q\frac{q}{h\nu}. \tag{3.35}$$

Responsivity is related to quantum efficiency through

$$R\frac{A}{W} = A\frac{q}{h\nu} = Q\frac{\lambda(\mu m)}{1.24\,(eV)}. \tag{3.36}$$

For a given quantum efficiency, responsivity increases linearly with wavelength.

Frequency Response
Limitations on the frequency response of the detector arises from two characteristic time constants. One is the internal time constant determined by the transit time of electrons and holes in the semiconductor; the other is the time

constant of the electrical circuit. The time constant τ_c associated with circuit parasitics is approximated by

$$\tau_c = (R_L + R_s)C, \quad (3.37)$$

where R_s is the internal series resistance of the diode, R_L is the external load resistance, and C is the capacitance associated with the junction depletion layer. Here, the variation of source series resistance with the intensity of optical input is neglected.

The intrinsic time constant τ_i of the device is determined by the transit time of optically generated electrons and holes through the n-layer thickness d_n, the p-layer thickness d_p, and the depletion layer d (Figure 3.15). The electric field and the drift velocities of electrons and holes in these regions are different. In most commonly used diodes with sufficiently high bias fields it is safe to assume that both electrons and holes move at their respective saturated drift velocities, which are almost equal. Therefore, the intrinsic time constant is approximately

$$\tau_i = \frac{d_p + d_n + d}{v_{\text{sat}}}. \quad (3.38)$$

This expression assumes that light absorption takes place in the entire volume of the diode. If the absorption occurs in a smaller volume than $d_p + d_n + d$, the time constant τ_i is reduced.

The two time constants above combine to give the bandwidth B of the photodiode,

$$B = \frac{1}{2\pi(\tau_i + \tau_c)}. \quad (3.39)$$

If a low doped i layer is placed between the p and n regions, the resulting p–i–n diode has considerably less junction capacitance, and τ_i is the dominant time constant determining its bandwidth.

Detector Gain

The gain Γ of the photodetector is defined as the ratio of the photocurrent I to the current that would have resulted from the charge directly generated by the absorbed optical power P_{abs}:

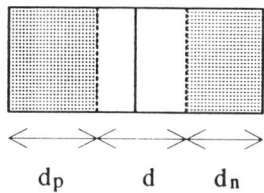

FIGURE 3.15 Dimensions used to calculate intrinsic time constant of photodiode.

$$\Gamma = \frac{I}{qQP_{\text{abs}}/h\nu}, \qquad (3.40)$$

where q is the electronic charge and Q is the intrinsic quantum efficiency of the detector.

In most diodes the gain Γ is close to unity, especially at low bias voltages. Two mechanisms are known to give rise to higher gain in photodetectors: one is the avalanche multiplication process, the other is charge injection. Since in the avalanche gain processes, the electron transit time is increased proportional to the gain of the diode, the bandwidth is reduced proportionally, and the gain–bandwidth product tends to remain constant [14].

Charge injection gain is a process that is caused by the difference in the transit times of electrons and holes. Figure 3.16 shows the contribution to current in the external circuit by either the electrons or the holes as they are swept out of the detector. The maximum gain that can be achieved with this process alone is unity, which occurs when no recombination occurs in the bulk or the surface of the semiconductor. If, however, the mobility of the electrons and the holes differ appreciably, a net charge is left in the material for a short period of time. During that time, charge of opposite polarity may be injected into the device to keep the neutrality, giving rise to an extra current that can flow until all of the slow carriers are swept out. It can be shown that in this case the maximum achievable gain is [14]

$$\Gamma = \frac{1}{2}\left(1 + \frac{\mu_n}{\mu_p}\right). \qquad (3.41)$$

The p–n and p–i–n diodes cannot be easily integrated with microwave monolithic circuits (MMICs) because of the required p doping (MMICs commonly use n-doped MESFET active devices). Schottky diodes on n-doped semiconductor are more attractive for integration with such high speed circuits. The time constant associated with Schottky diodes is in general smaller

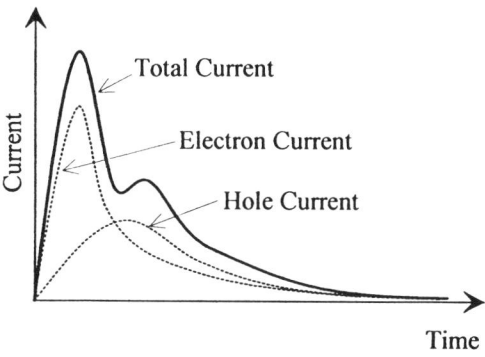

FIGURE 3.16 External circuit current as charge carriers are swept out.

than that of *p–n* diodes mainly because the *p*-doped drift region is missing. In Eq. (3.38) this corresponds to $d_p = 0$. Planar Schottky photodiodes have been reported with bandwidths reaching 100 GHz [15].

Surface Recombination
For diodes with a planar structure another important factor to be considered is the surface recombination of carriers, which leads to gain reduction. If surface recombination occurs over an area of width W, the gain expression representing this phenomenon is

$$\Gamma = \frac{1}{1 + 2S\tau_i/W}, \tag{3.42}$$

where S is the recombination velocity. This reduction in gain is often combined with an increase in bandwidth that tends to keep the gain–bandwidth product relatively constant. Surface recombination does however change the shape of the impulse response due to the fact that carriers generated near the surface are affected most. Recombination can be minimized by surface passivation which may be a layer of higher bandgap material grown on top the active region (such as AlGaAs over GaAs). The higher bandgap material is transparent to the radiation intended for detection. By having a lower refractive index than GaAs, AlGaAs can also serve as one part of a multilayer antireflection coating that allows more light to reach the active region.

3.4.2 Phototransistors

MESFETs and HEMTs have been used as high speed photodetectors compatible with MMIC processing [16]. The advantage of using transistors as opposed to diodes for photodetection is the internal current gain of transistors in the absence of the avalanche effect. One suggested mechanism for this amplification process is that the photogenerated carriers in the depletion layer primarily cause a current flow between gate and drain. This current in turn modulates the electrical potential on the gate that, through the transconductance of the device, modulates the drain current. In other words, the MESFET behaves as the combination of a photodiode and a transistor amplifier.

Some disadvantages of transistor photodetectors are (1) the presence of the gate serves to block part of the incident light and (2) high frequency transistors require precise processing (such as gate recess and mushroom profile), which can add large uncertainties to the response of the photodetector.

3.4.3 Photoconductors

One of the simplest structures used as a photodetector is the photoconductive gap. This structure consists of two ohmic contacts to a thin layer of doped semiconductor with an externally applied electric field. Charge transfer dynamics of photoconductors vary a great deal with the strength of the applied field. This is due to nonlinear dependence of carrier velocities on the electric field, which leads to the development of space charge layers (such as Gunn domains). The only analytical solutions available are for the low field regime where electron and hole mobilities μ_n and μ_p are assumed to be constant.

The intrinsic time constant of the photoconductive detector having an active layer thickness W in the low field regime is given by [14]

$$\tau_i = \frac{W}{2S} \frac{1}{(1 + W/2S\tau_0)}, \qquad (3.43)$$

where τ_0 is the volume lifetime and S is the surface recombination velocity. Photoconductive detectors have been reported with bandwidths exceeding 60 GHz [14].

3.4.4 MSM Detectors

Very fast, efficient, and easy to fabricate detectors can be made using the MSM structure, which consists of two Schottky contacts with a depleted light gathering gap on undoped semiconductor [17]. This detector is explained in some detail here due to its ease of integration on optoelectronic circuits. Subpicosecond electrical pulses have been obtained with this type of detector [18], and high responsivities have been reported [19]. The internal efficiency of the MSM photodetector is comparable with that of the PIN (P-insulating-N) diode. However, in MSM detectors with interdigitated electrodes (Figure 3.17), a large area of photoabsorption is masked by the electrical contacts, resulting in an overall responsivity reduction by as much as 50% unless transparent electrodes or backside illumination is used. The dark current of the MSM detector is much less than that of a photoconductive detector due to the fact that one of the electrodes is always reverse biased.

The responsivity of the MSM detector can be increased under a high applied electric field, which leads to internal current gain. Internal current gain

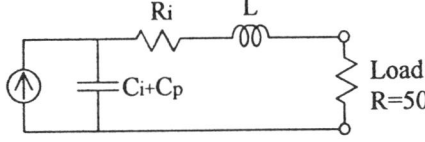

FIGURE 3.17 Interdigitated MSM photodetector with circular detection aperture and its equivalent circuit model.

can increase the responsivity by as much as a factor of 10 [20] at low frequencies and to a lesser extent at high frequencies. The suggested mechanism for this current gain is the possibility of hole injection from the anode into the channel due to accumulation of electrons near the anode.

Frequency Response of MSM Detectors

The response of an MSM detector to an optical impulse normally contains a fast leading edge and a relatively slow trailing edge (Figure 3.18). The trailing edge, which is caused by long carrier lifetimes, gives rise to increased responsivity at low frequencies. The result is a reduced 3-dB bandwidth (Figure 3.19). The bandwidth may be increased by carrier lifetime reduction through either ion implantation or low temperature material growth. Both of these techniques give rise to extra recombination centers that reduce carrier lifetime. The extra bandwidth achieved in this way is at the expense of reduced responsivity at all frequencies and does not boost the high speed performance of the device.

In order to see this mathematically, assume a simple impulse response approximated by zero risetime and an exponential decay, that is,

$$f(t) = e^{-\alpha t}, \quad \text{for } t > 0. \tag{3.44}$$

The magnitude of the transformed function in the frequency domain is

$$F(\omega) = \frac{1}{\sqrt{\alpha^2 + \omega^2}}. \tag{3.45}$$

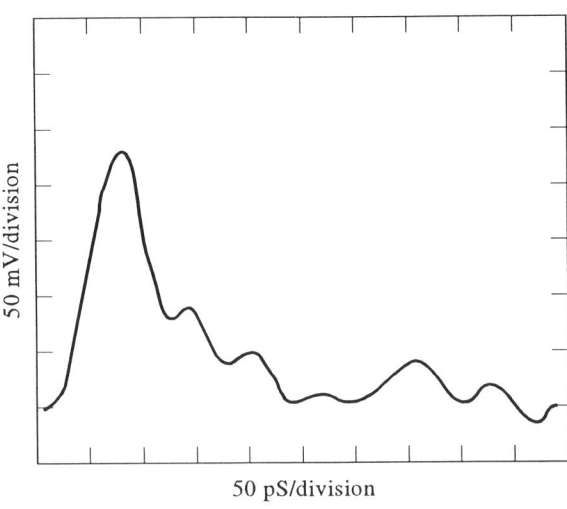

FIGURE 3.18 Measured response of a 2-μm-gap, 100-μm-diameter interdigitated MSM photodetector on GaAs to a 10-ps input optical pulse. Leading-edge slope is limited by oscilloscope response.

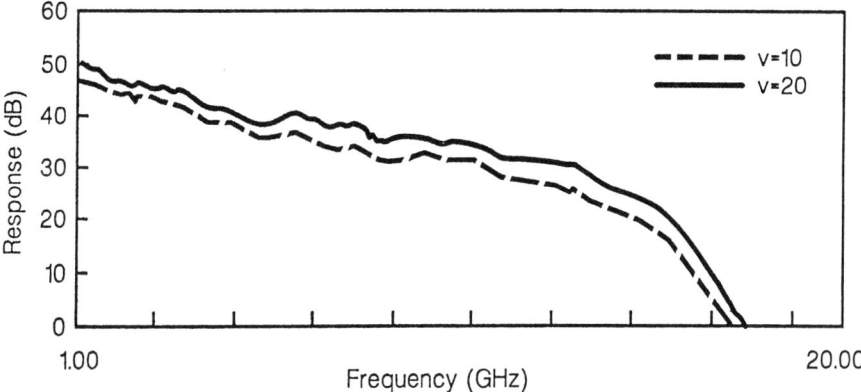

FIGURE 3.19 Measured frequency response of an MSM photodetector at two different bias points.

Reducing the tail of the response by reducing carrier lifetime corresponds to an increase in the decay constant α. In the frequency domain, a higher value of α corresponds to increased 3-dB bandwidth at the expense of reduced responsivity everywhere (Figure 3.20). In narrow-band applications where the slope of the response can be tolerated, this "flattening" of the response curve may not be desirable [21].

The gap capacitance C_g of an interdigitated MSM detector with a gap periphery L in the quasi-static approximation is given by

$$C_g = \epsilon_0(\epsilon_r + 1)\frac{K(k)}{K(k')}L, \qquad (3.46)$$

where

$$k = \tan^2\left(\frac{\pi}{4}\frac{W}{W+S}\right)$$

for finger width W and finger spacing S and $k' = \sqrt{1-k^2}$. The ratio of the elliptic integrals in Eq. (3.46) can be replaced by

$$\frac{K(k)}{K(k')} = \frac{\pi}{\ln\{2[(1+\sqrt{k'})/(1-\sqrt{k'})]\}}, \qquad 0 \le k^2 \le 0.5. \qquad (3.47)$$

Consider the MSM detector layout and its equivalent circuit model shown in Figure 3.17. Fingers and gaps are both 2.0 μm wide, and the diameter of the active circle is 100 μm. Linear periphery L of the detection gap is approximately half of the area of the circle divided by the width of the gap. For this structure $L = 1963$ μm, and the series capacitance is predicted to be

80 OPTICAL ACTIVE DEVICES

0.12 pF. In a 50-Ω system, this value of capacitance excluding the parasitics corresponds to a circuit bandwidth of 26 GHz. Based on Eq. (3.38), with the saturated velocity of GaAs taken to be 2×10^7 cm/s, charge transfer time constant is predicted to be 10 ps, corresponding to a frequency bandwidth of 16 GHz. This MSM would therefore operate in the transit time limited regime.

The optimum high speed design of an MSM detector would require that the time constants due to transit time and due to gap capacitance be equal [22]. The width of the gap and the thickness of the active region give control over the transit time, and the overall size of the device determines the total gap capacitance. A possible approach to a first-order optimum design would therefore be (1) from the required device bandwidth and Eq. (3.39) determine the time constants $\tau_c = \tau_i$, (2) from τ_i and the saturated carrier velocity of the medium determine the required gap, and finally (3) using Eq. (3.46),

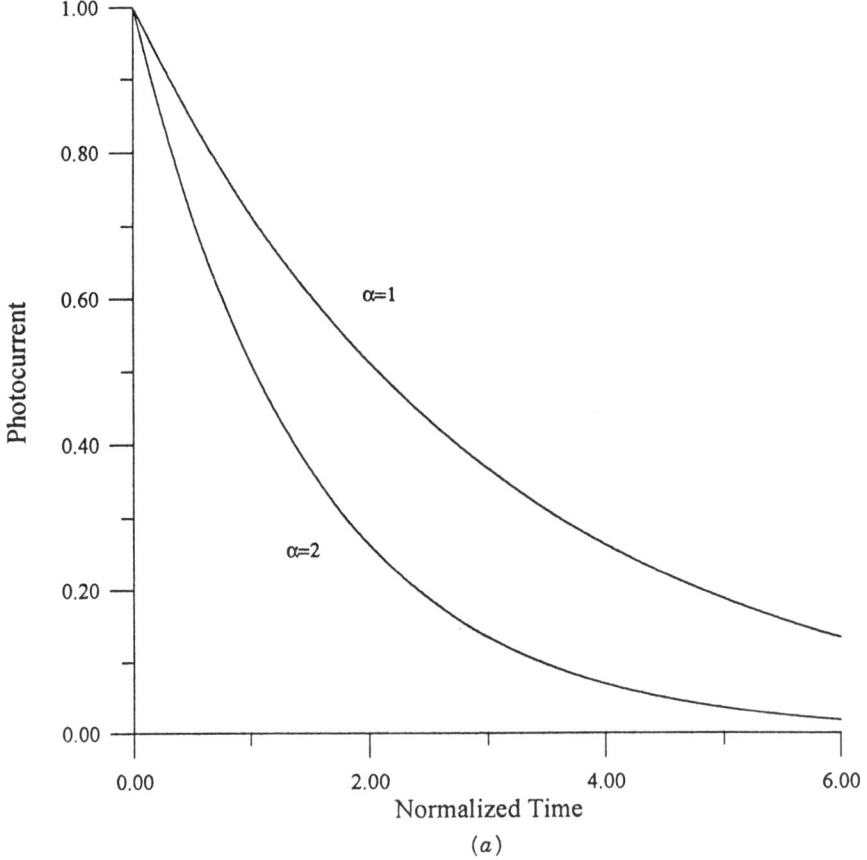

FIGURE 3.20 Effect of reducing the tail of impulse response on frequency domain transfer function.

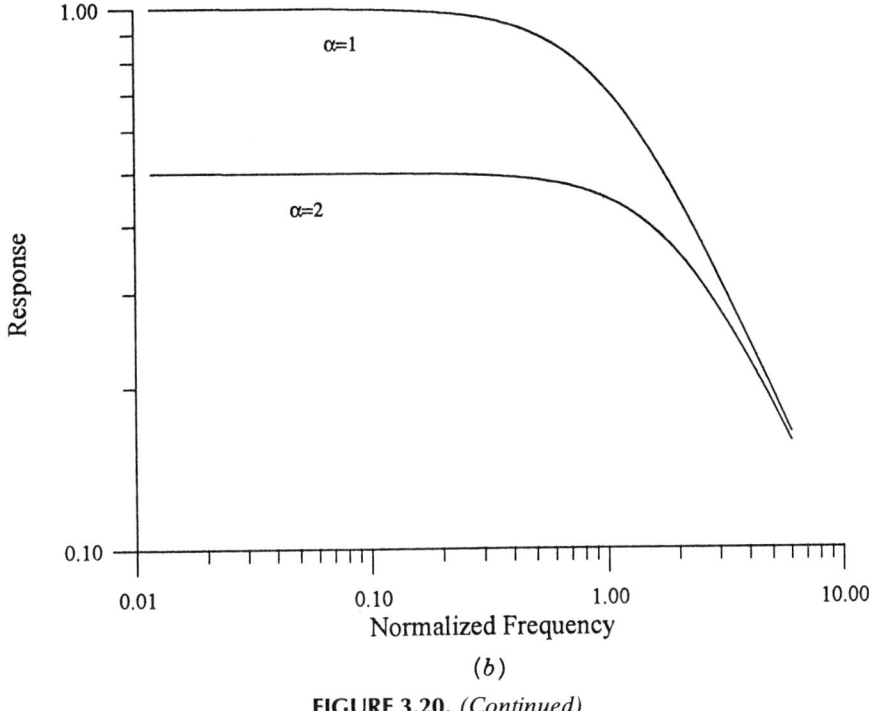

FIGURE 3.20. *(Continued)*

determine the gap periphery that leads to the τ_c that you started with. This procedure is shown in the following example

Example Design a circular aperture interdigitated MSM detector on InGaAs[4] with a detection bandwidth of 110 GHz. Assume equal gap and finger widths.

Note that 110 GHz bandwidth requires that $\tau_i = \tau_c \approx 0.7$ ps. Taking the saturated velocity of InGaAs to be 3.5×10^7 cm/s, the gap width and the active layer thickness should both be 0.25 μm or less.

In order to calculate the maximum gap capacitance, external load resistance may be taken as 50 Ω, and compared with that R_s may be ignored. The maximum allowed gap capacitance is $C = 0.7/50$ pF or 14 fF. Using Eq. (3.46), capacitance per unit length of a quarter micrometer gap is found to be 62 fF/mm. The total gap periphery of the detector should therefore be 226 μm. This leads to a total gap area of $226 \times 0.25 = 56.5$ μm², which is

---

[4]Due to the low Schottky barrier to InGaAs, such detectors normally employ a Schottky enhancement layer such as InAlAs, which also serves as a surface passivation layer and a reflection reducing coating.

half of the total area of the detector. The detector diameter is thereby found to be 12 μm. ∎

It should be pointed out that the 50-Ω external load for calculating τ_c was assumed as an example. Bandwidth and responsivity can be traded off by choosing other impedance levels. Best trade-offs at any given frequency are achieved through the use of a transimpedance amplifier that presents a small resistance to the detector, while being capable of driving higher impedance loads.

In the optimum design approach discussed here it is clear that light gathering capability and responsivity are traded off with detector speed. Smaller detectors require that the light be focused onto a smaller spot, and the thinner active layer absorbs less light and reduces responsivity.

Antireflection Coating

Optical reflection from the air–semiconductor interface causes a loss of signal in the detector. At normal incidence, over 30% of the incident power is reflected back at the GaAs surface. To reduce this reflection, an antireflection coating may be deposited on the surface. The antireflection coating is a quarter-wavelength layer of a material whose refractive index is the geometric mean of the indices on the two sides of the layer. If the refractive index of GaAs is taken to be 3.59, at the air–GaAs interface the required antireflection coating would have a refractive index of $\sqrt{3.59} = 1.89$. At the wavelength $\lambda = 850$ nm, the layer thickness should be $8500/4(1.89) = 1124$ Å. Deposited SiO_2 with a dielectric constant of 1.48 is a good approximation for most applications. The SiO_2 cannot completely eliminate surface reflections, but it can reduce the level to about 6% of the incident power. More precise coatings use silicon oxynitride, whose refractive index can be varied depending on its composition.

For very wide band applications, multilayer AR coatings are utilized. As was mentioned before, one layer of such a coating could play the dual role of also providing surface passivation for reduced surface recombination.

3.5 ELECTRO-OPTIC MODULATORS AND SWITCHES

Electro-optic modulators are normally referred to those employing Kerr and Pockels effects in materials. The dielectric constant of the material changes as a function of the applied electric field. When other electro-optic effects such as the Franz–Keldysh effect (the shift in the absorption band edge due to an applied electric field) are employed in modulators, the name *electro-optic* is normally not used.

The change in the refractive index η of an electro-optic crystal due to an applied electric field is small and is conventionally expressed in terms of a power series as

ELECTRO-OPTIC MODULATORS AND SWITCHES

$$\frac{1}{\eta^2} = \frac{1}{\eta_0^2} + rE + RE^2 + \cdots . \quad (3.48)$$

Here, r and R are the linear and the quadratic electro-optic coefficients, respectively. Linear electro-optic effect is not observed in liquids or in crystals possessing inversion symmetry. In such materials, the reversal of the applied electric field (from E to $-E$) should not change the refractive index due to the absence of a preferred direction. The only way this requirement can be satisfied is for the linear electro-optic coefficient to vanish. In general, the applied electric field in the equation above is a vector, the electro-optic coefficients are tensor elements, and the refractive index value depends on the polarization and propagation direction of light.

In a crystal with no particular symmetry restrictions there are three *principal directions* along which the **D** and **E** vectors of the optical field are parallel. There is a refractive index η_i associated with each of these directions. An optical beam propagating through this crystal with an arbitrary polarization direction has an energy density W associated with it that is given by

$$W = \frac{1}{2\epsilon_0} \sum_{i=1}^{3} \frac{D_i^2}{\eta_i^2}. \quad (3.49)$$

An external electric field applied to the crystal in an arbitrary direction changes the principal axes of the crystal, and the above equation has to be written in the more general form

$$W = \frac{1}{2\epsilon_0} \sum_{i=1}^{3} \sum_{j=1}^{3} \frac{D_i D_j}{\eta_{ij}^2}. \quad (3.50)$$

In Eq. (3.49) the refractive indices η_i formed a diagonal matrix with only three nonzero terms. In the more general expression (3.50) the index matrix is no longer diagonal but rather a symmetrical matrix with nonzero off-diagonal terms. It is possible to convert the symmetrical 3×3 index matrix to a six-element vector and rewrite Eq. (3.50) as

$$W = \frac{1}{2\epsilon_0} \left\{ \left(\frac{1}{\eta^2}\right)_1 D_1^2 + \left(\frac{1}{\eta^2}\right)_2 D_2^2 + \left(\frac{1}{\eta^2}\right)_3 D_3^2 \right. $$
$$\left. + 2\left(\frac{1}{\eta^2}\right)_4 D_2 D_3 + 2\left(\frac{1}{\eta^2}\right)_5 D_1 D_3 + 2\left(\frac{1}{\eta^2}\right)_6 D_1 D_2 \right\}. \quad (3.51)$$

The applied electric field E expressed in the coordinate system of the principal axes changes the six refractive indices according to

$$\Delta\left(\frac{1}{\eta^2}\right)_i = \sum_{j=1}^{3} r_{ij} E_j, \quad (3.52)$$

where the r_{ij} are the elements of the 6×3 linear electro-optic tensor.

84 OPTICAL ACTIVE DEVICES

Example Lithium niobate is one of the commonly used materials for electro-optic modulation. Linear electro-optic coefficients for this material at room temperature and at the wavelength of 633 nm are given by (in pV/m)

$$r_{13} = r_{23} = 8.6,$$
$$r_{22} = -r_{12} = -r_{61} = 3.4,$$
$$r_{33} = 30.8,$$
$$r_{51} = r_{42} = 28.$$

In order to make use of lithium niobate's large electro-optic coefficient r_{33}, both the polarization of the optical beam and the direction of the externally applied electric field should be in the z direction. For this purpose, the direction of propagation of light is chosen to be either in the x or y direction with electrodes placed on the two sides of the beam. ∎

It can be seen from the above example that despite the tensor relationship given in Eq. (3.52), crystal symmetries generally tend to confine the number of coefficients of interest to one or a few.

3.5.1 Phase Modulators

When light passes through an electro-optic material, the application of an electric field to the material causes a change in the speed of propagation of light that leads to a phase shift. In the presence of the linear electro-optic effect only and when the change in the refractive index is small, the phase shift $\Delta \phi$ is given by

$$\Delta \phi = \frac{2\pi L}{\lambda}(\eta_0 - \eta) \approx \frac{\pi L}{\lambda} \eta_0^3 r E. \tag{3.53}$$

Here, λ is the free-space wavelength of light and L is the length of the active region in the electro-optic material. The electric field required to cause 180° phase shift is called E_π and is obtained by setting $\Delta \phi = \pi$:

$$E_\pi = \frac{\lambda}{L \eta_0^3 r}. \tag{3.54}$$

There are direct trade-offs between the length of the modulator, the electro-optic coefficient, and E_π. So, in order to keep the applied electric field low, the length of the modulator can be increased. It was assumed in the derivation of the above equation that the potential on the electrode does not vary with location. In order to keep this assumption valid, care has to be taken at high speeds to keep the length of the modulator small compared to the wavelength of the electrical signal. If this condition is violated, standing waves or spatially varying electric field patterns will form on the electrodes that reduce the efficiency of the modulator. It is possible to avoid this phenomenon in long modulators by making electrodes "traveling-wave" structures. This concept is addressed next.

Example Let us find the maximum operation frequency for a 1-cm-long lithium niobate modulator.

The length of the modulator has to be small compared with the wavelength of the electrical signal, $L \ll \lambda_e$. Let's take the limit at $L = \frac{1}{10}\lambda_e$. The electrical wavelength is given by $\lambda_e = c/(f\sqrt{\epsilon_{\text{eff}}})$ [5] and the effective dielectric constant is approximately $\epsilon_{\text{eff}} = \frac{1}{2}(1 + \epsilon_r)$. Taking the dielectric constant of lithium niobate to be 45,

$$L = \frac{c}{10f\sqrt{23}} \quad \text{or} \quad f = 626\,\text{MHz}. \qquad \blacksquare$$

Traveling-Wave Modulators

At high frequencies, keeping the length of the electrode short compared with the electrical wavelength leads to very small modulators requiring unreasonably high values of E_π. In order to increase the efficiency of high speed modulators, it is possible to use long electrodes. But since local field variations need to be avoided, the long electrode should be fabricated as a controlled impedance transmission line with the proper termination (Figure 3.21). If the propagation speed of light in the optical waveguide and that of the microwave modulating signal on the electrical transmission line are equal, the structure can be as long as desired without any subsequent limitation on the operation frequency. This condition is normally not met in traveling-wave modulators. Denote the refractive index of the optical waveguide by η. The phase velocity of light in the waveguide is simply c/η. On the other hand the phase

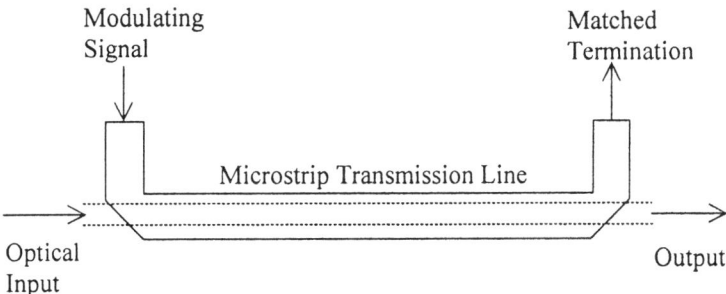

FIGURE 3.21 Traveling-wave electro-optic modulator.

---

[5] Electrical signals that travel along conductors placed on dielectric materials have a propagation speed between c/η_{air} and $c/\eta_{\text{dielectric}}$. The effective dielectric constant is defined to represent this speed. For a detailed discussion of effective dielectric constant see Chapter 4.

velocity of the electrical signal is given by $c/\sqrt{\epsilon_{\text{eff}}}$. These two numbers are, in general, not the same. Also, in a number of commonly used electro-optic crystals such as lithium niobate, the refractive index at optical wavelengths is significantly different from the index at microwave and radio frequency (RF) wavelengths.

The phase angle ϕ_o of the modulated signal on the optical beam after propagating through a device of length L is given by

$$\phi_o = \frac{2\pi L}{\lambda} = \frac{2\pi L f \eta}{c}, \tag{3.55}$$

where f is the modulation frequency. The phase angle ϕ_μ of the microwave signal propagating on the electrical transmission line is given by

$$\phi_\mu = \frac{2\pi L f \sqrt{\epsilon_{\text{eff}}}}{c}. \tag{3.56}$$

As the two signals drift with respect to each other, modulation efficiency is reduced through a cumulative dephasing effect. In fact after the two signals have drifted by 90 degrees, any further modulation begins to "undo" the previous modulation on the optical beam. Therefore an approximate upper limit on the operation frequency of the modulator is obtained by setting the difference between the two equations above to $\frac{1}{2}\pi$:

$$\Delta\phi = \frac{2\pi L f}{c}(\sqrt{\epsilon_{\text{eff}}} - \eta) = \tfrac{1}{2}\pi. \tag{3.57}$$

This leads to a maximum operation frequency given by

$$f_{\text{max}} = \frac{c}{4L(\sqrt{\epsilon_{\text{eff}}} - \eta)}. \tag{3.58}$$

Example Let's repeat the above example with the 1.0-cm lithium niobate modulator, this time assuming a traveling-wave structure:

$$f = \frac{3 \times 10^{10}}{4L(\sqrt{23} - 2.2)} = 2.9\,\text{GHz}.$$

Maximum operation frequency is raised by approximately a factor of 4. ∎

The dielectric constant of lithium niobate is much higher at microwave frequencies than in the optical range. This causes the electrical signal to travel much slower than the optical signal. In a number of other electro-optic materials (such as GaAs), the dielectric constant difference in the two regimes is a lot less pronounced. In those cases the electrical signal tends to travel faster than the optical signal because it partially travels in air. In either case, for ultra-broadband modulators there are techniques to equalize the velocities of the optical and microwave signals.

In cases where the electrical signal is faster than the optical signal, very broadband circuits are available to achieve synchronous propagation. Circuits that effectively increase the path length of the electrical signal include meander lines and periodically loaded transmission lines. These circuits can be made to have the same phase velocity retardation effect at both low and high frequencies and can operate from dc to the millimeter-wave range.

Achieving broadband synchronous propagation with a medium where the electrical signal is slower than the optical signal is unlike the previous case, in that the optical path length and its phase velocity are not allowed to vary. This is to minimize optical losses. The phase velocity of the electrical signal is the one to be modified. One technique available for increasing the velocity of the electrical signal is reducing ϵ_{eff} through an added layer of low dielectric constant material under the electrodes. Silicon dioxide is one candidate for this application [23]. A thick dielectric layer is needed to equalize a large phase velocity mismatch such as encountered with lithium niobate.

The phase velocity of the electrical signal may also be increased by circuit techniques, but this cannot be done over a very broad low-pass frequency range. A low-pass broadband signal such as a short pulse cannot be made to travel faster by any circuit design method. However, band-pass modulators with multioctave bandwidths have been fabricated. Basically, there are two circuit design methods for synchronizing a slow electrical signal with a fast optical beam. One is to use non-TEM signal propagation near the cutoff frequency where phase velocity is substantially higher than for a TEM structure. The other is to adjust for the gradual dephasing of the signals by periodically reversing the polarity of the electrical signal. It is clear that both of these schemes have the inherent restriction of not being effective at very low modulation frequencies.

Non-TEM structures that can be used to achieve phase velocity synchronization include rectangular waveguides, slotlines, and finlines. In general, any transmission line with a finite low frequency cutoff can perform this task. The phase velocity in a non-TEM transmission line increases as the operation frequency approaches the cutoff frequency f_c. The exact expression for the phase velocity v_p is given by [24]

$$v_p = \frac{v_{p\infty}}{\sqrt{1 - (f_c/f)^2}}, \qquad (3.59)$$

where $v_{p\infty}$ is the phase velocity at very high frequencies. The bandwidth that can be achieved by a modulator using a non-TEM transmission line is a function of the initial phase velocity mismatch between the electrical and the optical signals. In general, the larger the mismatch, the narrower the bandwidth.

The second approach for achieving synchronous propagation is to allow the signals to drift by close to 180°, at which point the polarity of the electrical signal is reversed [25]. This is shown schematically in Figure 3.22, where a half CPW (coplanar waveguide) transmission line is used. The length L_p of each line section is determined by setting $\Delta\phi$ of Eq. 3.57 to 180 degrees:

$$\Delta\phi = \frac{2\pi L_p f}{c}(\sqrt{\epsilon_{\text{eff}}} - \eta) = \pi, \qquad (3.60)$$

leading to

$$L_p = \frac{c}{2f(\sqrt{\epsilon_{\text{eff}}} - \eta)}. \qquad (3.61)$$

The bandwidth of this type of traveling-wave modulator decreases with the number of periods n_p. If the bandwidth is defined by a 90 degree additional phase drift from the center frequency, each section contributes $\pi/2n_p$ to this drift. Inserting this into Eq. (3.57), the following simple result is obtained:

$$2\Delta f = \frac{f}{n_p}. \qquad (3.62)$$

For example, a 10-section modulator has approximately a 10% bandwidth.

Optimized Designs

In the previous discussion, in order to estimate the frequency response of a traveling-wave modulator, frequency domain arguments were used. It is instructive to investigate the behavior of such modulators in the time domain. Consider a well-terminated traveling-wave modulator allowing a CW optical beam to pass. If the modulator is driven by a single electrical impulse and the transmission line is dispersionless (TEM), the resulting modulation on the optical beam is a time replica of the electrode pattern with proper scaling. The mapping between a position x along the transmission line and the corresponding time t on the modulated signal is given by

$$t = x\left(\frac{1}{v_\mu} - \frac{1}{v_o}\right), \qquad (3.63)$$

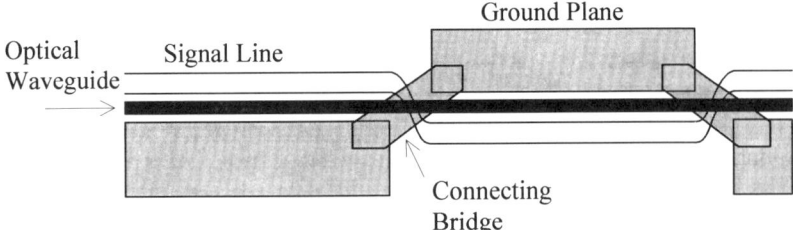

FIGURE 3.22 Phase equalization in a traveling-wave modulator by periodic phase reversals.

where v_μ and v_o are microwave and optical phase velocities, respectively. Therefore the impulse response of the modulator can be calculated using the given scaling. From this simple observation it follows that the frequency response of the modulator is the Fourier transform of the scaled electrode pattern. For example, a modulator with a single uniform section of transmission line has a frequency response in the form of sinc (x). Shortening the transmission line broadens the frequency response. Similarly, periodic polarity reversals on the transmission line shift the response to higher frequencies but lead to reduced bandwidths with increased number of periods.

Aperiodic phase reversals can give much broader bandwidths. An example is the Barker code modulator, whose phase reversal pattern is based on Barker bipolar sequences. Figure 3.23 shows the schematic diagram of a 13-bit Barker code modulator with a reported bandwidth in excess of 40 GHz [26]. Of course there are trade-offs among the length of the modulator, its bandwidth, and its dc response. For a detailed discussion see reference [27].

3.5.2 Amplitude Modulators

Electro-optic modulation changes the electrical path length of light by changing its propagation velocity. To first order, this effect results in a pure phase modulation. Since in present-day optoelectronic systems phase-sensitive detection is rarely utilized, it is desirable to modulate the amplitude or the intensity of the light rather than its phase.

One way to convert a phase modulation into an amplitude modulation is to place the electro-optic modulator between a polarizer–analyzer pair (Figure 3.24). The electro-optic effect creates a variable birefringence that can be used to rotate the plane of polarization of light depending on the applied modulation signal. The polarizer–analyzer pair converts the polarization rotation to an intensity variation. If the polarizer–analyzer pair is arranged in such a way that no optical power emerges when there is no phase delay, then the output intensity as a function of phase delay ϕ is simply a phasor addition given by

$$I_{\text{out}} = I_{\text{in}} \sin^2(\tfrac{1}{2}\phi). \tag{3.64}$$

Since phase modulation was seen to be proportional to the applied field (or the electric potential), intensity modulation resulting from this arrangement has the following dependence on the applied voltage V:

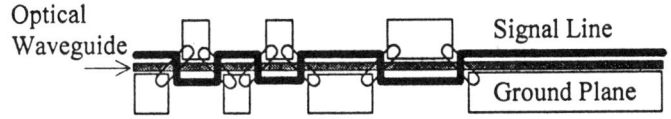

FIGURE 3.23 Schematic diagram of a 13-bit Barker code modulator.

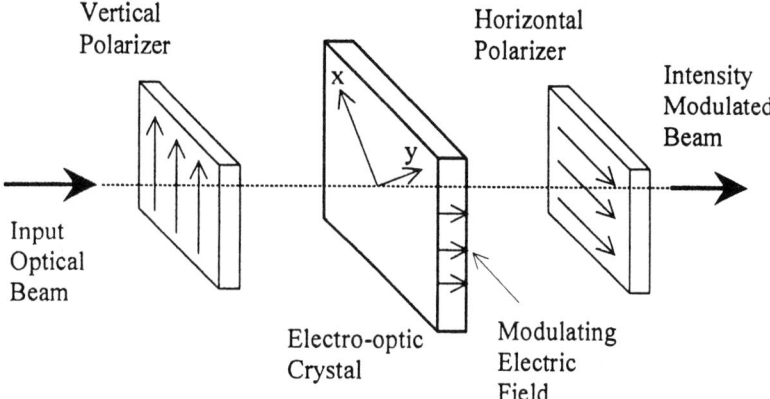

FIGURE 3.24 Converting phase modulation to amplitude modulation by a polarizer–analyzer pair.

$$I_{\text{out}} = I_{\text{in}} \sin^2\left(\frac{1}{2}\pi \frac{V}{V_\pi}\right), \quad (3.65)$$

where V_π is the magnitude of the electric potential that causes 180 degrees of phase shift.

A polarizer–analyzer arrangement is not compatible with monolithic integration. A similar arrangement that can generate amplitude modulation from phase-modulated light and lends itself to monolithic fabrication is the Mach–Zehnder interferometer.

Mach–Zehnder Interferometer

The Mach–Zehnder arrangement is shown in Figure 3.25. It consists of a power divider, two equal-length optical waveguides, and a power combiner. Light propagating through this structure divides, propagates in identical branches, and combines in phase and only suffers minor attenuation in the dividing and combining process. If, however, the electrical path length of one of the branches is altered, the subsequent addition does not reproduce the original intensity. In fact, if light in one branch undergoes an extra 180 de-

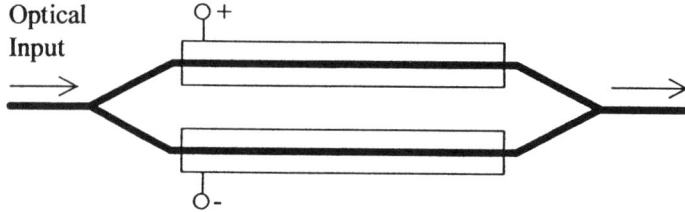

FIGURE 3.25 Schematic diagram of a Mach–Zehnder modulator.

grees of phase shift compared to the other branch, the output will be zero. This is very similar to the intensity modulator with crossed polarizers. The output as a function of the differential phase delay $\Delta\phi$ between the branches is obtained by phasor addition:

$$I_{\text{out}} = I_{\text{in}} \cos^2(\tfrac{1}{2}\Delta\phi). \tag{3.66}$$

3.5.3 Waveguide Switches

Waveguide switches are needed in fiber-optic telephone communications in order to switch an optical signal from one waveguide to another. Normally, an electrical signal is used to do the switching. A great variety of designs are available for switching single or multiple channels. They include methods employing beam steering, holographic techniques, and Bragg deflection [28]. Here, only three types of switches are selected for discussion, all of which fall in the category of electro-optic waveguide switches.

Some waveguide switches have close resemblances to amplitude modulators and may even be used as dual-purpose devices. Consider the amplitude modulators described in the previous section. When the electrical signal sets the transmission amplitude to zero, all the optical power is reflected back along the same optical channel. If the modulator is built such that unused power, instead of being reflected back, is coupled into a third optical waveguide, then the modulator can operate as a switch. The three integrated switch designs addressed below are all of this type.

Directional Coupler Switches

If two parallel identical optical waveguides are placed side by side at a close spacing, they can interact through the evanescent tails of their optical fields. An optical signal propagating in one waveguide will be periodically coupled completely to the other waveguide and then back to the first. The half period of this back-and-forth coupling is called the coupling length L_c, which is a function of the strength of the interaction between the two waveguides. This effect is utilized in optical directional couplers. If the propagation constants in the two waveguides are not identical, only a fraction of the optical power couples back and forth, and L_c is different. (See Section 4.9 for a more detailed discussion.)

Figure 3.26 shows two coupled waveguides with an interaction length $l = 7L_c$. All of the optical power propagating in waveguide A will have coupled into waveguide B by the end of the interaction section. Now consider slightly perturbing the refractive indices of the two waveguides by an applied electric field such that propagation constants in the two waveguides are different. The coupling picture changes. At the proper magnitude of the applied electric field E_s, the coupling length may be reduced from $\tfrac{1}{7}l$ to $\tfrac{1}{8}l$. In that case, the optical signal that started in waveguide A will still be in waveguide A after propagating through the interaction length. A digital signal alternating

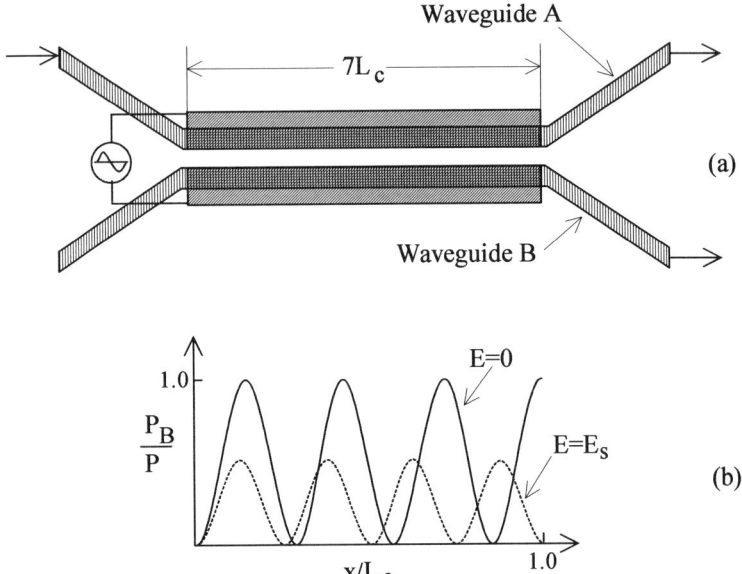

FIGURE 3.26 (a) Directional coupler electro-optic switch. (b) Intensity of optical signal in waveguide B as a function of position along interaction section.

between $E = 0$ and $E = E_s$ can switch the optical field between waveguide A and waveguide B. Also, an analog signal can modulate the intensity of the optical field in either waveguide.

High speed modulators and switches of this type have been fabricated utilizing both discrete and traveling-wave electrodes [29–31]. Directional coupler switches are characterized by high extinction ratio and moderate drive voltages but require precise control of the fabrication parameters.

Branching-Waveguide Switches

A passive branching waveguide is used as an optical power divider (see Section 5.3). It consists of a tapered section connecting the input optical waveguide to two output waveguides. If the structure is symmetrical, equal power is expected to be coupled into the output waveguides. The symmetry of the power divider can be changed by an external electric field applied through electrodes placed over the tapered section of the device (Figure 3.27). If the waveguide material is electo-optic, the refractive index of one side of the active section may be made slightly higher that the other. More of the optical power will propagate toward the side of the branch with a higher refractive index. The extinction ratio depends on the applied potential. Extinction ratios on the order of 10 dB can be achieved by voltages ranging from 6 to 40 V [32].

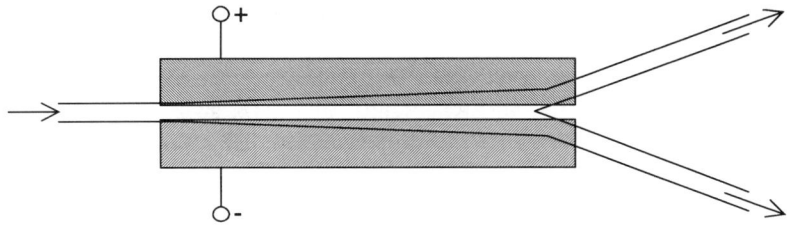

FIGURE 3.27 Branching-waveguide electro-optic switch.

TIR Switches

In the two intersecting optical waveguides shown in Figure 3.28, light propagating in either waveguide is expected to traverse the junction region with little scattering and to continue propagating in the same waveguide. This picture can be influenced by an external electric field inducing a refractive index change inside the waveguide junction. Consider a narrow region along the long axis of the junction whose refractive index is reduced by an applied electric field. Denote this change, as seen by the incident optical field, by $\delta\eta$. Any light that is incident on this region at an angle θ_i greater than the critical angle θ_c will be *totally internally reflected* (TIR) into the opposite waveguide. In this way, an applied potential can switch the light from one waveguide to the other. A switch utilizing this effect is thus called a TIR switch. The condition for total internal reflection is

$$\sin(\theta_i) \geq \frac{\eta - \delta\eta}{\eta}, \tag{3.67}$$

where η is the refractive index of the waveguide mode. This expression can be rewritten in terms of the angle θ_w between the waveguides. For a small $\delta\eta/\eta$ the angle θ_w is also small, and the following approximation holds:

$$\theta_w \leq 2\frac{\delta\eta}{\eta}. \tag{3.68}$$

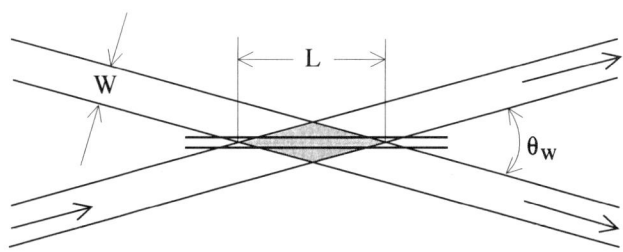

FIGURE 3.28 Waveguide switch based on total internal reflection.

94 OPTICAL ACTIVE DEVICES

For a given waveguide width, the overall length of the switch is determined by the refractive index change that can be achieved.

Example Determine the shortest length of a TIR switch with 6-μm-wide waveguides on an electro-optic material. Assume the achievable refractive index change is 0.5%.

Solution From the geometry shown in Figure 3.28, the length L is given by

$$L = \frac{w}{\sin(\theta_w/2)} \approx \frac{2w}{\theta_w}.$$

The largest allowed θ_w is $2 \times 0.005 = 0.01$ rad. The shortest length L is therefore 1200 μm. ∎

The TIR switches are in general shorter than coupled-waveguide types and are particularly applicable to integrated circuits employing quantum wells, where relatively large refractive index changes can be achieved by moderate applied fields [33].

For other types of optical waveguide switches see references 34 and 35.

3.5.4 Electro-Optic Polymers

The application of polymers and plastics to optical components is not very new. Plastic fibers have been used for short-distance transmission of optical information since the 1960s. The first electro-optic modulation in poled polymer films was reported in 1983 [36, 37]. Practical modulators appeared not long after that [38]. Amorphous polymer coatings are easily applied to various surfaces and thereby offer processing options that are not available with single crystals and epitaxial materials. Their role in both discrete and integrated optoelectronics is expanding, and they bring the advantage of lower cost to most electro-optic applications such as modulators. Their major weaknesses—power handling and operation at elevated temperatures—have improved considerably with recent advances.

Thin films of organic materials used to fabricate active devices do not in general manifest a linear electro-optic effect when initially applied to the surface. The molecules are disordered and the material possesses inversion symmetry. A molecular alignment process such as electrical poling is needed to activate the linear electro-optic response of these materials. Electrical poling is done by placing the material in a strong electric field ($\approx 10^8$ V/m) at an elevated temperature above its glass transition (50–200°C). After the temperature is reduced in the presence of the electric field, the material retains approximately a 10% molecular alignment, which is enough to give it a linear electro-optic coefficient comparable with lithium niobate. The molecular alignment is also the source of an induced optical birefringence that allows the definition of optical waveguides in the poling process.

Electo-optic polymers possess two important characteristics that make them particularly suitable for the fabrication of high speed modulators and switches. First is the fact that their dielectric constants do not vary substantially between microwave and optical frequencies. This leads to an easier synchronization of optical and microwave signals, leading to much broader bandwidths. Traveling-wave modulators without phase velocity compensation built on optical polymers can theoretically operate 10–15 times as fast as those built on lithium niobate. The second favorable factor for high speed operation is the lower dielectric constant of electro-optic polymers. In many cases, electrical transmission lines built on these materials can be much wider for the same electrical impedance. This could have the consequence of lower conductor loss.

A detailed description of electro-optic polymers is not in the scope of this book. The interested reader should consult references 38 and 39.

PROBLEMS

3.1 How much does a temperature rise of one degree affect the wavelength of a 200-μm-long laser with cleaved facets operating at 980 nm? Assume a refractive index of 3.5 and a linear expansion coefficient of $9 \times 10^{-6}/°C$.

3.2 Calculate the groove spacing needed for a first-order Bragg reflector to be used as the output coupler of a laser operating at 980 nm ($\eta = 3.5$). How much does a temperature rise of one degree affect the wavelength selected by this Bragg reflector if it has 50 grooves?

3.3 Referring to Figure 3.13, the step response of a laser diode is characterized by constants t_1, t_3, M_p, c, σ, and ω_n. Discuss the effects of each of these constants on the frequency response of the laser diode.

3.4 Design a traveling-wave phase modulator on lithium niobate ($r = 30$ pV/m) with a V_π of 6 V and the highest achievable bandwidth. The electrical waveguide is a microstrip line with a substrate thickness of 80 μm.

3.5 A traveling-wave modulator utilizing a slotline transmission line operates at 30 GHz. The velocity of the optical beam in the electro-optic material is faster than that of a TEM electrical signal by a factor of 1.8. Determine (a) the required cutoff frequency of the slotline and (b) the bandwidth of the modulator as defined by a 10% deviation from synchronous propagation.

REFERENCES

[1] J. Wilson and J. F. B. Hawkes, *Optoelectronics, an Introduction*, 2nd ed., Prentice-Hall International, London, 1989.

[2] A. Yariv, *Optical Electronics*, 3rd ed., Holt, Rinehart, and Winston, New York, 1985.

[3] K. Vahala, L. Chiu, S. Margalit, and A. Yariv, "On the Linewidth Enhancement Factor in Semiconductor Injection Lasers," *Appl. Phys. Lett.*, Vol. 42, p. 631, 1983.

[4] H. Kroemer, "A Proposed Class of Heterojunction Injection Lasers," *IEEE Proc.*, Vol. 51, pp. 1782–1783, Dec. 1963.

[5] G. C. Osbourn et al., "Applications of Multi Quantum Wells, Selective Doping and Superlattices," in *Semiconductors and Semimetals*, Vol. 24, Academic, New York, 1987.

[6] S. Simhony et al., "Vertically Stacked Multiple Quantum Wire Semiconductor Diode Lasers," *Appl. Phys. Lett.*, Vol. 59, No. 18, p. 2225, Oct. 1991.

[7] P. R. Selway et al., "Measurement of the Effect of Injected Carriers on the p-n Refractive Index Step in a Single Heterostructure Diode Laser," *Electron. Lett.*, Vol. 10, p. 453, 1974.

[8] G. H. B. Thompson, *Physics of Semiconductor Laser Devices*, Wiley, New York, 1980.

[9] G. P. Agrawal and N. K. Dutta, *Long-Wavelength Semiconductor Lasers*, Van Nostrand Reinhold, New York, 1986.

[10] T. H. Windhorn and W. H. Goodhue, "Monolithic GaAs/AlGaAs Diode Laser/Deflector Device for Light Emission Normal to the Surface," *Appl. Phys. Lett.*, Vol. 48, No. 24, pp. 1675–1677, June 1986.

[11] G. A. Evans et al., "Surface Emitting Second Order Distributed Bragg Reflector Laser with Dynamic Wavelength Stabilization and Far Field Angle of 0.25°," *Appl. Phys. Lett.*, Vol. 49, No. 6, pp. 314–315, Aug. 1986.

[12] D. Botez, L. M. Zinkiewicz, T. J. Roth, L. J. Mawst, and G. Peterson, "Low Threshold Current Density Vertical Cavity Surface Emitting AlGaAs/GaAs Diode Lasers," *IEEE Phot. Tech. Lett.*, Vol. 1, No. 8, pp. 205–208, Aug. 1989.

[13] G. F. Franklin, J. D. Powel, and A. Emami-Naeini, *Feedback Control of Dynamic Systems*, 3rd ed., Addison-Wesley, Reading, MA, 1994.

[14] H. Beneking, "On the Response Behavior of Fast Photoconductive Optical Planar and Coaxial Semiconductor Detectors," *IEEE Trans. Elect. Dev.*, Vol 29, No. 9, pp. 1431–1441, Sept. 1982.

[15] S. Wang and D. Bloom, "100 GHz Bandwidth Planar GaAs Schottky Photodiode," *Elect. Lett.*, Vol. 19, pp. 554–555, 1983.

[16] T. Sugeta and Y. Mizushima, "High Speed Photoresponse Mechanism of a GaAs MESFET," *Jpn. J. Appl. Phys.*, Vol. 19, No. 1, pp. L27–L29, Jan. 1980.

[17] W. C. Koscielniak, M. A. Littlejohn, and J. Pelouard, "Analysis of a GaAs MSM Photodetector with 0.1 Micron Finger Spacing," *IEEE EDL*, Vol. 10, No. 5, May 1989.

[18] D. Krokel, D. Grischkowsky, and M. B. Ketchen, "Optoelectronic Generation

and Detection of 350-fs Electrical Pulses," *IEEE CLEO Dig.*, Vol. 11, pp. 246–247, April 1989.

[19] C. S. Harder, B. Van Zeghbroeck, H. Meier, W. Patrick, and P. Vettiger, "5.2 GHz Bandwidth Monolithic GaAs Optoelectronic Receiver," *IEEE EDL*, Vol. 9, No. 4, April 1988.

[20] M. Ito et al., "Monolithic Integration of a Metal–Semiconductor–Metal Photodiode and a GaAs Preamplifier," *IEEE Electron. Dev. Lett.*, Vol. 5, No. 12, pp. 531–532, 1984.

[21] M. L. Riaziat, K. J. Weingarten, L. Generali, D. Gerstenberger, A. Drobshoff, and L. Ching, "All Optical Characterization of MMICs," *IEEE Proc. GaAs IC Symp.*, pp. 347–351, Oct. 1990.

[22] M. L. Riaziat, I. C. Pao, C. Yuen, and R. Marsland, "Realization and Performance of a Quarter Micron Feature Size InGaAs/InP MSM Photodetector," *IEEE IEDM Dig.*, pp. 191–194, Nov. 1991.

[23] C. H. Bulmer, W. K. Burns, and C. W. Pickett, "Linear 0–20 GHz Modulation with a 1×2 Directional Coupler," *IEEE Phot. Tech. Lett.*, Vol. 3, No. 1, Jan. 1991.

[24] S. Ramo, J. R. Whinnery, and T. Van Duzer, *Fields and Waves in Communication Electronics*, Wiley, New York, 1967.

[25] M. L. Riaziat, G. F. Virshup, and J. N. Eckstein, "Optical Wavelength Shifting by Travelling-Wave Electro-optic Modulation," *IEEE Phot. Tech. Lett.*, Vol. 5, No. 9, pp. 1002–1005, Sept. 1993.

[26] M. Nazarathy, D. W. Dolfi, and R. L. Jungerman, "Velocity Mismatch Compensation in Traveling Wave Modulators Using Pseudorandom Switched Electrode Patterns," *J. Opt. Soc. Am. A*, Vol. 4, No. 6, pp. 1071–1079, June 1987.

[27] D. W. Dolfi, M. Nazarathy, and R. L. Jungerman, "40 GHz Electro-optic Modulator with 7.5 V Drive Voltage," *Electron Lett.*, Vol. 24, No. 9, pp. 528–529, April 1988.

[28] H. Nishihara, *Optical Integrated Circuits,* McGraw-Hill, New York, 1989.

[29] M. Kondo et al., "Integrated Optical Switch Matrix for Single-Mode Fiber Networks," *IEEE J. Quant. Elect.*, Vol. 18, No. 10, pp. 1759–1765, Oct. 1982.

[30] E. Marcatili, "Optical Subpicosecond Gate," *Appl. Opt.*, Vol. 19, No. 9, pp. 1468–1476, May 1980.

[31] K. Kubota, J. Noda, and O. Mikami, "Traveling Wave Optical Modulator Using a Directional Coupler $LiNbO_3$ Waveguide," *IEEE J. Quant. Elect.*, Vol. 16, No. 7, pp. 754–756, July 1980.

[32] M. Haruna et al., "Electro-optic Branching Waveguide Switch with Low Drive Voltage," *Opt. Lett.*, Vol. 8, No. 10, pp. 534–536, Oct. 1983.

[33] J. Nayyer, Y. Suematsu, and K. Shimomura, "Analysis of Reflection Type Optical Switches with Intersecting Waveguides," *IEEE J. Lightwave Tech.*, Vol. 6, No. 6, pp. 1146–1152, June 1988.

[34] Y. Silberberg, P. Perlmatter, and J. Baran, "Digital Optical Switch," *Appl. Phys. Lett.*, Vol. 51, No. 16, pp. 1230–1232, Oct. 1987.

[35] M. Diemeer, J. Brons, and E. Trommel, "Polymeric Optical Waveguide Switch Using the Thermooptic Effect," *IEEE J. Lightwave Tech.*, Vol. 7, No. 3, pp. 449–453, March 1989.

[36] K. Singer, N. G. Kuzyk, and J. Sohn, "Second Order Nonlinear Optical Processes in Orientally Ordered Materials: Relationship Between Molecular and Macroscopic Properties," *J. Opt. Soc. Am. B*, Vol. 4, p. 968, June 1987.

[37] S. Lalama and A. Garito, "Origin of the Nonlinear Second Order Optical Susceptibilities of Organic Systems," *Phys. Rev. A,* Vol. 20, p. 1179, 1979.

[38] R. Lytel, G. F. Lipscomb, M. A. Stiller, J. I. Thakara, and A. J. Ticknor, "Organic Electro-optic Modulators and Switches," *Proc. SPIE*, Vol. 971, Aug. 1988.

[39] R. A. Hann and D. Bloor (Eds.), *Organic Materials for Nonlinear Optics*, Royal Society of Chemistry, London, 1989.

CHAPTER FOUR

Microwave Circuits and Integrated Transmission Lines

Low frequency circuit theory has been built around the assumption that all circuit dimensions are small compared with the electromagnetic wavelengths of interest. With this assumption, propagation phase delays are ignored. A very elegant and well-developed theory has resulted from this simplification. The opposite constraint has been used in ray optics, where all dimensions encountered are assumed to be much larger than the wavelength of light. This assumption ignores diffraction and interference effects associated with the wave nature of light. Microwave circuits operate in the regime where neither the approximation of low frequency circuit theory nor the approximation of ray optics holds valid. Simple interconnects of circuit theory have to be treated as transmission lines, and diffraction can never be ignored as it is in ray optics. Integrated optical components operate in the same microwave regime due to the fact that the dimensions of the components are comparable to the optical wavelengths used. It will be seen that some of the mathematical tools used to analyze microwave transmission lines are equally applicable to optical fibers and integrated optical waveguides.

4.1 MICROWAVE INTEGRATED CIRCUITS

Microwave integrated circuits are manufactured in either monolithic or hybrid form depending on the design approach, the application, and the frequency of operation. In a monolithic microwave integrated circuit (MMIC) active and passive components are processed on the same material, most commonly GaAs. The MMIC concept requires precise modeling and layout of the circuits. In order for these circuits to have acceptable yield, active and passive components have to be characterized accurately. Moreover, the reproducibility of their characteristics over time is essential.

100 MICROWAVE CIRCUITS AND INTEGRATED TRANSMISSION LINES

In hybrid microwave circuits, passive components that occupy large areas are processed on different substrates than the active components. Active components are selected and tested before combining with the rest of the circuit. Bond wires are used for connections among the various elements of the circuit. Variations in the characteristics of active components are compensated by manual tuning of some passive component parameters. The hybrid technique allows more control over the final characteristics of the circuit in the presence of diversity in component characteristics. Its drawbacks are its labor-intensive nature, relatively larger circuit sizes, and the inaccuracies and lower reliabilities associated with bond wires.

Hybrid and monolithic approaches are not competing technologies. Large and complicated hybrid circuits incorporate a number of monolithic circuits. Cost trade-offs and the final requirements of the circuit determine what fraction of its components should be manufactured by monolithic processing.

In this chapter, the discussion of microwave circuits is restricted to small-signal and large-signal circuits designed with the MESFET as the primary active device. Although any of the active elements discussed in Chapter 2 can be used as gain elements in microwave circuits, the emphasis on the MESFET was chosen based on current interest and level of activity in the industry.

4.2 MESFET CHARACTERISTICS

In small-signal circuit design, many of the characteristics of the gain element are conveniently expressed in terms of linear matrix parameters. Two-port scattering, impedance, and admittance parameters are among those commonly used for a MESFET. Scattering parameters are of particular interest at high frequencies due to the fact that they are often the directly measured parameters. Highly sophisticated two-port scattering parameter measurement equipment known as network analyzers (see Chapter 8) have been available for a long time. A major attraction of S-parameter characterization over other two-port parameters is that commonly used high frequency active devices are not likely to be unstable (break into oscillation) during S-parameter measurements. This is because of the reflectionless terminations used in such measurements. An extensive circuit theory has been developed based on S parameters for high frequency analog design. It is therefore instructive to investigate the scattering parameters of simplified MESFETs and to establish their interrelationships and the information they give about the operation of the device.

Small-signal equivalent circuits of both intrinsic and extrinsic FETs are shown in Figures 2.12 and 2.14. The input of a MESFET is approximated by a series RC combination. At low frequencies S_{11} is very closely an open circuit due to the very small gate capacitance. As the frequency is increased, S_{11} traces a constant resistance circle on the Smith chart (Appendix C). To a first approximation, S_{22} represents the parallel RC combination of the output circuit. At low frequencies it is nearly a constant resistance. With increased frequency it tends to trace a constant conductance circle on the Smith chart. Here, S_{12} is a measure of the isolation between the input and output of the device. It has a very small magnitude that increases with frequency. The term S_{21} represents the gain of the MESFET. Its magnitude is very high at low frequencies, given by $S_{21} = -2g_m Z_0$, where g_m is the transconductance and Z_0 is the termination impedance. As frequency is increased, S_{21} spirals toward the center of the Smith chart.

For the ideal MESFET of Figure 2.12 the value of S_{12} is zero at all frequencies. When this condition holds, the device is referred to as *unilateral*. Any parasitic feedback element such as gate-to-drain capacitance or source-to-ground inductance will keep the device from being unilateral. In most high frequency applications, device performance deviates considerably from its unilateral approximation. This is due to the fact that with increased frequency, even small parasitic feedback elements have significant effects.

Power gain obtained from a transistor depends on input and output impedances presented to it.[1] In a 50-Ω environment, small-signal power gain is simply $|S_{21}|^2$ of the device. By adding input and output matching networks, higher power gain values can be obtained up to a maximum given by

$$G_{\max} = \frac{|S_{21}|^2}{(1 - |S_{11}|^2)(1 - |S_{22}|^2)}. \tag{4.1}$$

The above expression is valid for unilateral devices only. The parameter $G_{\max}$ is called the *maximum available gain* of the device and is defined as the power gain when input and output of the device are simultaneously conjugate matched, that is, reflection coefficients of S_{11}^* and S_{22}^* are presented to the input and output of the FET, respectively.

If the device is not unilateral, matching at each port of the device affects the other. It may nevertheless be possible to obtain simultaneous input and output conjugate matching. Maximum available gain in this case is called MAG, which is a function of the stability of the device:

$$\mathrm{MAG} = \left|\frac{S_{21}}{S_{12}}\right| \left(k - \sqrt{k^2 - 1}\right). \tag{4.2}$$

In the above equation k is the so-called *stability factor* of the MESFET and is given by

$$k = \frac{1 + |S_{11}S_{22} - S_{12}S_{21}|^2 - |S_{11}|^2 - |S_{22}|^2}{2|S_{11}||S_{22}|}. \tag{4.3}$$

The significance of the stability factor is discussed further in the next sections.

---

[1] Some of the circuit parameters of the MESFET mentioned in this section were discussed in more detail in Section 2.5.

4.3 REACTIVELY MATCHED SINGLE-STAGE AMPLIFIERS

It is customary to have input and output impedances of a high speed amplifier matched to those of coaxial cables or rectangular waveguides.[2] For matching to coaxial cables or broadband matching, the impedance of choice is normally 50 Ω. This standard allows the cascading of devices without any concern for multiple reflections and instabilities caused by interconnect lines. Matching to controlled impedance levels provides for the efficient transfer of power from one device to the next without the need for tuning elements. Calibrated attenuators and filters also require controlled impedance levels for their proper operation. It is an important consideration in the design of matching networks to look at a range of frequencies rather than a single frequency. Normally, it is a particular bandwidth with an allowable passband ripple and a desired stopband isolation that needs to be designed. Equalization of gain versus frequency of an amplifier as well as out-of-band isolation is accomplished in the design of the matching networks. The MAG of the active device normally has a downward slope as a function of frequency. Gain equalization over a broad bandwidth is often achieved by doing partial matching at lower frequencies and reserving the highest degree of matching for the upper frequency end of the operation band.

A single-stage reactively matched FET amplifier consists of an active device and input and output matching sections (Figure 4.1). For a unilateral device (one with input and output impedances independent of the termination on the other side), it is relatively simple to design matching networks. In the following discussion this case is addressed first. In most high frequency applications, however, this assumption is far from applicable. The magnitude of the scattering parameter S_{12} is a good measure of nonunilateral behavior. Significant contributors to S_{12} are often parasitic parallel capacitances or series inductances that cause its value to increase with frequency. When the frequency is high enough, input and output matching networks affect each other and need to be arrived at simultaneously. The degree of added com-

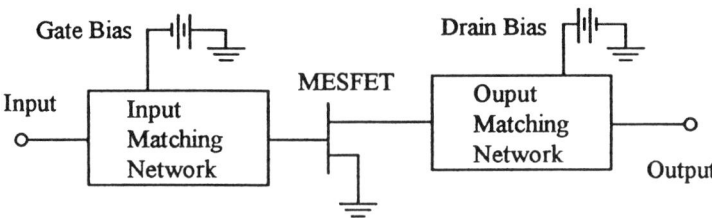

FIGURE 4.1 Common-source MESFET reactively matched amplifier.

---

[2]The concept of impedance matching in circuit theory is identical to reflection elimination in optics. Familiar methods such as Brewster's angle matching and antireflection (AR) coating can all be formulated in terms of impedance matching.

plexity to the circuit design in this case depends on the stability characteristics of the active device.

In the case of a unilateral FET, input and output matching networks may be designed independently. The input impedance of the MESFET is approximated by a series RC combination. The output is approximated by a parallel RC. Matching networks can be synthesized for both ports using either lumped or distributed circuit elements. In either case, desired characteristics of the passband and the stopband, such as bandwidth, ripple, and attenuation, determine the number of elements and the topology used.

4.3.1 Lumped-Element Matching

Lumped elements such as resistors, capacitors, and inductors can be used in high speed circuit design as long as their physical sizes are small compared to the guided electromagnetic wavelength used. The precision of monolithic integrated circuit fabrication makes the miniaturization of components possible so that lumped elements can be used even at millimeter-wave frequencies. Nevertheless it should be kept in mind that high speed lumped elements normally include parasitic components that need to be taken into account. For example, an integrated spiral inductor on a thin substrate has an unavoidable capacitance to ground which is imposed on it by its construction.

Series and parallel reactive elements (either lumped or distributed) can be used to match two real or complex impedances [1]. Figure 4.2a shows the general transformation of a load resistance R_L to an impedance Z by two reactive elements Z_1 and Y_1. Consider a more specific example where a 10-Ω load resistance is matched to a 50 Ω source. It is left as an exercise to the reader to show that $Z_1 = j20$ and $Y_1 = j0.04$ will perform the matching. Figure 4.2b has these values converted to an inductance and a capacitance at the frequency of 10 GHz. In Figure 4.3 the response of this matching network is plotted from 100 MHz to 50 GHz.

The bandwidth of the matching network can be increased by adding more elements. For example, a wider bandwidth can be achieved by matching the 10-Ω load first to 25 Ω and then to 50 Ω (Figure 4.2c). This approach doubles the number of elements in the network and adds more poles to the transfer function. The extra bandwidth is shown in Figure 4.3. Higher number of elements leads to more degrees of freedom and, at first glance, to a more involved task of optimization. Among the parameters of the matching network that are affected by the different combinations of element values are bandwidth, passband ripple, and stopband isolation. The required trade-offs among these parameters determine the choice of the matching network. There are a number of standard networks with various optimization criteria and trade-offs that one could choose from in most commonly encountered cases. The most noteworthy among these networks are Chebyshev (equal-ripple) and Butterworth (maximally flat) networks. A Chebyshev network has an optimum response in the sense that it gives the minimum passband

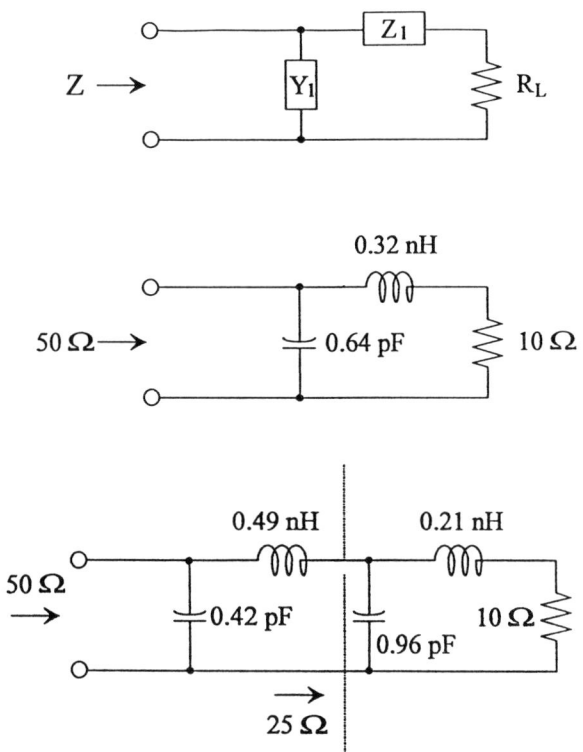

FIGURE 4.2 Multiple-section lumped-element matching networks.

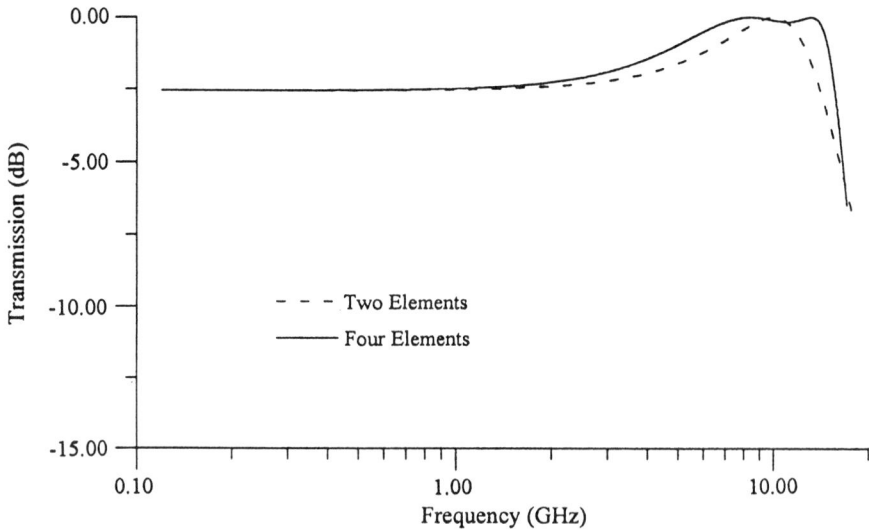

FIGURE 4.3 Extra bandwidth achieved by a four-element matching network vs. a two-element network.

ripple for any given stopband isolation or the maximum stopband isolation for a given passband ripple. A Butterworth network, on the other hand, yields a transfer function that has the highest number of zero derivatives with respect to frequency at the center of its passband. Since the parameters passband ripple and stopband isolation are more applicable to system design, the Chebyshev-type network is more commonly used.

In order to see more of the characteristics of the Chebyshev network, let us examine Chebyshev polynomials of the nth degree. The first three polynomials are $T_1(\omega) = \omega$, $T_2(\omega) = 2\omega^2 - 1$, and $T_3(\omega) = 4\omega^3 - 3\omega$. If ω is replaced by $\cos(\theta)$, the polynomials can be written as

$$T_n(\cos\theta) = \cos(n\theta). \tag{4.4}$$

This equation points to an important characteristic of Chebyshev polynomials. The magnitude of the nth-degree polynomial remains less than or equal to unity for $\omega < 1$. It is equal to unity at n points between $\omega = 0$ and $\omega = 1$. Beyond $\omega = 1$, its magnitude increases indefinitely. Chebyshev networks and polynomials share this characteristic, which is referred to as "equal ripple" in the passband. The response of the Chebyshev network is commonly written as [2]

$$|S_{21}|^2 = \frac{1}{1 + \epsilon^2 T_n^2(\omega/\omega_c)}, \tag{4.5}$$

where ω_c is the half power or the 3-dB point in the frequency response and n is the number of poles of the transfer function, which is equal to the number of reactive elements in a ladder network as well as to the number of ripples in the passband. The magnitude of the passband ripple r, in decibels is related to the parameter ϵ through $r = 10\log(1 + \epsilon^2)$.

Matching networks (or filters in general) with known transfer functions, such as the Chebyshev type, can be synthesized using sets of well-developed rules. Normally, filters and matching networks are designed in the same way initially, with equal source and load impedances and with normalized frequencies. The impedances and frequencies are scaled to the desired values after the completion of the synthesis. One common method of synthesis known as the *Darlington method* is described here as a step-by-step recipe:

- Describe the transfer function $T(\omega)$ (insertion loss in many microwave circuits)[3] as a function of normalized frequency. This could be a Chebyshev polynomial.
- Convert from the ω plane to the s plane by substituting ω with s/j.
- Compute the reflection coefficient from $|\rho(s)|^2 = 1 - |T(s)|^2$ (lossless network).

---
[3] The notation is not to be confused with Chebyshev polynomials.

- Extract $\rho(s)$ from $|\rho(s)|^2 = \rho(s)\rho^*(s)$ by assigning all the left-hand plane poles and zeros to $\rho(s)$.
- Find the input impedance of the network from

$$Z_{in}(s) = \frac{1 + \rho(s)}{1 - \rho(s)}.$$

- Determine element values by a continued fraction expansion of the input impedance function [3].
- Scale the elements to the proper frequency and the proper source and load impedances.

The last item needs further discussion on frequency scaling and impedance transformation: Frequency scaling from f_1 to f_2 is done by multiplying both L and C values by the ratio f_1/f_2. Any resistance in the network is left unchanged. Impedance transformation is done in two steps. First, changing the input and output impedances from the nominal 1-Ω level to Z is done simply by multiplying all impedances by Z. The resulting element values are RZ, LZ, and C/Z. Second, different input and output impedance values can be accomodated by using *LC transformers* as follows: A two-component combination (L-section) of a network can be easily converted to a three-component T or Π section. This conversion may be used to transform impedances. Figure 4.4 depicts some of the conversions and the resulting impedance transformations.

For specific network types such as Chebyshev or Butterworth, with a passband between 0 and 1, steps 1–6 have been calculated and tabulated in various textbooks [4]. Of particular interest is the low pass Chebyshev network shown in Figure 4.5. Element values for this network depend on the number of elements n and the allowed passband ripple r and are given by

$$g_k = 4\frac{a_{k-1}a_k}{b_{k-1}g_{k-1}}, \quad k = 2, 3, \ldots, n,$$

$$g_1 = \frac{2a_1}{p},$$

(4.6)

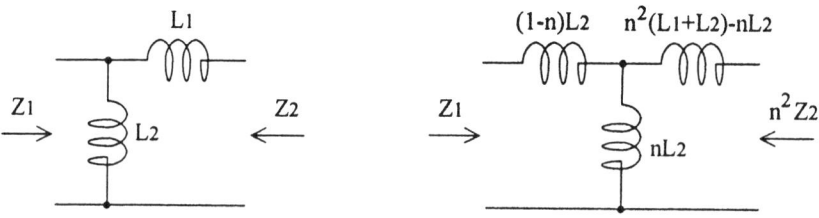

FIGURE 4.4 Example of transformation from L section to T section, yielding a lower impedance ($n < 1$).

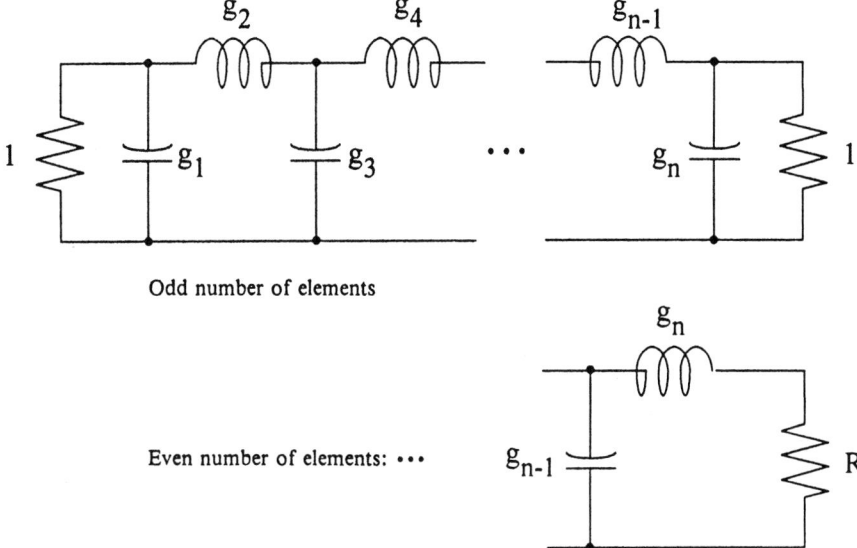

FIGURE 4.5 Prototype low-pass network.

where

$$a_k = \sin\left[\frac{(2k-1)\pi}{2n}\right],$$
$$b_k = p^2 + \sin^2\left(\frac{k\pi}{n}\right),$$
$$p = \sinh\left(\frac{B}{2n}\right),$$
$$B = \ln\left(\coth\frac{r}{17.37}\right),$$

(4.7)

and the passband ripple r is expressed in decibels. When the number of elements n is odd, source and load impedances are both unity. For even n, however, the source impedance Z is given by

$$Z = \tanh^2\left(\frac{B}{4}\right).$$

(4.8)

Example Let's repeat the matching problem of Figure 4.2 using a four-element low pass Chebyshev network.

Solution Since the number of elements is even, it is possible not to do any extra impedance transformation. The source-to-load impedance ratio of 0.2

translates into a passband ripple of 2.55 dB according to Eqs. (4.7) and (4.8). Next, the g values are found from Eqs. (4.6) and (4.7):

$$g_1 = 3.149, \quad g_2 = 0.803, \quad g_3 = 4.013, \quad g_4 = 0.630.$$

For establishing the passband edge at 10 GHz, the g values are divided by $\omega = 2\pi 10^{10}$. Starting at the 50-Ω side of the network, g_1 corresponds to the shunt capacitance, g_2 to the series inductance, etc. In order to normalize these values to 50 Ω, capacitance values are divided by 50 and inductance values are multiplied by 50. The resultant element values are

$$C_1 = 1.002 \text{ pF}, \quad L_2 = 0.639 \text{ nH}, \quad C_3 = 1.277 \text{ pF}, \quad L_4 = 0.502 \text{ nH}.$$

With a load impedance of 10 Ω, the impedance of the matching network and the reflection coefficient Γ are calculated at the 50-Ω port. Since the network is lossless, power transmitted to the load is $1 - |\Gamma|^2$. The plot of transmitted power versus frequency is shown in Figure 4.6. ∎

Low pass networks can be converted to band-pass types by adding resonating elements to either the series or the shunt branches. This causes the zero-frequency point of the transfer function to be shifted to the selected resonance frequency.

In synthesizing matching networks for the input and output of a MESFET, the matching needs to be done between a real source and a complex load or vice versa. Input and output ports of the MESFET are series and paral-

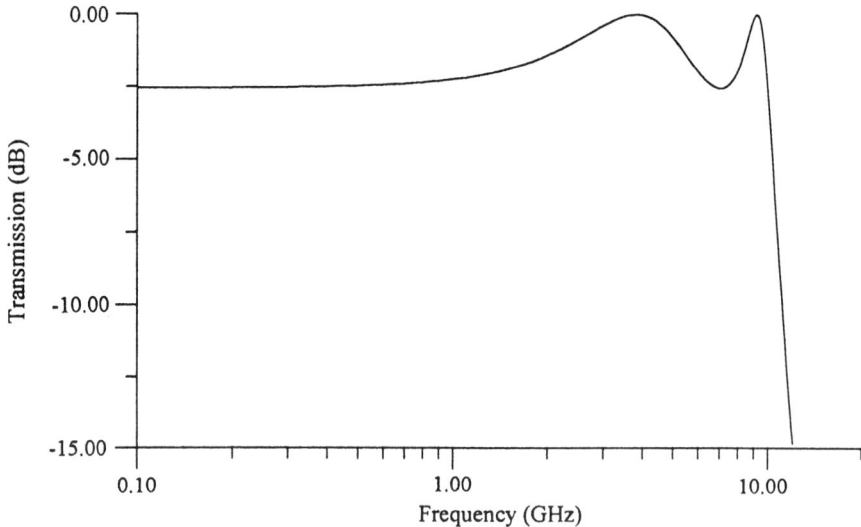

FIGURE 4.6 Response of a four-element Chebyshev low pass network.

lel *RC* combinations with other added parasitic elements. In the synthesis process, an attempt is made to absorb the reactive elements into the matching network. This may place severe restrictions on the characteristics of the matching network and the achievable gain–bandwidth product (see Section 4.3.2).

Detailed analytical procedures exist for designing matching networks based on desired frequency responses that allow for absorbing reactive parasitics [1, 4, 5]. These details are not discussed here partly due to the fact that in practice they are accomplished more accurately by automated extraction and simulation software.

4.3.2 Distributed-Element Matching

Distributed elements are sections of transmission line that can be interconnected to obtain matching. The choice between lumped and distributed circuit design depends on the particular circuit at hand, but in general higher frequencies make the use of distributed components more attractive for integrated circuits. Traditionally, the Smith chart was used for designing distributed-element matching in microwave circuits. This is still a powerful tool especially when non-TEM transmission lines are used. With TEM and quasi-TEM transmission lines (such as stripline and microstrip), sections of line may be used in a fashion similar to lumped-element capacitors and inductors. It is possible to convert matching networks from lumped-element to distributed-element by substituting stubs for each element of the ladder. This allows the application of well-developed synthesis routines mentioned above to distributed element circuit design. This technique is known as the *commensurate-length* design. Another approach to distributed-element synthesis is the use of tapers and transmission line transformers. This is a very powerful and elegant technique for which simple lumped-element analogs do not exist. Both of these techniques are discussed in this section.

Commensurate-Length Design

Transmission line sections may be substituted for the reactive elements of an *RLC* network to yield a similar frequency response. This transformation (known as Richards' transformation) assumes that the transmission lines are dispersionless (impedance and propagation constant independent of frequency) and the cutoff frequency is zero. The building blocks used in Richards' transformation are transmission lines of equal length l (hence the name commensurate-length). The corresponding electrical length is $\theta = \beta_0 l = \pi\omega/2\omega_0$, where β_0 is the propagation constant. At the frequency ω_0 the line is a quarter wavelength long. It is instructive to investigate the characteristics of the line under the frequency transformation $\Omega = \tan(\pi\omega/\omega_0)$. The input impedance of a shorted section of transmission line of characteristic impedance Z_L is $j\Omega Z_L$. An open-circuited section of characteristic admittance Y_C has the input admittance $j\Omega Y_C$. In the Ω domain, a short stub behaves as an inductor, while an open stub acts as a capacitor. It is therefore possible to convert a lumped-element RLC network to a distributed-element circuit by the following substitutions:

$$L \to Z_L, \quad C \to Y_C, \quad R \to R. \tag{4.9}$$

The transformation involves no approximation. Once the substitution between lumped and distributed elements is made, the transfer function replicates exactly.

The mapping between ω and Ω is not one to one. The zero-to-infinity range in Ω corresponds to the zero-to-ω_0 range in ω. This means that the frequency response of the commensurate-length distributed-element network repeats periodically. The period is $2\omega_0$. This is shown graphically in Figure 4.7. Note that strictly speaking there are neither low pass nor high pass circuits in these distributed networks.

At high frequencies where the use of distributed elements are practical, it is difficult to physically realize a circuit consisting only of short and open stubs. Since there are no zero electrical length interconnects available, direct connection of transmission line stubs could be topologically impossible. For this reason, another circuit element known as the unit element (UE) was introduced in the commensurate-length design approach. The unit element is a two-port section of transmission line with the same electrical length as the open and short stubs (90° at ω_0). Any number of these unit elements may be inserted before and after the network without altering the amplitude response. Through a set of identities known as Kuroda's transformations, it is possible to change the order of a unit element and a capacitive or inductive element [3, 4]. In this way Kuroda's transformations allow the unit elements to be moved into the network to the desired locations where they are needed to facilitate the physical realization of the network. Kuroda's transformations are listed in Figure 4.8, and the method used to derive them is discussed briefly in the next section.

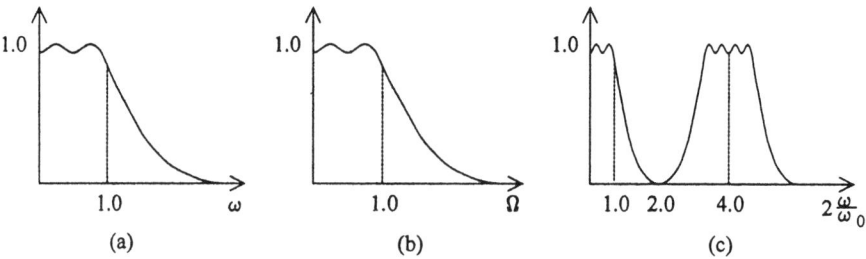

FIGURE 4.7 (*a*) Transfer function of a lumped-element prototype network. (*b*) Transfer function of the corresponding commensurate-length distributed network in Ω domain. (*c*) Transfer function of the distributed network in ω domain.

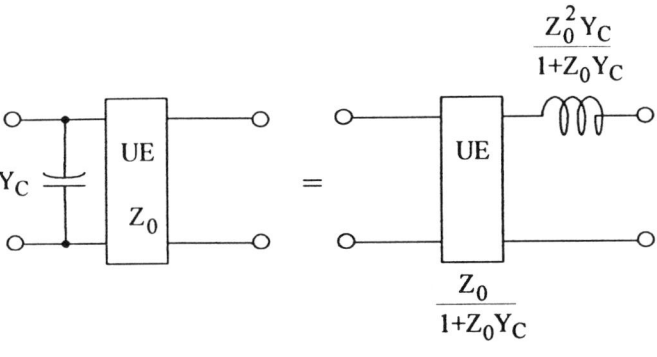
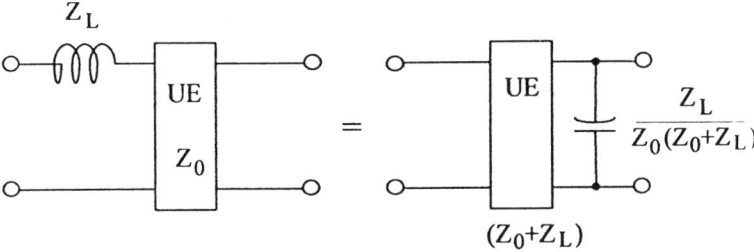

FIGURE 4.8 Kuroda's transformations used to move unit elements to locations of interest in the network.

Distributed Transformers

A section of transmission line whose length is $\frac{1}{4}\lambda$ serves as an impedance inverter; that is, if it is connected to a load impedance Z_L, the impedance seen at its input is Z_0^2/Z_L, where Z_0 is the impedance of the transmission line section. (This will be shown in the discussion that follows.) The input impedance of this combination is matched to the source impedance Z_s if $Z_s = Z_0^2/Z_L$, or

$$Z_0 = \sqrt{Z_s Z_L}. \tag{4.10}$$

A section of transmission line satisfying this condition is known as a *quarter-wave transformer*.

Example The impedance of a medium for unguided plane-wave propagation is defined as $\sqrt{\mu/\epsilon}$. For free-space propagation this value is close to 377 Ω. For nonmagnetic dielectrics it is $377/\sqrt{\epsilon_r} = 377/\eta$. Design an antireflection coating for green light ($\lambda = 550$ nm) impinging on glass with a refractive index of 1.5.

Solution A simple antireflection coating is a quarter-wave transformer that matches the impedance of free space to that of a dielectric. The impedance of the required coating is

$$377/\eta_0 = \sqrt{(377)(377/\eta)},$$

where η_0 is the refractive index of the coating section. It follows that

$$\eta_0 = \sqrt{(1)(\eta)}.$$

This is the familiar requirement that the refractive index of the coating should be the geometric mean of the indices on both sides. In this case, $\eta_0 = \sqrt{1.5} = 1.22$. The thickness of the coating is a quarter of the wavelength in that medium, which is

$$\frac{\lambda}{4\eta} = \frac{550}{4(1.22)} = 113 \text{ nm}. \qquad \blacksquare$$

In order to investigate the performance of a quarter-wave transformer, consider the impedance transformation properties of a section of transmission line. The input impedance of a transmission line of length l terminated in a load impedance Z_L is given by

$$Z = Z_0 \frac{Z_L \cos\theta + jZ_0 \sin\theta}{Z_0 \cos\theta + jZ_L \sin\theta}, \qquad (4.11)$$

where $\theta = \beta l$ is the electrical length of the line. For $\theta = \frac{1}{2}\pi$, $Z = Z_0^2/Z_L$, as stated before. If the impedances are normalized with respect to Z_0, the transformer section inverts the normalized load impedance. The quarter-wave transformer can also be identified as the "unit element" discussed in the previous section. The impedance inversion property of this element is what is used to derive Kuroda's transformations given in Figure 4.8.

As might be expected, a quarter-wave transformer has a very limited bandwidth. Figure 4.9 shows the transformer bandwidth for two different ratios of load-to-source impedances. The bandwidth is narrower for higher transform ratios. It is clear that the bandwidth may be increased by using two transformer sections or matching to an intermediate impedance value first. This is analogous to adding extra sections to the lumped-element matching networks discussed before. This concept may be carried a step further to multisection quarter-wave transformers. A multisection transformer is shown in Figure 4.10. In a multisection transformer in general, there are multiple reflections in every section, and waves propagate in both directions throughout the structure. To simplify the problem, it is safe to assume that when multiple sections are used, the mismatch between two consecutive sections is very small. Only the first-order reflections need to considered, and multiple reflections can be ignored. The reflection coefficient Γ of the whole structure may be considered as the sum of the first-order reflections ρ_i at each stage, with the proper phase factor. For the structure shown in Figure 4.10 this is given by

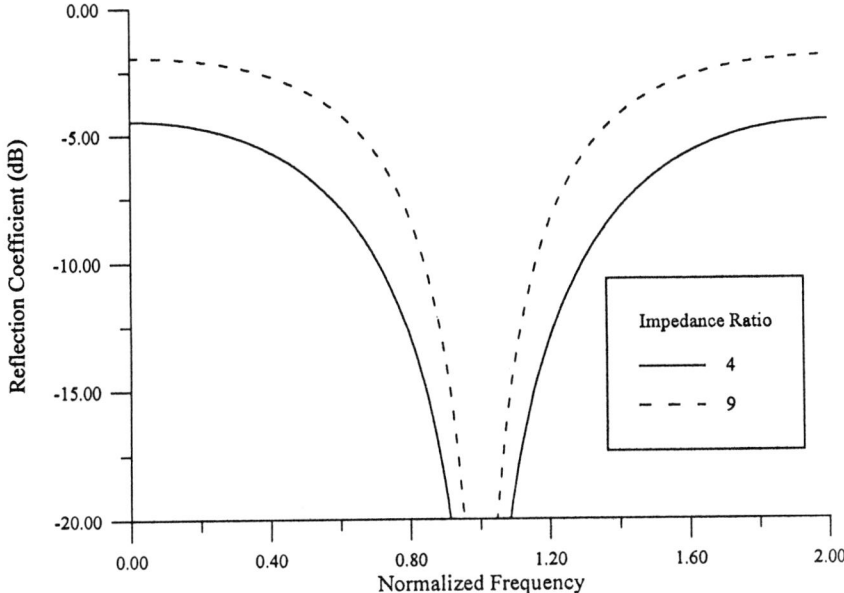

FIGURE 4.9 Bandwidth of quarter-wave transformer for different transform ratios.

$$\Gamma = \sum_{i=0}^{N} \rho_i e^{-j2i\theta}, \qquad (4.12)$$

where θ is the electrical length of each section. It is taken to be the same for all sections. If the network is assumed to be symmetric in the sense that $\rho_i = \rho_{N-i}$, the above summation may be written as the following cosine series:

$$\Gamma = 2e^{-jN\theta} \sum_{i=0}^{N/2} \rho_i \cos(N - 2i)\theta. \qquad (4.13)$$

It is clear that ρ_is may be selected to tailor the response of the multisection transformer to network requirements.

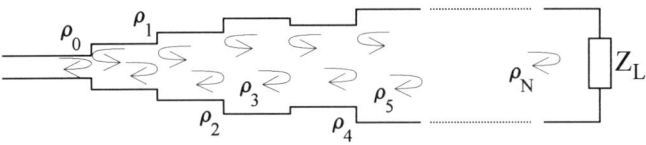

FIGURE 4.10 Multisection transformer and reflections at each section. Electrical lengths of all sections are equal.

Chebyshev Transformer

Designing a multisection transformer with given reflection coefficient characteristics involves equating expression (4.13) to the desired response over frequency. It should be kept in mind, however, that the cosine series nature of this expression implies a periodic response in $\theta = \omega l \eta / c$.[4] The period is $2\omega_0$ in frequency domain or π in θ domain. A Chebyshev response with the center of the passband at ω_0 and one edge of the passband at ω_1 may be written as

$$\Gamma(\omega) = T_n \frac{\cos(\omega l \eta / c)}{\cos(\omega_1 l \eta / c)}, \qquad (4.14)$$

where T_n is the nth-order Chebyshev polynomial. This response with extra amplitude and phase factors can be set equal to the reflection coefficient Γ of Eq. (4.13):

$$\Gamma = A e^{-jN\theta} T_N \frac{\cos(\theta)}{\cos(\theta_1)} = 2 e^{-jN\theta} \sum_{i=0}^{N/2} \rho_i \cos(N - 2i)\theta. \qquad (4.15)$$

At zero value of θ, the reflection coefficient is simply $(Z_L - Z_0)/(Z_L + Z_0)$. From this observation, the value of A is easily determined:

$$A = \frac{Z_L - Z_0}{(Z_L + Z_0) T_N(\sec \theta_1)}. \qquad (4.16)$$

The constant A is the maximum magnitude of Γ in the passband, which is, by definition, the passband ripple $r = 20 \log(A)$. For a given network of order N, the ripple is determined by the source-to-load transform ratio and the desired bandwidth.

The other constants in the expansion determine the reflection coefficients between every two sections. From these constants the impedance of each section of the transformer is determined. Numerically, the determination of the constants requires the expansion of $T_N[\cos(\theta)/\cos(\theta_1)]$ in a cosine power series. In order to illustrate this, a few of these expansions are given below:

$$T_2(\cos \theta \sec \theta_1) = \sec^2 \theta_1 \cos(2\theta) + \sec^2 \theta_1 - 1,$$
$$T_3(\cos \theta \sec \theta_1) = \sec^3 \theta_1 \cos(3\theta) + 3 \sec \theta_1 (\sec^2 \theta_1 - 1) \cos \theta. \qquad (4.17)$$

Example Design a 10-GHz third-order Chebyshev transformer to match a 100-Ω load to a 50-Ω source. Assume an ϵ_{eff} of 8 for the transmission line. The required bandwidth is 2.0 GHz.

---

[4] For TEM transmission lines, η is the refractive index of the medium at the operation frequency. For quasi-TEM transmission lines it is customary to replace η with $\sqrt{\epsilon_{\text{eff}}}$. For non-TEM transmission lines it is best to leave the definition of the electrical length as $\theta = \beta l$, where β is the propagation constant.

REACTIVELY MATCHED SINGLE-STAGE AMPLIFIERS   115

Solution   The length of each section is found from the center frequency:

$$\frac{\omega_0 l \sqrt{\epsilon_{\text{eff}}}}{c} = \frac{\pi}{2}, \qquad \frac{(2\pi)10^{10} l \sqrt{8}}{3 \times 10^{11}} = \frac{\pi}{2}, \qquad l = 2.65 \text{ mm}.$$

The edge of the passband $\omega_1 = 9$ GHz. Therefore, $\theta_1 = \frac{9}{10}(\pi/2) = 0.45\pi$. Using this, we can find the passband ripple A:

$$A = \frac{100 - 50}{(100 + 50) T_3[\sec(0.45\pi)]} = \frac{50}{150 T_3(6.39)} = 3.25 \times 10^{-4},$$

which is -70 dB.

In order to find the other physical parameters of the transformer, we use third-order cosine series expansion:

$$2\rho_0 \cos(3\theta) + 2\rho_1 \cos(\theta) = A T_3(\sec \theta_1 \cos \theta).$$

Using identities (4.17), it follows that $\rho_0 = \frac{1}{2}A(\sec^3 \theta_1) = 0.042$ and $\rho_1 = \frac{3}{2}A \sec \theta_1 (\sec^2 \theta_1 - 1) = 0.124$. Next, from these reflection coefficients and the initial line impedance of 50 Ω, the other line impedances are calculated:

$$Z_1 = \frac{1 + \rho_0}{1 - \rho_0} Z_0 = 54.4\,\Omega, \qquad Z_2 = \frac{1 + \rho_1}{1 - \rho_1} Z_1 = 69.8\,\Omega.$$

The transformer consists of two sections of transmission line with the impedances of 54.4 and 69.8 Ω and the physical length of 2.65 mm each.   ∎

Transmission Line Tapers

In a multisection transformer consider taking the following limit: Let the length of each section become very small and allow the number of sections to increase. As the length l diminishes, the frequency ω_0 becomes infinite, and the response loses its periodicity. In the limit, the transmission line will have a smooth taper instead of discrete sections. At each point x along the taper the local characteristic impedance of the taper is $Z(x)$. Using the same "small reflections" approximation used above, we find the incremental contribution of point z to the reflection coefficient to be

$$d\Gamma = \frac{Z(x + dx) - Z(x)}{Z(x + dx) + Z(x)} = \frac{dZ/dx}{2Z}dx = \frac{1}{2}\frac{d}{dx}(\ln Z)\,dx. \qquad (4.18)$$

The reflection coefficient at the input of the taper is the sum of the individual contributions with the proper phase factor. In the continuous limit, it is given by the following integral:

$$\Gamma = \int_0^L e^{-j2\beta x} \frac{1}{2}\frac{d}{dx}(\ln Z)\,dx, \qquad (4.19)$$

where L is the physical length of the taper and $\beta x = \omega x/c$ is substituted for θ in Eq. (4.13). As expected, Γ is a function of frequency ω.

The impedance of the taper as a function of x is found by inverting Eq. (4.19), which may be recognized as a Fourier integral. The limits of the integral may be extended to $\pm\infty$ if $Z(x)$ is assumed to be constant outside the taper region. The inversion of the integral gives

$$\frac{d}{dx}(\ln Z) = \frac{2}{\pi} \int_{-\infty}^{\infty} e^{j2\beta x}\Gamma(\beta)\,d\beta. \qquad (4.20)$$

For a given reflection coefficient as a function of frequency, this expression finds an impedance taper as a function of position. Of particular interest is the Chebyshev taper whose explicit description will not be given here but may be found using the above Fourier transform. For a given passband ripple and stopband isolation, the Chebyshev taper offers the shortest length. See reference [2] for a more detailed discussion.

4.3.3 Generalized Scattering Parameters

We started this chapter by investigating methods of matching the input and output of an active device such as the MESFET to transmission line impedances. In the previous sections, one-port lumped-element and distributed-element matching were discussed. These discussions are directly applicable to the design of input and output matching networks for two-port active devices as long as the device is assumed to be unilateral, that is, S_{12} of the device is zero. The unilateral assumption is far from valid at higher microwave and millimeter-wave frequencies. The input impedance of the transistor is affected by the output termination, and vice versa.

In practice, simultaneous input and output matching networks are designed in two steps. First, matching networks are synthesized separately for the input and output assuming isolation between the two networks. Second, the unilateral assumption is removed, and the two networks are reoptimized for a simultaneous match using numerical optimizers. It is however possible, if desired, to solve for simultaneous matching with coupled equations. To achieve this, *generalized S parameters* of the device are used instead of the measured parameters.

Note that S parameters of a device are normalized to standard transmission line impedances. They are defined with the assumption that no signal is reflected back toward the device at any port. Generalized S parameters (denoted by S') allow this assumption to be relaxed. They incorporate extra parameters Γ_i representing the reflection coefficient of arbitrary loads presented to the various ports of the device. The derivation of generalized S parameters is done by a matrix renormalization procedure. The scattering matrix is said to be renormalized with respect to arbitrary termination impedances or reflection coefficients. If the diagonal matrix of the reflection coefficients is denoted by $\boldsymbol{\Gamma}$ and the identity matrix by $\mathbf{I}$, the renormalized scattering matrix $\mathbf{S}'$ is given by [6]

$$\mathbf{S}' = (\mathbf{I} - \mathbf{S})^{-1} \cdot (\mathbf{S} - \boldsymbol{\Gamma}) \cdot (\mathbf{I} - \mathbf{S} \cdot \boldsymbol{\Gamma})^{-1} \cdot (\mathbf{I} - \mathbf{S}). \tag{4.21}$$

This expression is general and valid for linear multiport networks. In the particular case of a two-port network terminated in a load with a reflection coefficient Γ_L, the generalized S_{11} is found to be

$$S'_{11} = \frac{S_{11} - \Gamma_L D}{1 - S_{22}\Gamma_L}, \tag{4.22}$$

where D is the determinant of the S matrix.

4.3.4 Conditional Stability

Conditional stability of an active device means that some passive reactances, when presented to the device, cause it to oscillate. In other words, some impedance combinations at either the input or the output of a two-port device may cause the magnitude of either S'_{11} or S'_{22} to become greater than unity. If for a particular load impedance the magnitude of S'_{11}, for example, is greater than 1, a standing wave established on the input transmission line can grow into an unwanted oscillation. A MESFET capable of this behavior is called *potentially unstable*.

Looking at the input of a MESFET, it is possible to find the values of load reflection coefficient Γ_L that could make the device unstable. Solving the equation

$$|S'_{11}| = \left| S_{11} + \frac{S_{12}S_{21}\Gamma_L}{1 - S_{22}\Gamma_L} \right| = 1 \tag{4.23}$$

gives the instability boundary for load values presented to the device. The solution is a circle on the Smith chart (Figure 4.11), known as the *stability circle*. The radius r_L and the center C_L of the circle are found to be

$$r_L = \left| \frac{S_{12}S_{21}}{|S_{22}|^2 - |D|^2} \right|, \quad C_L = \frac{(S_{22} - DS^*_{11})^*}{|S_{22}|^2 - |D|^2}, \tag{4.24}$$

where D is again the determinant of the S matrix. On one side of this boundary, the operation of the device in unstable. In most cases, the center of the Smith chart is where the original S parameters were measured and falls within the stable region. If this is the case, the region inside the stability circle represents load impedances that make the device unstable. The overlap area of the stability circle with the Smith chart is the region of passive loads to be avoided in circuit design [7].

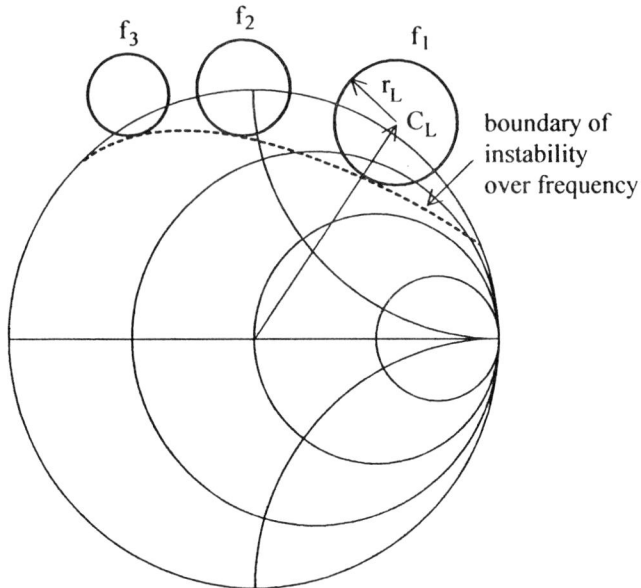

FIGURE 4.11 Stability circle and its overlap with the Smith chart as a function of frequency.

Active devices for which the stability circle falls completely outside the Smith chart are called unconditionally stable. In order for this condition to hold, the magnitude of C_L minus the radius r_L should be greater than unity. This quantity is defined as the stability factor k that we alluded to earlier in this chapter:

$$k = |C_L| - r_L. \qquad (4.25)$$

Stability circles are defined for both the input and the output terminations of the device at every frequency. In amplifier design over a broad bandwidth it is necessary to monitor the location of the stability circles at different frequencies in order to assure stable operation.

Equation (4.2) above gave the MAG of the device in terms of the value of the stability factor. Notice that this expression is valid for $k \geq 1$ only. For $k < 1$, MAG is not defined. There are a number of other parameters that are used in the absence of MAG to quantify the gain available from a transistor. One of the more commonly used parameters is MSG, or the *maximum stable gain*, which is defined simply as the ratio S_{21}/S_{12}. The MSG is the maximum available gain of the device once enough loss has been added to make it unconditionally stable.

4.3.5 Gain–Bandwidth Limitation

In a single-stage amplifier, the gain at any frequency is a function of the input and output match, which is characterized by the reflection coefficient Γ at either port. It is not possible to realize a high degree of matching over an arbitrarily broad frequency range. (Chebyshev networks offer some of the best-trade-offs.) For matching to a parallel RC combination there is a restriction on the magnitude of Γ over a given bandwidth. This restriction is known as Fano's limit and is stated here without proof [8]:

$$\int_0^\infty \ln|1/\Gamma|\, d\omega \leq \frac{\pi}{RC}. \tag{4.26}$$

Fano's limit is applicable to reactive lossless matching.

Consider a circuit that is designed to operate between frequencies f_1 and f_2. Outside this range, Γ may be close to unity, in which case $\ln|1/\Gamma| \approx 0$. Furthermore, if we assume that the reflection coefficient is constant in the operating range, the above expression simplifies to

$$|\Gamma| \geq \exp\frac{-1}{2\,\Delta f(RC)}. \tag{4.27}$$

Let us examine a reactively matched single-stage amplifier consisting of an active device for which the slope of MAG with frequency may be ignored. Assume that the input port of this device is perfectly matched. The power gain delivered by this arrangement to the output load is $(\mathrm{MAG})(1 - |\Gamma|^2)$, where Γ represents the degree of matching between the output of the device and the load impedance. By substituting the minimum value of Γ over any given bandwidth from Eq. (4.27), we find the gain–bandwidth product of this hypothetical case to be given by

$$\mathrm{GBW} = (\mathrm{MAG})\left[1 - \exp\frac{-1}{\Delta f(RC)}\right]. \tag{4.28}$$

This is only due to output matching. Another factor should be added to this expression to represent the GBW limitations due to the input match. The general conclusion to be made here is that over a given bandwidth the maximum gain out of a reactively matched amplifier is not only limited by the MAG of the active device but also by the RC time constants of the input and output of the device.

4.3.6 Noise Matching

The highest gain of an active device is realized when the matching networks achieve simultaneous conjugate match to the input and output of the device.

Similarly, the lowest noise figure is realized if the input impedance presented to the device is Γ_{opt}. These two requirements are in general not satisfied simultaneously. Single-stage reactively matched amplifiers need to find a trade-off between the input return loss and the noise figure of the amplifier (noise figure is insensitive to output loading conditions). The trade-off is made quantitative using contours of equal noise figure and contours of equal return loss on the Smith chart. For more detail see references [9–11]. Simultaneous gain and noise matching using feedback is discussed in Section 4.5.

4.4 BALANCED CONFIGURATION

Identical pairs of poorly matched devices or amplifiers can be combined in a balanced configuration in order to achieve matching by the elimination of reflected signals. This is a popular and simple method of achieving a *lossy match*.

Consider a matched 3-dB directional coupler that is a two-way power divider. It is shown in Section 4.7 that if the coupler is lossless its two outputs are 90° out of phase, that is, $S_{21} = jS_{31}$ and $S_{24} = jS_{34}$. They all have magnitudes of $1/\sqrt{2}$, and the matrix is reciprocal, that is, $S_{ij} = S_{ji}$. Imagine two identical active devices with input reflection coefficients Γ connected to ports 2 and 3 of the directional coupler (Figure 4.12). If a signal of unit amplitude is incident on port 1 of the coupler, the reflected signal to the same port will be

$$S'_{11} = S_{21}\Gamma S_{12} + S_{31}\Gamma S_{13} = 0. \qquad (4.29)$$

On the other hand, the signal arriving at port 4 of the coupler is

$$S'_{12} = S_{21}\Gamma S_{42} + S_{31}\Gamma S_{43} = \Gamma. \qquad (4.30)$$

The input port appears matched, and the reflected power is diverted to port 4, where it is dissipated. A matched broadband amplifier may be constructed

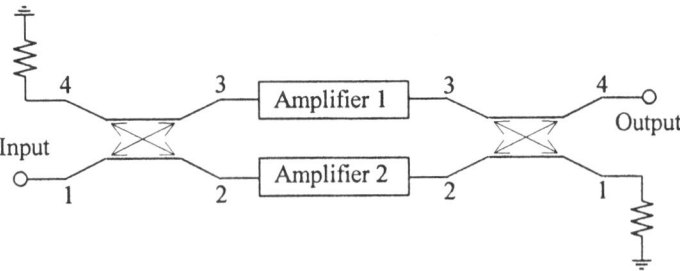

FIGURE 4.12 A balanced amplifier consists of individual amplifiers and two directional couplers.

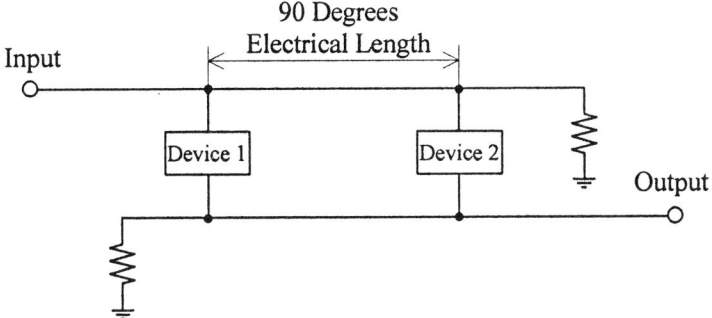

FIGURE 4.13 Schematic diagram of a traveling-wave balanced amplifier.

in this way using two poorly matched amplifiers in each arm. Note that the gain of the resulting *balanced amplifier* is the same as that of each individual amplifier, but the power handling capability is doubled. Also, due to the fact that the directional couplers are terminated in resistive loads, Fano's gain–bandwidth limitation does not apply.

Instead of using two separate directional couplers, a variation of the balanced amplifier uses a structure similar to a single branchline coupler with added terminations (Figure 4.13). This configuration is a traveling-wave balanced amplifier and consists of impedance transformers placed between the devices both on the input line and the output line of the amplifier. There are various ways of analyzing this arrangement. It is left as an exercise to show that the 90° section transforms the impedances such that at the input of the combination the impedance is Z_0, the impedance of the transmission line. Another way to look at this picture is to note any shunt reactive element giving rise to a small reflection on a transmission line may be completely matched by an identical element placed 90° away from it on the same transmission line. Reflections from elements that are 90° apart will be 180° out of phase at the input and therefore cancel. The presence of the matched termination is necessary because the reflections add in phase at the termination and are actually dissipated there.

The input matching obtained with a traveling-wave balanced amplifier is more efficient than those using directional couplers. This leads to the possibility of obtaining higher gain–bandwidth products.

4.5 FEEDBACK AMPLIFIERS

Input and output matching networks for an FET can be designed for the optimization of either gain or noise. In single-stage reactively matched amplifiers, in general, the two cannot be achieved simultaneously; that is, if the lowest noise figure is desired, input return loss needs to be compromised and vice versa. The introduction of feedback can help to achieve both goals at the same time [12]. Multiloop feedback can achieve shaped gain response over frequency with input and output return loss constraints [13]. Feedback is also a tool to control the stability of the active device. The general discussion of feedback control of active devices is beyond the scope of this book [14]. Here, the study is limited to specific examples of how feedback is used in high speed circuit design.

Consider a two-port (three-terminal) device such as a MESFET. There is an input terminal (gate), an output terminal (drain), and a common terminal (source). Feedback elements may be added to this device either in parallel configuration or in series. Parallel feedback forms a second path between the input and output terminals, while series feedback elements are placed between the common terminal and ground (Figure 4.14).

In the classical sense, the process of feedback adds a portion of the output signal to the input either in phase or out of phase. In-phase addition is called positive feedback, and out-of-phase addition is called negative feedback. Positive feedback increases the gain and destabilizes the device. Negative feedback stabilizes the device at the expense of the resulting gain. A more stable device has lower magnitude S_{11} and S_{22}, which makes it easier to achieve broadband matching. An FET in common-source operation has an inherent 180° phase shift (direction of the current source in the equivalent circuit). Therefore a simple resistive path between the input and output provides negative feedback. In high speed circuits, the notions of positive and negative feedback are modified slightly to account for more general phase angles between the input and the feedback signal. If vector addition of input and feedback increases the magnitude of the input signal, we are dealing with positive feedback and vice versa.

The use of reactive elements in the feedback loop gives rise to frequency-dependent feedback. This is important for gain equalization. For example, an inductor in the parallel feedback loop has the effect of providing a high degree of negative feedback at lower frequencies and less at higher frequencies. This results in the flattening of the gain slope.

Multiple loops of parallel and series feedback can in principle be designed

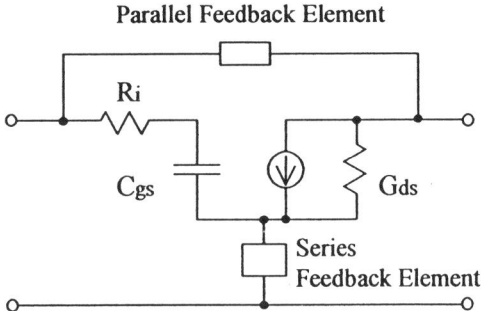

FIGURE 4.14 Intrinsic FET small-signal model and added series and parallel feedback elements.

to cause the points of optimum noise match and optimum gain match to coincide [12]. Simultaneous gain and noise matching is made possible in this way. This is a powerful tool for designing low noise amplifiers. Some authors have reported using multiple-stage feedback amplifiers where each stage is designed to achieve a different goal. For example a three-stage amplifier may have the first stage optimized for low noise, the second stage for flat gain, and the third stage for power [15].

A simple rule for obtaining the new linear parameters of a two-port device after introducing feedback is as follows: The admittance of the parallel feedback path is added to every Y parameter of the device. Similarly, the impedance of the series feedback path is added to every Z parameter of the device. This process may be repeated to account for multiloops of parallel and series feedback.

Example At 5 and 10 GHz calculate the S matrix of the intrinsic MESFET with the following parameters: $C_{gs} = 1.0\,\mathrm{pF}$, $r_i = 5\,\Omega$, $g_m = 0.1\,\mathrm{S}$, $C_{ds} = 0.2\,\mathrm{pF}$, and $G_{ds} = 5\,\mathrm{mS}$. Then add a resistive parallel feedback path of 200 Ω, and recalculate S parameters at the same frequencies.

Solution At 5 and 10 GHz the admittance parameters of the MESFET are $Y_{11} = 4.8 + j30.6$ mS and $18.0 + j57.2$ mS, $Y_{12} = 0$, $Y_{21} = -97.6 + j15.3$ mS and $-91.0 + j28.6$ mS, and $Y_{22} = 5 + j6.3$ mS and $5 + j12.6$ mS, respectively. When converted to an S matrix, the magnitudes of the relevant S parameters are $S_{11} = 0.867$, 0.834 and $S_{21} = 3.89$, 1.99 at the respective frequencies. The addition of the feedback loop is taken into account by adding the value of the feedback admittance (5 mS) to every element of the admittance matrix. When converted to S parameters, the new values are $S_{11} = 0.50$, 0.69 and $S_{21} = 2.41$, 1.55, respectively.

The introduction of the feedback loop is seen to have improved the input match and flattened the gain slope (from 5.8 to 3.8 dB). ∎

One obvious shortcoming of feedback is the reduction of output to input isolation. Other factors to be kept in mind are the restrictions placed on the feedback path by device biasing requirements. For example, in a parallel inductive feedback loop, normally there needs to be a dc blocking element to allow independent biasing of input and output terminals. This can add undesirable parasitics to the loop. Similarly, a series feedback path normally requires a parallel inductor to provide a dc path to ground. This is to avoid dissipating dc power in the feedback resistor.

4.6 DISTRIBUTED AMPLIFIERS

The gain–bandwidth limitation of Section 4.2.1 can be avoided by networks involving more than a single gain element. An example is the traveling-wave balanced amplifier that uses two gain elements. This concept may be expanded to the case where multiple active devices are used. The resulting distributed amplifier can achieve bandwidths that are much broader than those of most other types of amplifiers.

The basic idea of a distributed amplifier is to incorporate the input and output capacitances of the active device in artificial, or lumped-element, transmission lines. A lumped-element low-pass transmission line consists of series inductors L and parallel capacitors C in a ladder network (Figure 4.15). This transmission line has a "characteristic impedance" of $\sqrt{L/C}$ and an upper cutoff frequency of $\omega_c = 2/\sqrt{LC}$ [16]. The distributed amplifier substitutes the parallel capacitors of this structure with active device capacitances and the inductors with high impedance sections of transmission line. This idea was first implemented using triodes in 1948 [17] but did not gain widespread use. It was rediscovered in the 1980s as a design particularly suitable for use in microwave integrated circuits [18].

Let us refer to the distributed-amplifier schematic shown in Figure 4.16. It consists of a string of three-terminal (two-port) elements with gain, connected together to create input and output artificial transmission lines. If the input of the active device is purely capacitive, the signal propagates along the input artificial transmission line without significant attenuation and gets dissipated in the matched termination of the line. Meanwhile, on the output transmission line, the signal grows as it propagates toward the output by in-phase addition of output currents of each active device. A higher number of active devices leads to higher amplifier gain. For practical amplifiers, however, signal attenuation on the input transmission line is significant. After propagating through a few sections, the input signal becomes weak enough such that the succeeding active devices cannot provide any additional power to the output line. The maximum gain achievable by a distributed amplifier is limited in this fashion. The signal is a decaying wave on the input transmission line and a growing wave on the output transmission line. The decay rate of the signal is dominated by resistive losses in the active devices. In common FET distributed amplifiers the loss per unit section of the gate line is much higher than that of the drain line. Gate line loss is the major factor that limits the amplifier gain. This process is of course frequency dependent.

At low frequencies the decay rate is low on the gate line (S_{11} closer to unity) and the growth rate is high on the drain line. At high frequencies

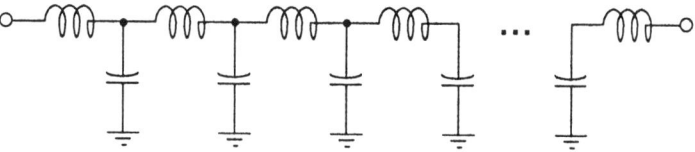

FIGURE 4.15 Artificial low pass transmission line made of series inductors and shunt capacitors.

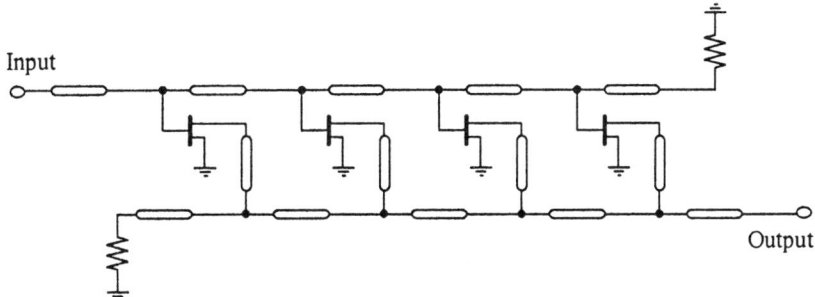

FIGURE 4.16 Schematic diagram of a distributed amplifier using four gain elements.

the reverse is true, leading to lower gain. However, more signal is dissipated in the input and output line terminations at low frequencies. This helps to flatten the gain slope of the amplifier. At the high frequency end of the amplifier's operation band f_c, its behavior is analogous to a 100% directional coupler with gain. At this frequency in a properly designed amplifier the power dissipated in either termination is negligible [19]. The maximum gain of the amplifier at this frequency is the MAG of the individual constituent gain elements.[5] Therefore, the maximum achievable *flat* gain over the operation band of the amplifier is MAG at f_c.

By concentrating on the operation of the device at f_c, the optimum number of gain elements can be determined. The dependence of the gain of the distributed amplifier on the number of gain elements is shown in Figure 4.17. The gain initially increases as more gain elements are added to the amplifier. This trend eventually reverses due to the fact that the signal on the input transmission line is a decaying wave. When the input signal decays to

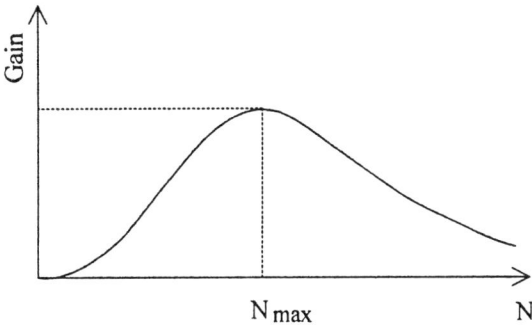

FIGURE 4.17 Distributed-amplifier gain as a function of total gate periphery or total number of devices.

---

[5]This follows from the definition of MAG, which is the gain when all of the available input power is delivered to and all of the available output power is extracted from the active device.

a critical level, the power gain of the last active element cannot compensate for the loss of the last section of the drain transmission line. Adding any more devices to the amplifier beyond this point reduces the overall gain. The number of gain elements, N_{max}, that can provide the highest flat gain is the number of gain elements required to achieve MAG at f_c. Therefore N_{max} is equal to MAG divided by the average gain contribution of each active device. A rough estimate of the value of N_{max} may be found by assuming that the device is unilateral and that all devices are presented with the same input and output impedances:

$$N_{max} \approx \frac{2}{(1-|S_{11}|^2)}(1-|S_{22}|^2). \qquad (4.31)$$

In many cases, this number is determined by losses on either the input or the output transmission line, whichever is dominant. Take the case where the input losses are dominant. The factor involving S_{22} may be ignored. To a first approximation, the input of the FET is r_i in series with C_{gs}. Thus

$$N_{max} \approx \frac{\omega^2(r_i+Z_0)^2 C_{gs}^2 + 1}{2\omega^2 r_i C_{gs}^2 Z_0}. \qquad (4.32)$$

The efficiency of the distributed amplifier for a fixed input signal is proportional to gain divided by current consumption. The highest efficiency is therefore achieved before the gain peak, or at a total periphery that is slightly below N_{max}. Total gate periphery is often used instead of number of active devices. This may be favorable if the gate width of each active device is not known before the amplifier is designed or if devices of unequal gate width are to be used.

Example An FET with a gate width of 150 μm has $C_{gs} = 0.127$ pF and $r_i = 9\, \Omega$. Determine the number of such FETs in a distributed amplifier required to provide the maximum flat gain up to 20 GHz, based on input losses. Input and output impedances of the amplifier are matched to 50 Ω.

Solution This is a case where Eq. (4.32) may be utilized. Plugging in the numbers, $N_{max} = 8.2$ is obtained. If the transistor gate width is fixed at 150 μm, eight transistors may be used. Since this number is approximate, in practice the gate width of each transistor is taken to be a variable to be determined in the automatic design optimization step. ∎

Implicit in the previous discussion was the requirement that at f_c the propagation constants on the gate and the drain lines be equal. This is to assure that signals arriving at the drain line add in phase. This requirement may be quantified in terms of physical parameters if the input and output lines are considered as periodically loaded transmission lines. In periodically loaded transmission lines, the phase change per section may be calculated from the following relationship [16]:

$$\cos \phi = \sqrt{\frac{Z_{oc}}{Z_{oc} - Z_{sc}}}, \qquad (4.33)$$

where Z_{oc} and Z_{sc} are the input impedances of a section when the output is either open circuited or short circuited, respectively. It is left as an exercise to show that using this expression the following solution for equal propagation constants along the gate and drain lines may be obtained:

$$\frac{Z_0^g}{Z_0^d} = \frac{Z_{in}}{Z_{out}}, \qquad (4.34)$$

where the Z_0 are the characteristic impedances of the unloaded transmission lines and Z_{in} and Z_{out} are the input and output impedances of the active device. Losses on the unloaded transmission lines are relatively small and their characteristic impedance ratios are real. On the other hand, the ratio of the input to output impedances of a MESFET is, in general, complex. A section of transmission line connecting the drain of the MESFET to the drain line (Figure 4.16) serves partially to adjust the Z_{in}/Z_{out} ratio to be real at f_c.

Early distributed amplifiers used identical sections and identical active devices. Later, it was recognized that nonidentical sections offered higher design flexibility and, in general, higher performance. There are analytical design criteria to yield flat gain over the operation band when identical sections are used [20]. In practice, and particularly with nonidentical section designs, gain flattening is done by numerical optimization. Monolithic distributed amplifiers have been fabricated as broadband low noise amplifiers [21], as medium power amplifiers [22], as combined feedback and distributed amplifiers [23], and as ultrabroadband amplifiers [24].

4.7 FET OSCILLATORS

In the previous sections of this chapter the versatility of three-terminal devices in amplifier design and their advantages over two-terminal devices were shown. As sources of microwave signal, the advantages of three-terminal devices over Gunn and IMPATT-type oscillators are not as easily recognized. Nevertheless, three-terminal devices do manifest distinct and important advantages. Among them are the possibility for monolithic integration, high efficiency in CW operation, lack of a threshold current, and ease of stabilization and compensation, to name a few.

In two-terminal oscillators, the application of proper bias voltage or current causes the device to exhibit negative resistance. The design process mainly addresses the issue of efficient power coupling to the load as well as the universal issues of stability, noise, power output, and thermal design.

In a transistor oscillator the negative-resistance component is normally not present and has to be introduced by feedback. In this section, the subject of oscillator design by feedback is discussed in some detail.

4.7.1 Oscillation Condition

An oscillator circuit consists of a source of microwave power and a load impedance to which the power is delivered. The source has a nonlinear, amplitude-dependent impedance $Z_S(A)$, while the load impedance Z_L is assumed to be linear and power independent. Both of these impedance values are as measured at the same arbitrarily chosen reference plane between the source and the load. Applying Kirchhoff's law at this plane, we get

$$I[Z_S(A) + Z_L] = 0, \qquad (4.35)$$

where I is the current at the reference plane. If we choose the nontrivial solution of $I \neq 0$, it follows that $Z_S(A) + Z_L = 0$. This is the *oscillation condition*. The real part of this solution is

$$R_S(A) = -R_L, \qquad (4.36)$$

meaning that the power is stabilized at the level where the negative resistance of the source equals in magnitude to the resistance of the load (plus any parasitic resistances). The oscillation condition is often expressed in terms of the source and load reflection coefficients Γ_S and Γ_L as

$$\Gamma_S(P)\Gamma_L = 1. \qquad (4.37)$$

Again, since $|\Gamma_L| < 1$, it follows that $|\Gamma_S| > 1$. The real part of the oscillation condition determines the oscillation amplitude. As the oscillation is built up from noise, the negative resistance causes the amplitude to increase. With the rise in amplitude, the negative resistance $R_S(P)$ diminishes until its value equals R_L, where the amplitude reaches steady state. The oscillation frequency is determined by the imaginary part of the oscillation condition, which sets the source and load reactances to be equal and opposite.

It is intuitively appealing to look at the oscillation condition graphically. Figure 4.18 shows two plots in the complex impedance plane. The first plot is the load impedance $Z_L(\omega)$ as a function of frequency. The second plot is the negative of the source impedance $\overline{Z}_S(A) = -Z_S(A)$ as a function of amplitude. The point of intersection of these two curves specifies the frequency and amplitude that satisfy the oscillation condition. Yet, any intersection point does not necessarily guarantee stable oscillation. In order for steady state oscillation to be sustainable, another condition known as the *stability condition* also needs to be satisfied. The stability condition requires that small disturbances to either frequency or amplitude decay with time. The necessary and sufficient condition for stability is that the angle ϕ between $\partial \overline{Z}_S(A)/\partial A$

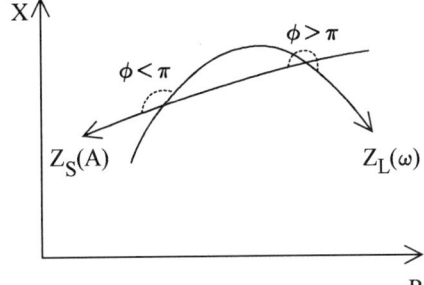

FIGURE 4.18 Graphical representation of oscillation condition and stability condition for oscillators.

and $\partial Z_L(\omega)/\partial \omega$ lie between zero and π [25]. Conditions where ϕ is close to either one of these two extremes represent noisy oscillations on the verge of instability. The lowest noise condition is where $\phi \approx \frac{1}{2}\pi$.

4.7.2 Stabilization of FET Oscillators

A free-running oscillator is subject to short-term and long-term frequency and phase drift, making it unusable in most applications. Popular methods of stabilizing the frequency of an oscillator include employing a high-Q resonant cavity or phase locking the oscillator to a stable reference signal.

Cavity Stabilization and DROs

Single-frequency (or mechanically tunable) FET oscillators commonly use high-Q resonant cavities as narrow-band filters to achieve frequency stability. Such narrow-band filters are used either in the feedback loop or at the output to filter the outgoing signal. Metallic cavities resonating at lower microwave frequencies are relatively large, and their physical structures are not in accord with integrated circuit technology. Low loss dielectric materials with high permittivities can form compact, high-Q resonators that may be integrated with microwave circuits. Materials developed specifically as dielectric resonator ceramics demonstrate low dielectric constant variation with temperature. The main constituent in most such materials is titanium oxide (TiO_2) due to its low loss and high dielectric constant. Other materials such as barium oxide, zirconium, and tin are added to titanium oxide to achieve temperature stability [26]. Dielectric resonators are designed such that the thermal expansion of the resonator compensates the change in permittivity, keeping the resonance frequency constant to a first order.

Dielectric resonators are commonly used as cylindrical cavities with evanescent field coupling to nearby microstrip lines. These resonators are typically a few millimeters in diameter and a similar height, achieving unloaded Q values of 5000–10,000. Dielectric resonator oscillators (DROs) are common in the frequency range of 2–20 GHz. Figure 4.19 shows two implementations of common-source MESFET DROs.

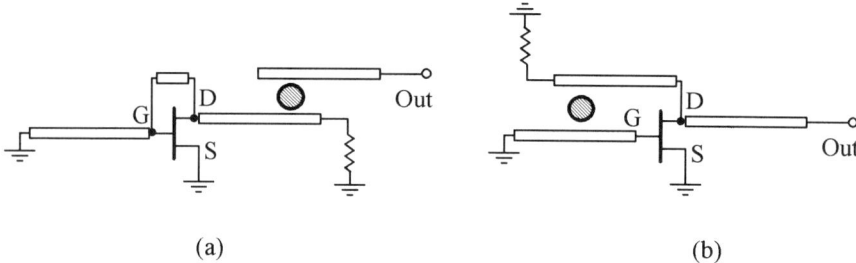

FIGURE 4.19 Two DRO configurations: (*a*) dielectric resonator as output bandpass filter and (*b*) in the feedback loop.

The oscillation frequency of the DRO can be mechanically tuned by placing a conducting plate at an adjustable distance to the top surface of the cylindrical cavity.

Phase Locking and Delay Line Stabilization

There is a class of oscillators known as voltage-controlled oscillators (VCOs) that, instead of being stabilized, employ a controllable reactance, or other means to externally tune the output frequency. A VCO is useful as a generator of frequency-modulated carriers. It can also be externally stabilized by phase locking to a stable oscillator, as shown in Figure 4.20*a*. Alternatively, it may be "phase locked" to its own delayed signal by delay line stabilization (Figure 4.20*b*). At the core of both of these approaches is a phase or frequency discriminator. The role of the discriminator is to generate an error signal proportional to the phase difference or the frequency difference of its two inputs. The output of the discriminator is fed back to the VCO in order to correct for the difference.

A simplified model of the discriminator is a black box whose output error signal as a function of time $E(t)$ is proportional to the phase difference between the two inputs. Referring to the configuration shown in Figure 4.20*b* the error signal is given by

$$E(t) \propto (\Delta\phi - \omega\,\Delta t). \tag{4.38}$$

The desired frequency of oscillation ω_0 is set by the proper choice of the phase shifter setting $\Delta\phi$ with respect to the delay time, that is,

$$\Delta\phi = \omega_0\,\Delta t - 2n\pi, \tag{4.39}$$

where n is an integer. Once this value is plugged into Eq. (4.39), the result is

$$E(t) \propto (\omega_0 - \omega)\,\Delta t. \tag{4.40}$$

This means that the sensitivity of the discriminator to small variations in ω increases linearly with the delay time. Delay time cannot be increased indefinitely. Limitations arise from the loss and bandwidth of the delay line

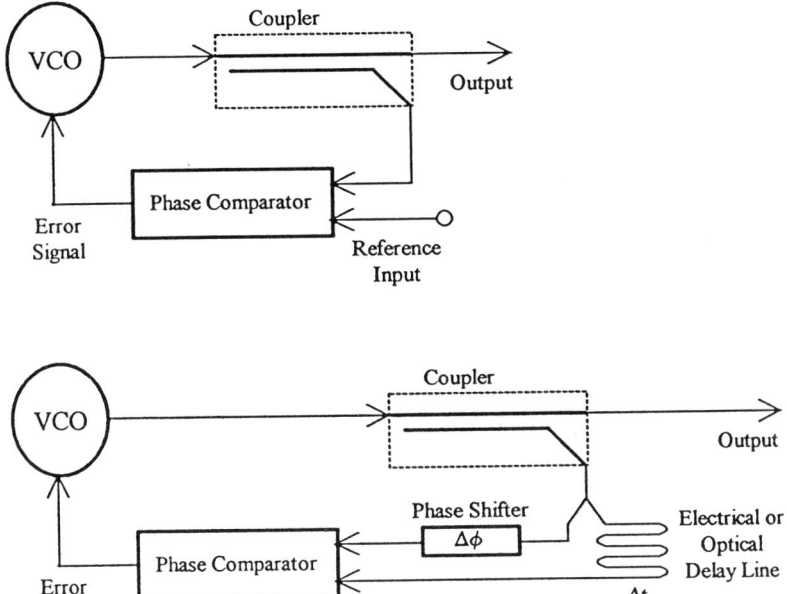

FIGURE 4.20 (a) Phase-locked VCO and (b) delay line stabilized VCO. Long delay times are possible with fiber optics.

itself as well as the problem of mode hopping. Some authors have shown that the problem of delay line loss for long delay times can be overcome by using fiber-optic delay lines [27]. The added complexity to the circuit is of course the presence of optical transmit/receive modules and the phase noise associated with them. Mode hopping problem arises from the fact that for any phase and delay time setting, oscillation is possible at many different frequencies (modes). The separation between oscillation modes narrows with increased delay time. When mode separation is less than the bandwidth of the delay loop filter, mode hopping can occur. This problem is particularly noticeable when the center frequency of the oscillator is being tuned.

4.8 INTEGRATED TRANSMISSION LINES

In microwave integrated circuits every "interconnect" whose length is not negligible compared to the wavelength needs to be treated as a transmission line and requires careful control of the parameters associated with it. Nevertheless, transmission lines should not be thought of as merely unintentional byproducts of high speed. Most microwave integrated circuits use transmission lines as circuit elements to achieve effects not easily realized otherwise (such as tapers and transformers). Also combinations of transmission line sections give rise to a variety of *distributed components*, one of the most well known being the directional coupler. The basics of some distributed components are discussed in Section 4.9.

The transmission line whose structure most resembles a low frequency "interconnect" is *microstrip*. Microstrip is also the most commonly used transmission line in microwave integrated circuits. Other transmission lines compatible with integrated circuit technology include *stripline, coplanar waveguide* (CPW), *coplanar strips* (CPSs), *slotline,* and *finline* (Figure 4.21). The characteristics of some of these transmission lines are discussed here, with the major emphasis on CPW.

4.8.1 Microstrip

Microstrip consists of a single conductive strip on one side of a dielectric substrate with the other side completely covered by a conducting ground plane. In hybrid microwave circuits, the substrate is made of a low loss dielectric material selected for particular trade-offs among loss, thermal conductivity, thermal expansion coefficient, cost, and so on. Table 4.1 lists some commonly used substrates and some of their physical characteristics. In a monolithic integrated circuit there is less flexibility in the choice of substrate, as it is selected based on the active devices that can be fabricated on it. Both GaAs and InP provide good low loss substrate materials for transmission lines as well as for many active devices. Due to its high loss, silicon is normally not considered a good transmission line substrate.

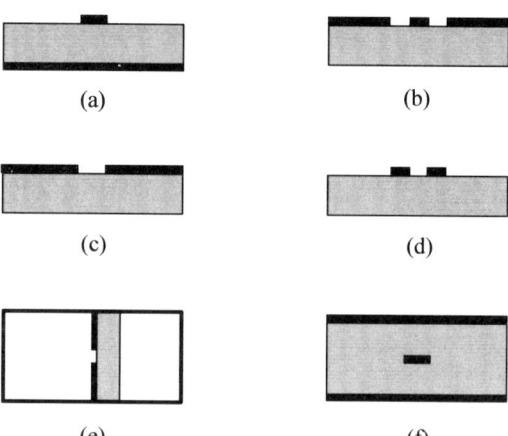

FIGURE 4.21 Cross sections of some transmission lines suitable for planar processing: (*a*) microstrip, (*b*) coplanar waveguide (CPW), (*c*) slotline, (*d*) coplanar strips (CPS), (*e*) finline, and (*f*) stripline.

TABLE 4.1 Materials Commonly Used as Transmission Line Substrates and Their Approximate Properties

Material	ϵ_r	Loss $10^4 \tan \delta$	Thermal Conductivity (W/cm–K)
Alumina	9.5	1.5	0.3
Sapphire	9.4, 11.6	1.0	0.4
Fused silica	3.8	1.0	0.01
Beryllia	6.7	4.0	2.5
Duroid	2.2	2.5	0.02
GaAs	12.5	≥ 6.0	0.44
InP	12.2	≥ 6.0	0.68
Silicon	12	≥ 50	1.4

Quasi-TEM Mode and Characteristic Impedance

Since the electromagnetic field guided by the microstrip line does not propagate in a single dielectric, microstrip is not capable of supporting a classical TEM mode. The microstrip field pattern, at any finite frequency, requires longitudinal field components in order to satisfy the boundary conditions at the air–dielectric interface. Strictly speaking, this is not a pure mode. However, in practice it is treated as one, and since it has a zero cutoff frequency and the field pattern resembles that of a TEM mode, it is called the quasi-TEM mode. Associated with the quasi-TEM mode is an *effective relative permittivity* ϵ_{eff} that defines its phase velocity: $v_p = c/\sqrt{\epsilon_{\text{eff}}}$. The value of ϵ_{eff} depends on the geometry of the microstrip line as well as the frequency. To a first-order approximation it is taken to be the average between the relative dielectric constants of air and the substrate:

$$\epsilon_{\text{eff}} \approx \tfrac{1}{2}(\epsilon_r + 1). \tag{4.41}$$

Microstrip characteristic impedance (referred to the quasi-TEM propagation) is found either by numerical analysis or by empirical formulas. For a microstrip line of width w and substrate height h, a useful empirical formula that gives the characteristic impedance is [28]

$$Z_0 = \frac{119.9}{\sqrt{2\epsilon_r + 1}} \left\{ \ln\left[4\frac{h}{w} + \sqrt{16\left(\frac{h}{w}\right)^2 + 2}\right] - \frac{1}{2}\left(\frac{\epsilon_r - 1}{\epsilon_r + 1}\right)\left(\ln\frac{\pi}{2} + \frac{1}{\epsilon_r}\ln\frac{4}{\pi}\right)\right\}, \tag{4.42}$$

which is valid for $w/h < 3.3$.

Example Calculate the characteristic impedance of a microstrip of width 75 μm over a 100-μm-thick GaAs substrate. For GaAs $\epsilon_r = 12.5$.

Solution Using Eq. (4.42), the value of Z_0 is found to be 54.8 Ω. This is close to the 50-Ω microstrip line dimensions used in many GaAs integrated circuits. ∎

Microstrip Loss

Losses in microstrip transmission lines are categorized into conduction loss, substrate (dielectric) loss, and loss by radiation.

Attenuation coefficient α as a function of signal amplitude along the line may be written as $dV/dx = -\alpha V$. For conduction loss this is given by

$$\frac{dV}{dx} = -\alpha_c V = -I\frac{dR}{dx} = -\frac{V}{Z_0}\frac{dR}{dx}, \tag{4.43}$$

or

$$\alpha_c = \frac{1}{Z_0}\frac{dR}{dx} = \frac{1}{Z_0 \sigma A}, \tag{4.44}$$

where σ is metal conductivity, $A = w\delta_s$ is the cross-sectional area of the current flow, and δ_s is the skin depth. The attenuation coefficient due to conductive loss may also be written in the following forms:

$$\alpha_c = \frac{1}{Z_0 w \sigma \delta_s} = \frac{R_s}{Z_0 w} = \frac{\sqrt{f}}{Z_0 w}\sqrt{\frac{\pi\mu_0}{\sigma}}, \tag{4.45}$$

where R_s is the surface resistance of the metal. The expressions given here for conductive loss do not take into account the effect of surface roughness. If the rms surface roughness is significant compared with the skin depth, conductive loss can be considerably higher than estimated here [28].

Associated with an electric field E applied to a dielectric, there is a stored energy density $\epsilon' E^2$ and a power dissipated per unit volume $\omega\epsilon'' E^2$, where ϵ' and ϵ'' are the real and imaginary parts of the dielectric's complex permittivity. The stored energy density represents the transmitted power density in a traveling wave. The ratio of the dissipated power to the transmitted power is therefore proportional to $\omega\epsilon''/\epsilon' = \omega\tan\delta$, where $\tan\delta$ is the loss tangent of the dielectric. This proportionality is valid for dielectric filled or partially filled transmission lines such as microstrip. In most commonly used microstrip substrates, transmission loss at low frequencies is dominated by conductor loss. Note, however, that conductor loss increases as the square root of frequency, while dielectric loss increases proportional to frequency. It follows that as the frequency increases, eventually dielectric loss will begin to dominate.

Radiation loss from a uniform microstrip line tends to couple energy more into surface waves supported by the substrate than into free space (see the same discussion under Coplanar Waveguide). Free-space radiation is more pronounced in resonant sections of microstrip and at abrupt discontinuities.

Dispersion

The effective dielectric constant for microstrip is dependent on the impedance and the frequency. At low frequencies it is approximately $\frac{1}{2}(\epsilon_r + 1)$ more valid for higher impedance lines. At very high frequencies, due to interaction with substrate modes, all of the energy will propagate in the dielectric so that the effective dielectric constant approaches ϵ_r. This transition leads to dispersion at intermediate frequencies.

The same effect can be expressed more quantitatively in terms of coupled-mode theory. When two modes with propagation constants k_1 and k_2 interact weakly in a transmission structure, their propagation constants are modified. Coupled-mode theory predicts the modified propagation constants to be [29]

$$k' = \left[\tfrac{1}{2}(k_1 + k_2) \pm \sqrt{\left[\tfrac{1}{2}(k_1 - k_2)\right]^2 \pm C^2} \right], \qquad (4.46)$$

where C is the mode coupling coefficient. If one of these modes is a substrate mode with its own dispersive behavior, its frequency-dependent propagation constant gives rise to dispersion in the other, which could be the microstrip quasi-TEM mode. In such cases, the coupling coefficient, which is a function of field overlap, also becomes frequency dependent, thereby enhancing the dispersive behavior of microstrip.

For design purposes, empirical formulas exist that predict the dispersive behavior of microstrip. One such formula given by Edwards and Owens [28] is

$$\epsilon_r - \epsilon_{\text{eff}}(f) = \frac{\epsilon_r - \epsilon_{\text{eff}}(0)}{1 + (h/Z_0)^{1.33}(0.43f^2 - 0.009f^3)}, \qquad (4.47)$$

where h is in millimeters and frequency is in gigahertz. Such empirical formulas may also be used to estimate the value of the coupling coefficient in Eq. (4.46).

4.8.2 Coplanar Waveguide

Coplanar waveguide was first proposed by Wen [30] as a transmission line appropriate for nonreciprocal applications such as isolators and circulators. In such applications the transmission line is deposited directly on the nonreciprocal substrate. Due to the circular field polarization it generates, CPW was considered a suitable choice. The second major application of CPW came with surface contacting probes for high speed on-wafer measurements. This application is discussed in detail in Chapter 8. In the early 1980s it was realized that since CPW requires the processing of only one side of the wafer (as opposed to both sides in the case of microstrip), it reduces the cost of integrated circuits considerably. Also, the ease of access to ground that it offers simplifies the design and enhances the performance of some high speed circuits. This combination made CPW a serious contender in high speed integrated circuit design. The first monolithic integrated circuit using CPW was reported in 1986. Presently, as its behaviour is analyzed further and more empirical and numerical models become available, CPW circuits are being developed by an increasing number of manufacturers.

In optoelectronics, applications of CPW are expanding. One attractive feature of CPW is the controllable high electric field that is available in the gaps. Optical waveguides may be fabricated in these gaps to take advantage of the electro-optic effect (see the discussion of optical modulators in Chapter 3).

A CPW in its ideal form consists of a conducting strip with two semi-infinite side conductors on the surface of an infinitely thick dielectric substrate (Figure 4.22). Practical realizations of CPW deviate from this definition in a number of ways. First, in most cases, the substrate cannot be considered infinitely thick. Second, the side conductors have finite widths. Third, the finite substrate may have a conducting backplate. Variations in these three parameters lead to five practical cases with different propagating modes and dispersion characteristics that will be discussed individually.

Transverse confinement of the fields as well as the presence of a top cover, which occur in packaged circuits, require additional analysis not discussed here [31, 32]. The discussions presented in this section are intended to help the reader gain a general understanding of the behavior of the CPW and to be able to avoid some commonly encountered pitfalls in working with these transmission lines. For more quantitative information about various CPW parameters, some found by numerical simulation, see references [33–35].

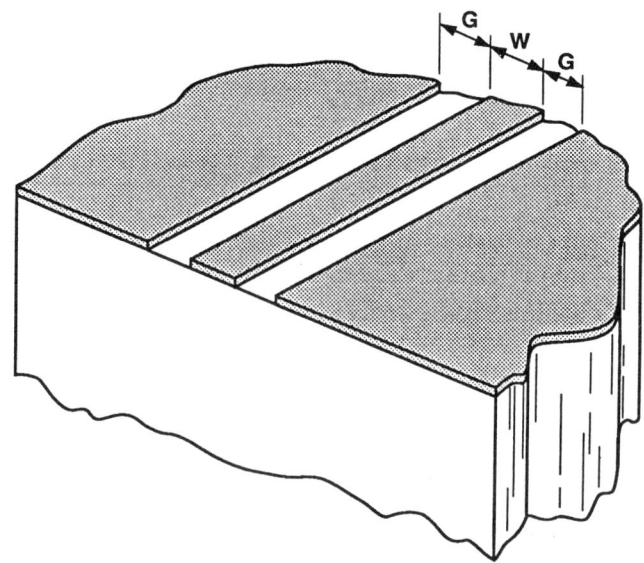

FIGURE 4.22 "Ideal" CPW has a semi-infinite substrate and wide ground planes.

Infinite Substrate

Guided modes of the ideal CPW structure have been studied by a number of authors [30, 36]. The ideal CPW structure is conveniently considered as a pair of coupled slotlines (complimentary structure to CPS) with odd and even normal modes. The odd and even refer to the symmetry of the electric fields in the slots with respect to the center conductor. The odd and even modes are what we refer to as the CPW mode and the slotline mode, respectively.

The CPW guided mode propagates at the interface between two dielectric media, with a phase velocity v_g that exceeds that of a TEM wave in the higher dielectric constant material (v_d). This condition gives rise to radiation from the guided CPW wave into the substrate. For constructive interference, the radiated wave and the guided CPW wave should have the same propagation constant along the direction of the CPW transmission line. This requirement restricts the propagation direction of the radiated wave to a semicone of angle θ given by

$$\cos\theta = \frac{k_g}{k_d} = \left(\frac{\epsilon_{\text{eff}}}{\epsilon_r}\right)^{1/2}, \tag{4.48}$$

where k_g and k_d are the propagation constants of the guided and radiated waves, respectively. Energy transfer from the CPW mode into the substrate causes attenuation of the guided wave. The attenuation constant α as calculated by Rutledge et al. [37] is given by

$$\alpha = f(\epsilon_r)\left(\frac{1}{\lambda_d}\right)^3 \frac{(W+2G)^2}{K(k)K'(k)} \tag{4.49}$$

in units of inverse length. Here, $k = W/(W+2G)$, K and K' are complete elliptic integrals of the first and second kind, and

$$f(\epsilon_r) = \left(\frac{\pi}{2}\right)^5 \frac{1}{\sqrt{2}} \frac{(1-1/\epsilon_r)^2}{\sqrt{1+1/\epsilon_r}}. \tag{4.50}$$

For GaAs substrates, this expression reduces to

$$\alpha = \frac{47.4}{K(k)K'(k)}\left(\frac{W+2G}{\lambda_d}\right)^2, \tag{4.51}$$

in units of decibels (per λ_d).

A 50-Ω CPW on GaAs has a k value close to 0.5, for which $K(k)K'(k) = 3.44$ and varies slowly with k. Substituting this in Eq. (4) gives

$$\alpha = 13.8\left(\frac{W+2G}{\lambda_d}\right)^2. \tag{4.52}$$

For $W + 2G \leq \lambda_d/20$, attenuation due to radiation is less than 0.034 dB per dielectric wavelength, which is easily acceptable for integrated circuit applications.

Finite Substrate without Backside Metalization

When substrate thickness is comparable to dielectric wavelength, reflections from the backplane air–dielectric interface should also be taken into account. The problem is that of a conductor-backed dielectric slab (Figure 4.23a). Surface wave modes of this structure are odd TE modes and even TM modes

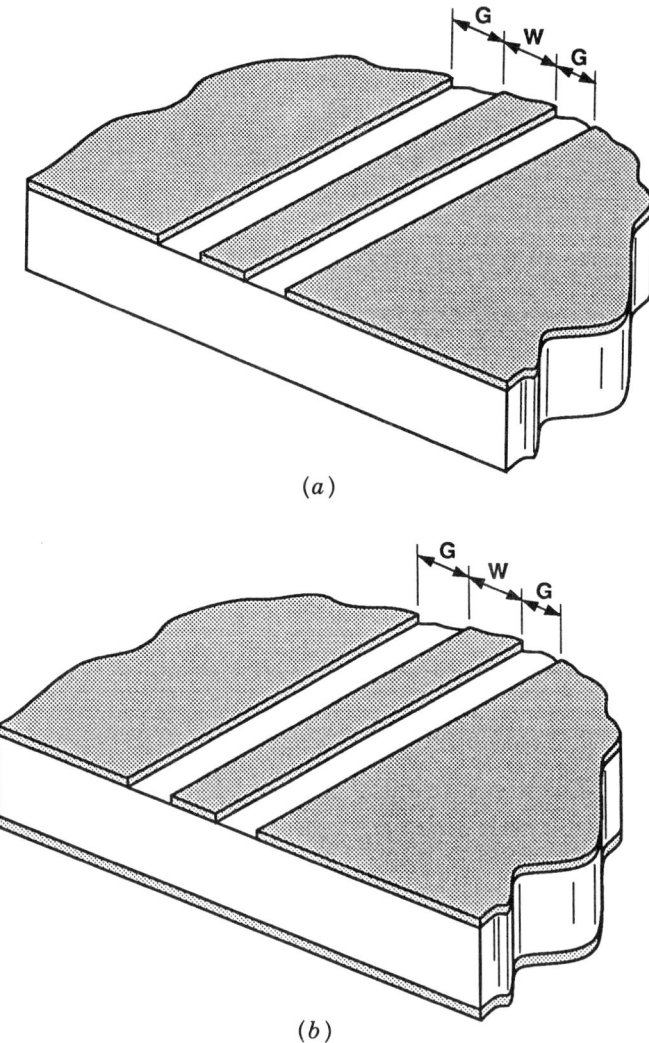

FIGURE 4.23 Four practical implementations of CPW.

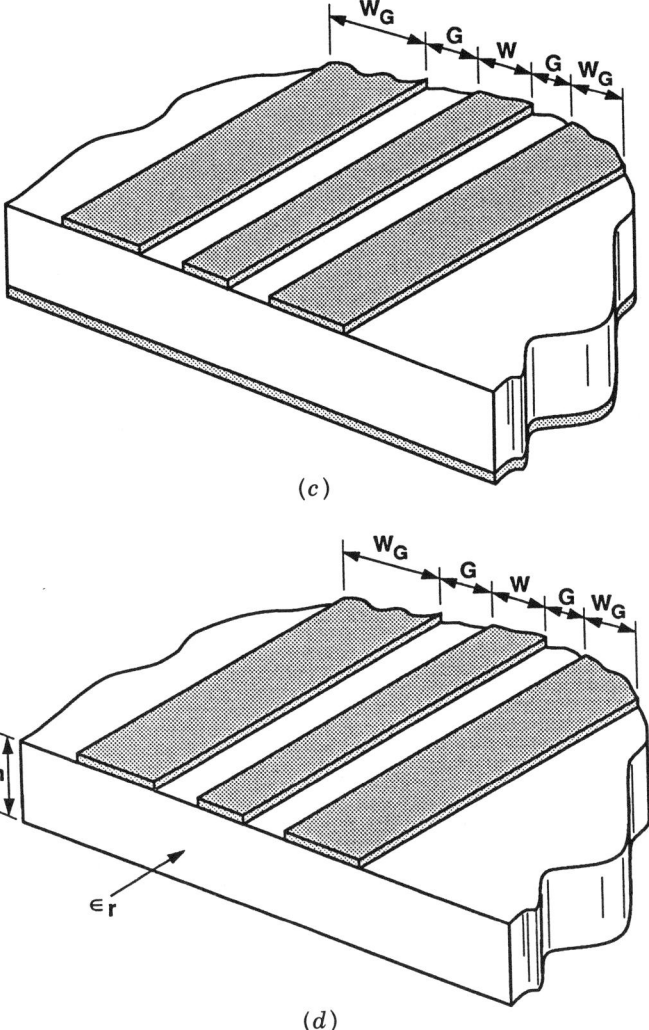

FIGURE 4.23. *(Continued)*

of a dielectric slab with twice the thickness and no metalization [38], where *odd* and *even* refer to the symmetry of the transverse field component. These modes (sometimes referred to as grounded slab modes) are similar to the ones encountered in the case of microstrip lines. The interaction of the CPW mode with the surface waves, however, is different from that of microstrip due to the difference in their field distributions. Most CPW structures have less field overlap with surface wave modes than microstrip and interact weakly with them. Despite this weak interaction, in frequency ranges where the propagation constants of the surface waves approach or exceed that of the CPW mode, considerable dispersion and radiative losses can occur.

The TE and TM modes of the conductor-backed dielectric slab are shown in Figure 4.24. For comparison, the normalized propagation constant range for the CPW mode, corresponding to $\epsilon_{\text{eff}}/\epsilon_r$ ranging from 0.48 to 0.64 is also shown. This range corresponds to the values most commonly encountered in monolithic circuits on GaAs. At low frequencies where the phase velocity of the surface wave mode is much higher than the CPW mode, only weak dispersion effects are observed. As the frequency increases, the phase velocity of the surface wave mode approaches and eventually becomes less than that of the CPW mode. Close to the point of intersection the two modes are synchronous (i.e., phase matched) and interact strongly, resulting in a highly dispersive behavior near this frequency. Above this point the surface wave mode mainly contributes to loss in the CPW mode, which keeps increasing with frequency. For wide-band integrated circuit applications it may be advantageous to select the substrate thickness such that the operating frequency range is below this intersection point. For the lowest order TM mode this requirement is satisfied if $h < 0.15\lambda_d$ (Figure 4.24a).

The next mode that can be excited as the frequency is increased is the TE_0 mode, whose cutoff frequency is at $2h \approx 0.25\lambda_d$. This mode will be synchronous with the CPW mode near $2h = 0.35\lambda_d$ (Figure 4.24b). A conservative guideline to avoid all potential problem points is to operate below the cutoff frequency of the TE_0 mode, i.e., $h < 0.12\lambda_d$. This guideline is also valid for microstrip since the same substrate modes are involved.

Finite Substrate with Backside Metalization

In many applications of CPW microwave integrated circuits, the substrate has a conducting surface on the backside, due to either intentional metalization or mounting in a metal housing. Backside metalization of the substrate together with wide topside ground planes (Figure 4.23b) give rise to parallel-plate waveguide modes. What is significantly different in this case compared to the unmetalized substrate is the presence of a zero-cutoff TEM mode (Figure 4.25). The phase velocity of this mode is smaller than the CPW mode and therefore constitutes a source of power loss for the CPW in the absence of lateral confinement. It should be kept in mind that lateral confinement of this structure gives rise to dispersion effects associated with rectangular waveguide modes. The functional dependence of attenuation coefficient α due to radiation in the TEM parallel-plate mode is given by [39]

$$\alpha = \frac{\pi^2}{2h} \frac{Z_0}{\eta_d} \frac{(W+2G)^2}{(\lambda_d)^2} \sqrt{1 - \frac{\epsilon_{\text{eff}}}{\epsilon_r}}, \qquad (4.53)$$

for $W + 2G$ small compared to the wavelength. Here, Z_0 is the CPW impedance, and $\eta_d = 377/\sqrt{\epsilon_r}$ is the wave impedance of the dielectric material.

Radiation from the CPW mode is much less than from the slotline mode, mainly due to the fact that in the CPW mode the electric fields (or the equivalent magnetic currents) point in opposite directions, causing partial cancellation and a second-order dependence of the radiated power on $(W + 2G)$.

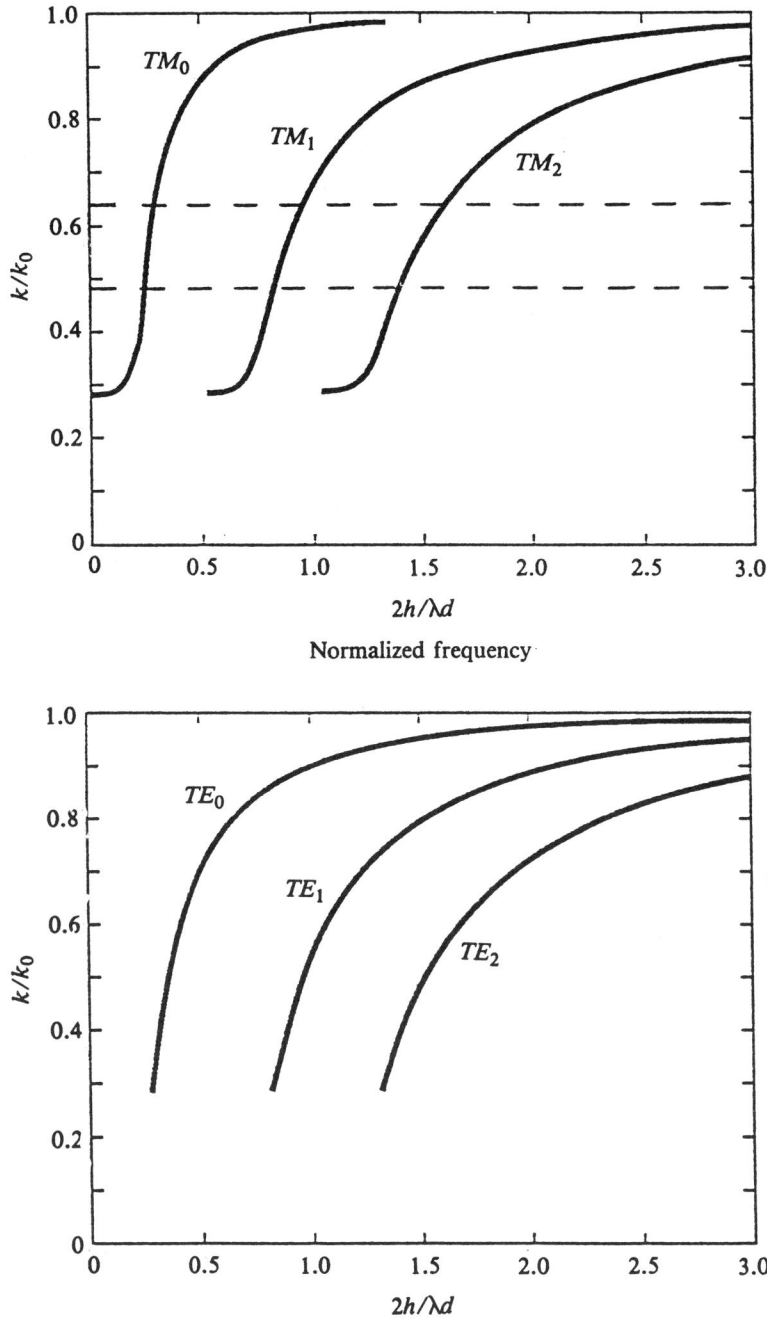

FIGURE 4.24 (a) TM and (b) TE surface wave modes of conductor-backed dielectric slab.

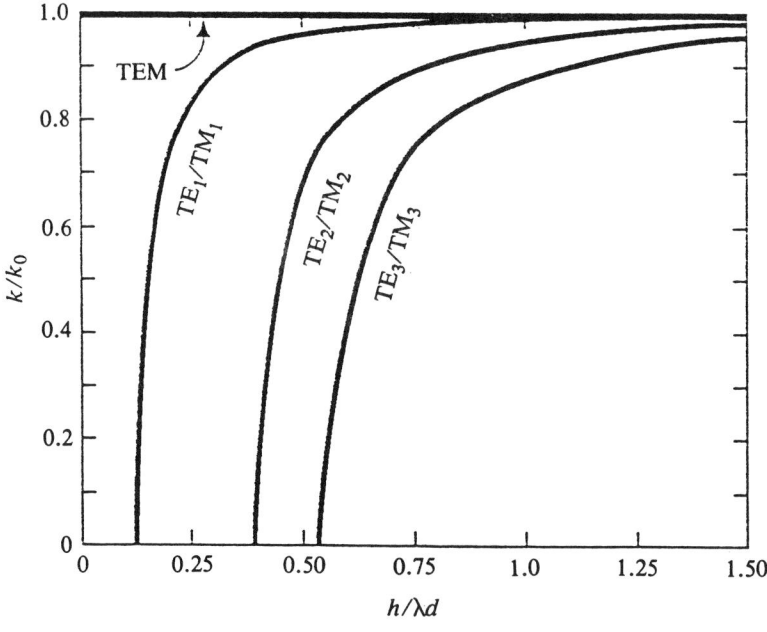

FIGURE 4.25 Parallel-plate waveguided modes for substrate metalized on both sides.

For a 50-Ω CPW line on GaAs with $W + 2G = \frac{1}{4}h \leq \frac{1}{20}\lambda_d$, attenuation is less than 0.17 dB per guide wavelength. This value also gives a qualitative estimate of the maximum level of cross talk and interference, through coupling with the parallel-plate TEM mode, that can be expected in a CPW integrated circuit with backside metalization.

The next higher order modes of the parallel-plate waveguide are TE_1 and TM_1 modes, which will be synchronous with the CPW mode near $h = 0.125\lambda_d$. Therefore the previously established criterion of $h < 0.12\lambda_d$ also avoids the higher order parallel-plate waveguide modes.

Finite Substrate and Finite Ground Planes Without Backside Metalization
A large area of substrate material is exposed in this case without any metallization on either side (Figure 4.23d). This portion of the substrate supports full slab modes including zero-cutoff TE_0 and TM_0 modes. The first substrate mode to start interacting with the CPW mode is TE_0, whose characteristics are given in Figure 4.26. Dispersion caused by the presence of TE_0 and TM_0 modes is unavoidable at any frequency, but the region where TE_0 has a phase velocity close to that of the CPW mode can be avoided if the criterion $h < 0.1\lambda_d$ is met ($k/k_0 < 0.48$ in Figure 4.26). This is a slightly more stringent requirement than encountered in other cases and justifies avoiding large areas of exposed substrate in close proximity to a CPW line.

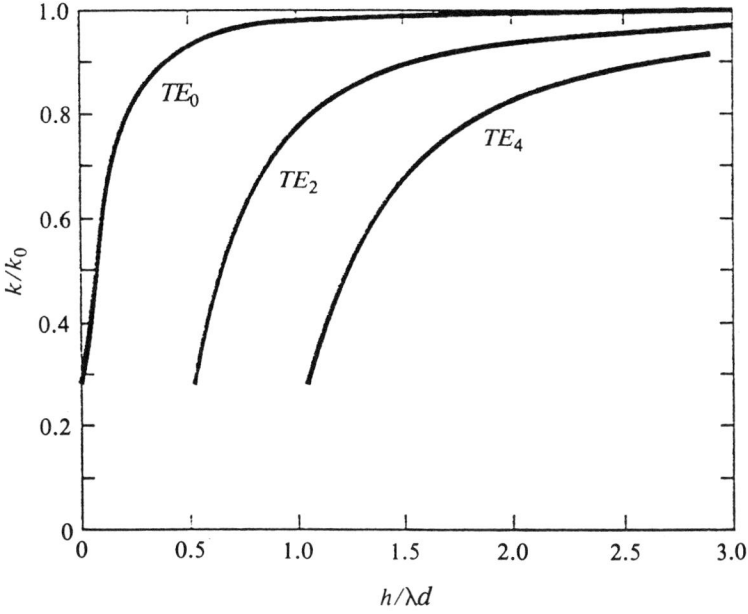

FIGURE 4.26 Dielectric slab even TE modes (no metalization on either side).

Finite Substrate and Finite Ground Planes with Backside Metalization
Finite ground planes on finite substrate with backside metalization behave as microstrip lines coupled together and to the center conductor (Figure 4.23c). The propagation modes on this structure are not as simple as the odd and even modes of the ideal CPW. In fact, none of the normal modes of this structure allow both ground planes to remain at zero potential [40]. This can cause significant problems in a circuit designed with CPW.

Three coupled microstrips have three normal modes of propagation. Each mode is represented by a three-dimensional vector, with elements representing relative potentials on the three lines [41]. The relative potentials remain the same as the mode propagates along the line. This is not the case if the transmission line is excited with three arbitrary voltages. For example, the excitation

$$V_0 = \begin{pmatrix} 0 \\ 1 \\ 0 \end{pmatrix}$$

represents a signal on the center conductor with the side conductors at zero potential. This is not a normal mode of the structure, which means the relationship among the three potentials on the line will change as the excitation propagates. The three normal modes are given by [40]

$$V_1 = \begin{pmatrix} -1 \\ 0 \\ +1 \end{pmatrix}, \quad V_2 = \begin{pmatrix} a \\ 1 \\ a \end{pmatrix}, \quad V_3 = \begin{pmatrix} -b \\ 1 \\ -b \end{pmatrix}. \quad (4.54)$$

Here, V_1 is the slotline mode that corresponds to the well-known even mode of the CPW on semi-infinite substrate. This is a non-TEM mode orthogonal to V_0. The slotline mode is normally suppressed in most applications [42, 43]. The other two modes, V_2 and V_3, are the microstrip mode and the CPW mode, respectively. These modes involve parameters a and b that are functions of the physical dimensions of the line. In general $a \neq b$ and the phase velocities associated with these modes are different. The $\begin{pmatrix} 0 \\ 1 \\ 0 \end{pmatrix}$ excitation is a combination of V_2 and V_3:

$$\begin{pmatrix} 0 \\ 1 \\ 0 \end{pmatrix} = \frac{1}{a+b} \left[b \begin{pmatrix} a \\ 1 \\ a \end{pmatrix} + a \begin{pmatrix} -b \\ 1 \\ -b \end{pmatrix} \right]. \quad (4.55)$$

Since the phase velocities v_2 and v_3 associated with modes V_2 and V_3 are different, the phase angle between the constituent modes of V_0 will drift along the transmission line and finite potentials will appear on the ground planes.

It is desirable for most applications to design the CPW in a way that would allow only the single propagating signal V_0. Although V_0 is not a normal mode, if the ground plane width W_g and the substrate thickness h are chosen to be larger than the linewidth $W + 2G$, the CPW mode V_3 will approach V_0, that is, the value of b becomes much smaller than 1.

In order to demonstrate this numerically, three line geometries were chosen for which the normal modes were calculated [40]. The parameters h and W_g were varied among the three lines, while ϵ_r, W, and G remained fixed at 12, 50 μm, and 25 μm, respectively.

For the first line, substrate thickness $h = 100\,\mu$m, with a ground plane width $W_g = 200\,\mu$m. The value of b for the coplanar mode was calculated to be 0.156, corresponding to a signal level on the ground plane 16 dB below the center conductor. For the second line the substrate thickness was increased to 400 μm, keeping all the other parameters the same. The value of b was reduced to 0.122, or 18 dB below the center conductor. For the third line all parameters were the same as the second line except the ground plane width, which was increased from 200 to 400 μm. The value of b was further reduced to 0.06, or 24 dB below the center conductor.

The above analysis shows that in the CPW mode the potential on the ground planes can be kept close to zero if the criteria of wide ground planes and thick substrate compared to $W + 2G$ are met. It should also be pointed out that once the signal V_0 has been excited on the transmission line, the maximum potential on the ground plane appears at a distance equal to half the beat wavelength between microstrip and CPW modes. For the three lines considered, this distance varies from 8.7 to 10.5 mm at the frequency of 30 GHz. In MMIC applications, where uninterrupted propagation distances are generally much less than this, problems encountered are not due to the fact that V_0 excitation is not a pure mode, but rather, due to unintentional excitation of the microstrip mode at the launch point. This is likely to occur at transitions between microstrip and CPW [40].

Dispersion Effects

The presence of surface wave modes, especially the zero-cutoff modes that exist in all finite substrate cases, contribute to not only power loss in the CPW mode but also to its dispersion.

Coupled-mode theory as expressed in Eq. (4.46) is a useful tool for analyzing CPW dispersion. The coupling coefficient C in Eq. (4.46) is a function of the field overlap integral I [44], given by

$$I = \int (\mathbf{E}_1 \times \mathbf{H}_2) \cdot d\mathbf{s}. \tag{4.56}$$

This integral is evaluated over the cross section of the waveguide. Subscripts 1 and 2 correspond to the two interacting modes.

Tables 4.2–4.4 list overlap integrals calculated for the interactions of microstrip and CPW with surface wave and parallel-plate modes. These integrals are normalized with respect to the incident power and give a comparison between the strengths of these interactions in various structures. The highest value of overlap is between the microstrip and TE_0 modes. This value was selected as the normalizing reference for all the cases considered. The overlap integral is of course frequency dependent where non-TEM modes are involved. The overlap integrals have been evaluated at frequencies close to the intersection point of the propagation constants of the substrate mode with that of CPW or microstrip.

The first structure listed is a 50-Ω microstrip line on a 100-μm substrate with a relative dielectric constant of 12.5. Field overlaps with TM_0 and TE_1 surface wave modes are given in Table 4.2. The overlap with TM_0 is selected as unity for reference.

The next structure is a 50-Ω coplanar waveguide without backside metalization. Two different values of $W + 2G$ are considered. The results are given in Table 4.3.

TABLE 4.2 Normalized Field Overlap Integrals for 50-Ω Microstrip on 100-μm Substrate

Mode	Overlap
TM_0	1.00
TE_1	0.857

TABLE 4.3 Normalized Field Overlap Integrals for 50-Ω Coplanar Waveguide on 400-μm Substrate

$W + 2G$ (μm)	Mode	Overlap
100	TM_0	0.353
100	TE_1	0.119
50	TM_0	0.213
50	TE_1	0.062

Finally, a 50-Ω coplanar waveguide with backside metalization is considered. The overlap integrals for its interaction with parallel-plate modes are given in Table 4.4.

It can be seen that coplanar waveguide field overlaps depend strongly on the value of $W + 2G$. Practical coplanar waveguides with $W + 2G \ll h$ interact less than microstrip lines with the first two TE and TM surface wave modes. In this respect the main difference between microstrip and coplanar waveguide is the degrees of freedom in their physical layouts. On a given substrate material microstrip has two degrees of freedom: linewidth W and substrate thickness h. Changing both of these while keeping line impedance constant only results in a frequency scaling and does not change the field overlaps. With CPW, on the other hand, $W + 2G$ can be varied independent of line impedance in order to minimize field overlap with surface wave modes and therefore to minimize dispersion. Some experimental results have been reported for the dispersion of CPWs [45]. As an extension of these results CPWs with $W + 2G = \frac{1}{8}h$ were fabricated on GaAs that showed no measurable dispersion in the frequency range of 10–100 GHz.

In summary, for coplanar waveguides, dispersion due to interaction with surface wave and parallel-plate modes in the substrate strongly depends on the ground-to-ground spacing $(W + 2G)$. If this spacing is small compared to dielectric wavelength and substrate thickness, both dispersion and radiation loss are minimized. In finite ground plane cases, the condition of small ground-to-ground spacing compared to substrate thickness is also necessary

TABLE 4.4 Normalized Field Overlap Integrals for 50-Ω Grounded Coplanar Waveguide on 400-μm Substrate

$W + 2G$ (μm)	Mode	Overlap
100	TEM	0.224
100	TM_1	0.559
100	TE_1	0.180
50	TEM	0.134
50	TM_1	0.371
50	TE_1	0.095

to avoid the excitation of the microstrip mode and to minimize the deviation of the CPW mode from the "odd" mode of the ideal line.

Limitations on the reduction of $W + 2G$ come from conductor loss, which imposes a minimum-width requirement on the center conductor. If substrate thickness is to be kept large compared with this dimension, it is inevitable that at high frequencies the operation of a CPW will not be below the cutoff frequency of surface wave modes. However, as was shown here, the interaction of the CPW mode with surface wave modes is negligible for integrated circuit applications if ground-to-ground spacing is small compared with dielectric wavelength.

4.8.3 Other Transmission Lines

Slotlines and *finlines* are non-TEM transmission lines used in very specific applications. The physical difference between the two is that a finline has no integrated ground planes and is placed in the electric field plane of a rectangular waveguide. Circuits using finlines are often referred to as *E-plane circuits*. An important characteristic of a finline is that it uses integrated circuit technology in a rectangular waveguide environment. The E-plane circuits are naturally shielded, and their electrical transitions to rectangular waveguides are simple. The most useful frequency range for E-plane circuits is 30–100 GHz where dimensions and tolerances are suitable for their fabrication [46].

In the upper millimeter-wave range (above 100 GHz) various types of dielectric guiding are sometimes used that have more relaxed dimensional tolerances than rectangular waveguides and less radiation losses than planar open waveguides. Dielectric slab waveguides supporting low loss modes have been in existence for many years, and their propagation modes have been analyzed thoroughly [38]. In order for the dielectric slab waveguide to be compatible with other waveguide structures, it is often fabricated as a slab deposited on a conducting ground plane. In this form it is referred to as an *image guide*. One major difference between millimeter-wave dielectric waveguides and optical dielectric waveguides such as the optical fiber is the way radiation is coupled into and out of these waveguides. Millimeter-wave dielectric guides are almost always connected to some form of a conductor-guided line. This is not the case with optical fibers.

4.9 PASSIVE COMPONENTS

Passive components are non–power generating devices used to manipulate and control wave propagation and distribution. Historically, most microwave passive components were designed for use with guided waves while optical passive components used free-space propagation. Presently, these domains overlap. There are quasi-optical components for use in microwaves and waveguide components in fiber optics. Passive components described in

148 MICROWAVE CIRCUITS AND INTEGRATED TRANSMISSION LINES

this section are microwave junctions and directional couplers usable with planar transmission lines.

4.9.1 Microwave Junctions

Scattering of electromagnetic waves by an arbitrarily shaped structure is an involved mathematical problem. However, any such structure enclosed in conductive walls with a finite number of viewing ports (transmission lines) forms a microwave junction whose scattering characteristics may be studied in detail by symmetry arguments alone. Any multiport linear microwave junction is fully characterized by a scattering matrix. Additional assumptions about the junction place mathematical restrictions on the scattering matrix. For example, any reciprocal junction (devoid of any dc bias) has a symmetrical S matrix; while a lossless junction has a unitary S matrix (Appendix C). The combination of lossless and reciprocal assumptions represents a large class of commonly encountered junctions. Any port of the junction that is reflectionless (matched) places a zero on the diagonal of the S matrix. In Appendix C it is shown that based on these facts, a lossless reciprocal three-port junction cannot be matched at all ports. This is an important rule particularly applicable to power combiners and power dividers. Appendix D shows that a matched five-port junction with rotational symmetry serves as a four-way equal power divider. The list is extensive, particularly for rectangular waveguide junctions (magic-T's, turnstile junctions, etc.) [47]. One useful junction that can be fabricated using integrated transmission lines is a four-port ring-type junction informally known as the *rat-race junction*. The schematic diagram of this junction is shown in Figure 4.27.

A signal incident on the junction at port 1 is divided equally into counterpropagating waves along the ring. At ports 2 and 4, the waves arrive in phase, while at port 3 they arrive out of phase. Port 3 is thus isolated from port 1 ($S_{13} = S_{31} = 0$). Similarly, ports 2 and 4 are isolated ($S_{24} = S_{42} = 0$). Note also that signals arriving at the output ports are 180° out of phase ($S_{12} = -S_{14} = 1/\sqrt{2}$).

The choice of the characteristic impedance Z_r of the ring is not obvious but can easily be derived from the admittance matrix of the junction [48]. First, the scattering matrix of the matched junction is formed based on the behavior described above:

$$[S] = \frac{1}{\sqrt{2}} \begin{bmatrix} 0 & 1 & 0 & -1 \\ 1 & 0 & 1 & 0 \\ 0 & 1 & 0 & 1 \\ -1 & 0 & 1 & 0 \end{bmatrix}. \tag{4.57}$$

This is then converted to the Y matrix through $[Y] = [Y_0]([Z_0] - [S])([Z_0] + [S])^{-1}$. This particular S matrix cannot be converted to Y as it stands and requires the addition of transmission line sections to each terminal. A con-

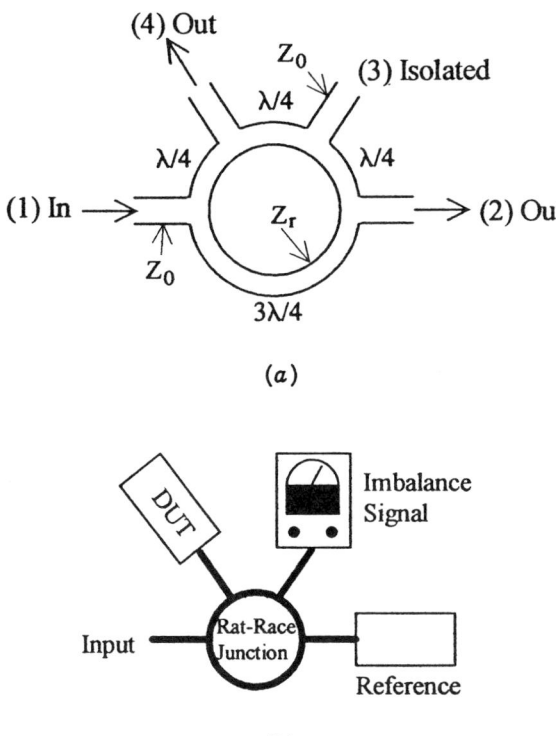

FIGURE 4.27 (*a*) Ring-type "rat-race" junction. (*b*) Equivalent circuit for reflectionless loads at ports 2 and 4.

venient choice is to add a quarter-wavelength section of transmission line to each terminal. This is represented by multiplying the entire S matrix by j. Not only can the new matrix be easily converted to Y, but the resulting normalized Y parameters have the same numerical values as the S parameters:

$$[Y] = \frac{jY_0}{\sqrt{2}} \begin{bmatrix} 0 & 1 & 0 & -1 \\ 1 & 0 & 1 & 0 \\ 0 & 1 & 0 & 1 \\ -1 & 0 & 1 & 0 \end{bmatrix}. \tag{4.58}$$

This is of course a coincidence and not in general true.

Let us try to interpret the structure of this Y matrix. First, the zero off-diagonal elements imply that there are no direct connections between ports 1 and 3, and similarly between ports 2 and 4. Second, Y parameters are short-circuit parameters, but we added quarter-wavelength sections between the ring and the terminals. So, the inactive terminals present zero admittance to the ring. This is equivalent to not being connected to the ring at all. In view of this, $Y_{12} = jY_0/\sqrt{2}$ implies a quarter-wavelength connection between ports 1 and 2, and $Y_{14} = -jY_0/\sqrt{2}$ suggests a three-quarter-wavelength connection between ports 1 and 4. Finally, the normalizing admittance of the ring is $Y_0/\sqrt{2} = 0.014$, corresponding to a characteristic impedance of $Z_r = 70.7\ \Omega$ for 50-Ω terminals. It can be seen that once the S matrix of a junction is conceived from its desired characteristics, the conversion to either Y or Z parameters can aid in the synthesis of the junction using TEM transmission lines. This concept will be demonstrated in the next section.

The rat-race junction may be used as a bridge network. If power at port 1 is fed to loads at ports 2 and 4, any identical reflections of these loads will still not couple to port 3. The output at port 3 is proportional to any mismatch between the two reflections. So, this junction may be used to monitor small changes in the impedance of one of the loads.

4.9.2 Directional Couplers

A four-port junction that is of particular interest in both microwaves and optics is the directional coupler. A directional coupler is a four-port reciprocal junction that, in its idealized form, is lossless and reflectionless with the following general S matrix:[6]

$$[S] = \begin{bmatrix} 0 & S_{12} & S_{13} & 0 \\ S_{12} & 0 & 0 & S_{24} \\ S_{13} & 0 & 0 & S_{34} \\ 0 & S_{24} & S_{34} & 0 \end{bmatrix}. \quad (4.59)$$

Ports 1 and 2 are the main ports; ports 3 and 4 are the coupled ports. Ports 1 and 4 as well as 2 and 3 are isolated. This means that power incident on port 1 is sampled at port 3 and power incident on port 2 is sampled at port 4. A directional coupler is characterized by the ratio of its sampled power to the incident power in decibels. A 3-dB directional coupler is an equal power divider for which $|S_{12}| = |S_{13}| = 1/\sqrt{2}$ and $|S_{21}| = |S_{24}| = 1/\sqrt{2}$. If the lossless condition of the junction is explicitly enforced, the S matrix is required to be unitary, that is, $[S]^t[S]^* = [I]$. One of the equations arising from this requirement is

$$S_{12}S_{13}^* + S_{24}S_{34}^* = 0. \quad (4.60)$$

If the structure of the directional coupler possesses reflection symmetry such that $S_{12} = S_{34}$ and $S_{13} = S_{24}$, it follows that $S_{12} = jS_{13}$. This means that in a symmetrical coupler if the incident power is equally divided between the two outputs, then the two outputs are 90° out of phase. Directional couplers

---
[6] Note that the discussion here is valid for optical as well as microwave directional couplers.

of this type are used in heterodyne balanced detection in order to suppress local oscillator noise (Chapter 7).

The physical design of a directional coupler depends on the application. There are numerous approaches to coupler design, some of which are described below.

Branch Line Couplers

Consider two parallel transmission lines that are weakly coupled by a signal path not appreciably disturbing the propagation in either line. Some of the signal from the first line is coupled to the second line where it will propagate in both directions. If instead of one signal path there are two signal paths placed a quarter of a guided wavelength apart, as shown in Figure 4.28, propagation in the second line will be possible only in one direction, namely the same direction as the original signal. In the reverse direction the signals coming from the two coupling paths will be out of phase and cancel. This simple structure forms a directional coupler known as a branch line coupler.

A branch line coupler can be designed using the procedure described under the ring-type rat-race junction. The desired S matrix of the junction is

$$[S] = \frac{1}{\sqrt{2}} \begin{bmatrix} 0 & 1 & -j & 0 \\ 1 & 0 & 0 & -j \\ -j & 0 & 0 & 1 \\ 0 & -j & 1 & 0 \end{bmatrix}, \qquad (4.61)$$

as discussed earlier. The conversion of this matrix to an impedance matrix results in

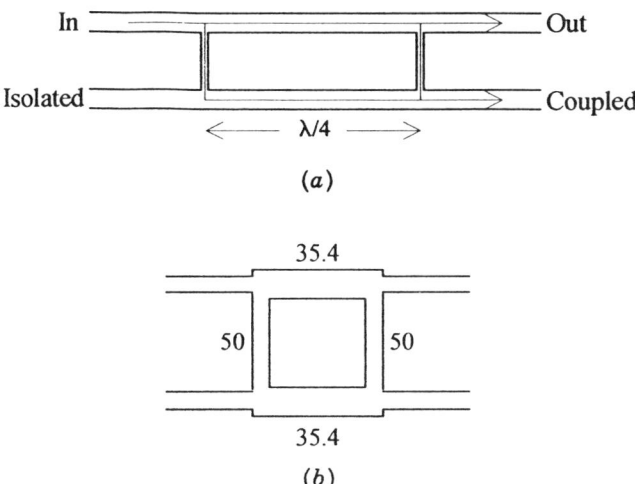

FIGURE 4.28 (a) Simple directional coupler consisting of two weakly coupling paths between two transmission lines. (b) Branch line directional coupler with 50-Ω ports.

$$[Z] = jZ_0 \begin{bmatrix} 0 & 0 & \sqrt{2} & 1 \\ 0 & 0 & 1 & \sqrt{2} \\ \sqrt{2} & 1 & 0 & 0 \\ 1 & \sqrt{2} & 0 & 0 \end{bmatrix}. \tag{4.62}$$

It is left as an exercise to show that the obtained impedance matrix represents the branch line directional coupler shown in Figure 4.28 consisting of $\frac{1}{4}\lambda$ transmission line sections with the indicated impedances. Finally, branch line couplers can be made of multiple sections for increased bandwidth. Figure 4.29 shows one such implementation.

Distributed Coupling

While branch line couplers are common as elements in integrated circuits, most stand-alone couplers employ continuous or distributed coupling between two transmission lines. Distributed coupling is achieved by placing the transmission lines close enough to each other that some of their fringing fields overlap. The strength of the coupling is measured by the coupling capacitance per unit length for planar waveguides and by coupling coefficients in general.

Distributed electromagnetic coupling between two transmission lines is commonly analyzed by coupled-mode theory [49, 50]. An intuitive rendition of this approach is to represent signal propagation along the coupled transmission lines as the superposition of normal modes known as odd and even modes. In the odd mode of propagation, the electric field vectors on the two lines are oppositely directed, while in the even mode they point in the same direction. The odd and even modes have different mode impedances. In homogeneous transmission lines such as stripline their phase velocities are the same. In transmission lines with inhomogeneous dielectrics such as microstrip and dielectric waveguides, the normal modes propagate at different phase velocities.

In practical planar transmission line couplers two design approaches are common. One is the backward coupler or the *contradirectional* coupler, which relies on the impedance difference between the odd and even propagation modes. The other is the forward or *codirectional* coupler, whose design is based on the velocity difference between the two normal modes.

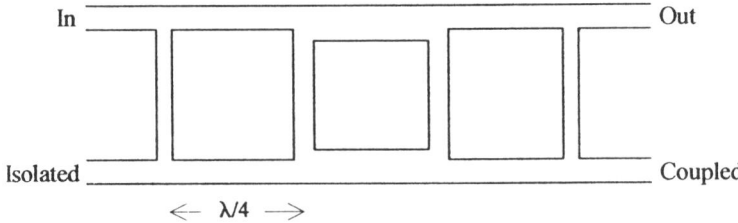

FIGURE 4.29 Schematic diagram of a multisection branch line coupler.

Backward Directional Coupler

Consider a section of coupled transmission lines of electrical length $\beta l = \theta$. If the coupled lines are identical, the odd and even normal modes are very well defined, as shown in Figure 4.30. A signal of amplitude A incident on port 1 of the coupled section is the sum of the odd and even modes each of amplitude $\frac{1}{2}A$. Outside the coupled section, the odd and even modes see the uncoupled impedance of the transmission lines Z_0. The presence of the coupled section is seen by each mode as a transmission line section of a different impedance (Z_{0e} for the even mode and Z_{0o} for the odd mode). This gives rise to transmitted and reflected portions of each mode. The reflection and transmission coefficients are denoted by Γ_o, Γ_e and T_o, T_e, respectively, for the odd and even modes.

The calculation of these coefficients for the normal modes is identical to the standard derivation of the S parameters of a transmission line section. For the even mode, the coefficients are

$$\Gamma_e = \frac{j(Z_{0e}^2 - Z_0^2)\sin\theta}{2Z_{0e}Z_0\cos\theta + j(Z_{0e}^2 + Z_0^2)\sin\theta}, \qquad (4.63a)$$

$$T_e = \frac{2Z_0 Z_{0e}}{2Z_0 Z_{0e}\cos\theta + j(Z_0^2 + Z_{0e}^2)\sin\theta}. \qquad (4.63b)$$

For the odd mode Z_{0e} is replaced by Z_{0o}. It is important to note that if $Z_{0o}Z_{0e} = Z_0^2$, then $\Gamma_o = -\Gamma_e$ and $T_o = T_e$. Combining these values as shown in Figure 4.30 results in $b_1 = b_3 = 0$, $b_2 = AT_e$, and $b_4 = A\Gamma_e$. It can be seen that the coupled port is port 4 rather than port 3, hence the name backward coupler. The coupled power is proportional to $|\Gamma_e|^2$, which is a maximum for $\theta = \frac{1}{2}\pi$. This maximum value of coupling may be expressed in decibels as

$$P_c = -20\log\left|\frac{Z_{0e} - Z_{0o}}{Z_{0e} + Z_{0o}}\right|. \qquad (4.64)$$

As can be seen, the directional coupler consists of a section of coupled transmission lines with the electrical length of $\frac{1}{2}\pi$. The level of coupling required determines the relationship between Z_{0e} and Z_{0o}. For example, a 6-dB coupler requires that $Z_{0e} = 3Z_{0o}$. Since the uncoupled transmission line impedance is the geometric mean of the odd and even impedances, these impedances would have to be 28 and 84 Ω, respectively, for 50-Ω termina-

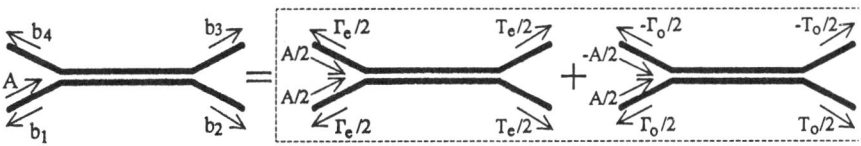

FIGURE 4.30 Odd and even mode analysis of the directional coupler.

tions. From the values of the odd and even impedances the physical dimensions and the proximity of the coupled lines may be determined. This is normally done by numerical analysis or the use of published curves such as those given in reference [51] and shown in Figure 4.31 for microstrip. The physical layout of the coupler discussed here is shown in Figure 4.32.

Note that in the analysis of the backward directional coupler it was tacitly assumed that the phase velocities of the odd and even modes are the same. This assumption is needed to define the $\frac{1}{2}\pi$ electrical length of the coupled section without ambiguity. When using transmission lines with inhomogeneous dielectrics such as microstrip, this assumption is not valid. Methods have been devised to minimize this phase velocity mismatch. Examples are

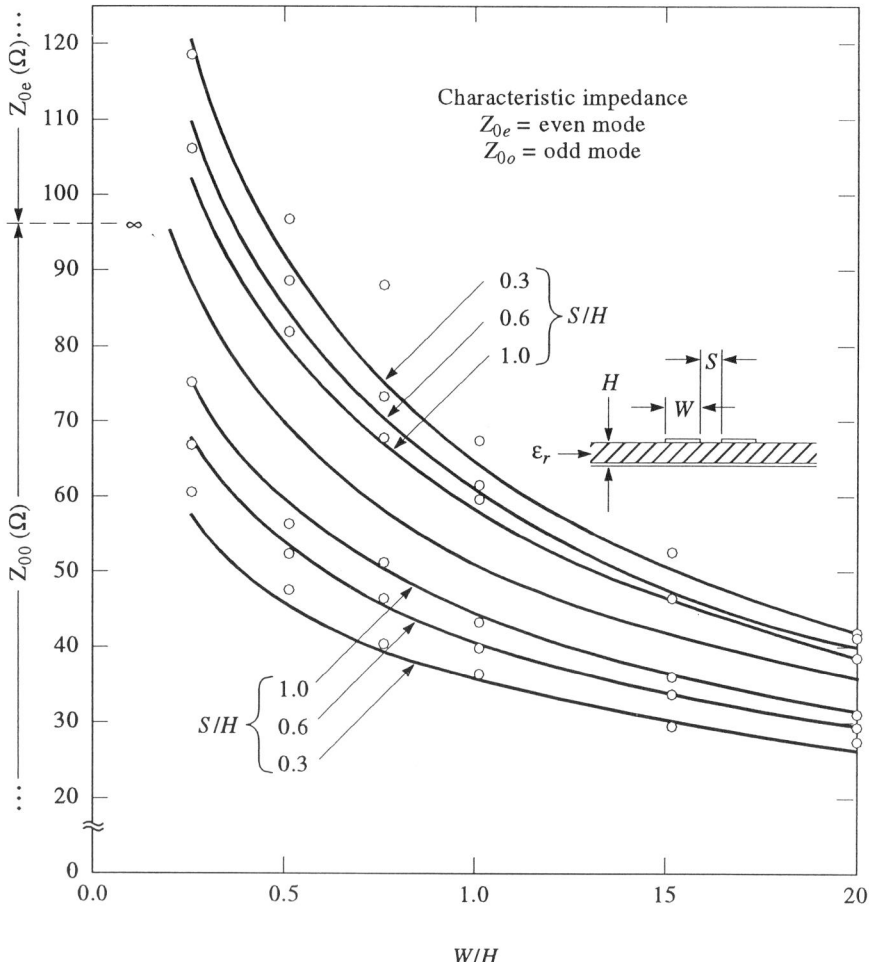

FIGURE 4.31 Odd and even mode impedances for microstrip. (Courtesy of Bryant and Weiss [51].)

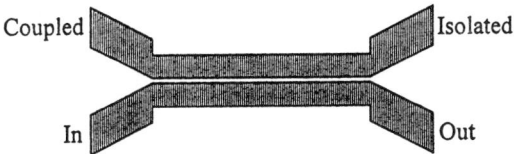

FIGURE 4.32 Physical layout of 6-dB directional coupler using microstrip.

dielectric deposition over the coupled section and adding equalizing corrugations at the coupling gap.

Forward Directional Coupler

Coupled transmission lines with inhomogeneous dielectrics such as microstrip and optical fiber demonstrate a distinct phase velocity deviation between the odd and even modes. If the coupled section is long enough, a beating pattern between the two modes will be observed. As the odd and even modes drift in and out of phase, the electromagnetic energy is periodically transferred to one line or the other. If the waveguides are labeled A and B, electromagnetic power as a function of distance z is given by

$$P_A = \sin^2 Kz, \qquad P_B = \cos^2 Kz. \tag{4.65}$$

The constant K is called the *coupling coefficient*. Lower values of K correspond to a more gradual power transfer between the lines. Complete power transfer as described above only occurs if the uncoupled waveguides have identical phase velocities. When nonidentical waveguides with a phase velocity mismatch $\Delta\beta$ are coupled, the power transfer is incomplete and given by [52]

$$P_A = \frac{\sin^2\{Kz[1 + (\Delta\beta/2K)^2]^{1/2}\}}{1 + (\Delta\beta/2K)^2}, \qquad P_B = 1 - P_A. \tag{4.66}$$

The *coupling length* L_c for the coupled section is defined as the length required for maximum energy transfer from one line to the other. This is one half of the beat wavelength. For identical lines L_c is given by

$$L_c = \frac{\pi}{2K} = \frac{\pi}{|\beta_{odd} - \beta_{even}|}. \tag{4.67}$$

In order to design a forward coupler, the coupling coefficient and the length of the coupled section need to be specified. The coupling coefficient is determined from tabulated data or by numerical analysis. If the transmission lines used are identical, the length of the coupler is selected as the fraction of L_c that gives the desired coupling. A more broadband approach is to use nonidentical waveguides of coupling length L_c. The coupling level is chosen by

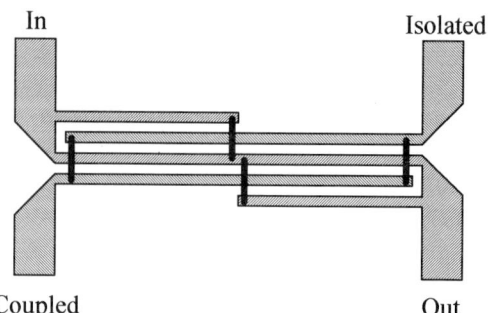

FIGURE 4.33 Schematic layout of the "Lange" coupler.

selecting the difference in the propagation constants of the waveguides [53]. In either case, the transition points between coupled and uncoupled waveguides should be reflectionless. This is in contrast with backward couplers where the transitions are assumed to be abrupt.

Interdigitated Couplers

The concept of an octave bandwidth interdigitated directional coupler was empirically discovered in 1969 by Lange [54]. This coupler, whose schematic diagram is shown in Figure 4.33, is more compact and achieves a higher bandwidth than the conventional parallel coupled microstrip. The approximate design rule is that the main section's length is $\frac{1}{4}\lambda$ at the lowest frequency of the operation band and that the length of each small finger is $\frac{1}{4}\lambda$ at the highest frequency of the band. Note that the length of the coupled section in conventional couplers is $\frac{1}{2}\lambda$ at midband.

The Lange coupler achieves its broad bandwidth by effectively compensating for the phase velocity difference between the odd and even modes. The procedure for designing the Lange coupler as well as other interdigitated couplers are described elsewhere [55, 56].

PROBLEMS

4.1 Consider an intrinsic MESFET with $r_i = 5\,\Omega$, $C_{gs} = 0.12$ pF, $g_m = 85$ mS, $G_{ds} = 0.02$ mS, and $C_{ds} = 0.08$ pF. Calculate S_{11}, S_{22}, and S_{21} at 1, 5, and 10 GHz. Is this a good amplifier? What characteristics should the S parameters of a better amplifier have?

4.2 Plot the magnitudes of $T_3(\omega/\omega_c)$ and $1/[1 + T_3^2(\omega/\omega_c)]$ over the range $0 < (\omega/\omega_c) < 2$.

4.3 Design input and output matching networks for the intrinsic MESFET model of Figure 2.12 with the following parameter values: $g_m = 50$ mS, $C_{gs} = 0.65$ pF, $R_i = 5\,\Omega$, $C_{ds} = 0.1$ pF, and $G_{ds} = 5$ mS.

The frequency range of interest is 2.0–3.0 Ghz. Plot the resulting S_{21} for this frequency range.

4.4 In a balanced amplifier determine the roles of finite isolation and finite directivity of the directional couplers on gain and return loss.

4.5 Calculate the new Y parameters of a two-port network after having added a parallel feedback path of admittance Y_f.

4.6 Calculate the new Z parameters of a two-port network after having added a series feedback path of impedance Z_f.

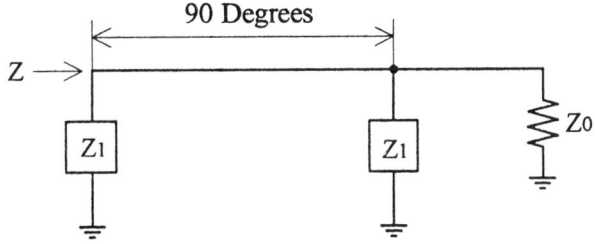

4.7 In the figure above calculate the impedance Z assuming the characteristic impedance of the transmission line to be Z_0.

4.8 Derive the conditions for synchronous propagation on the gate line and the drain line of a distributed amplifier. Take each line to be a periodically loaded transmission line.

4.9 For a delay line stabilized oscillator derive an expression relating oscillation mode spacing to delay time Δt and phase shifter setting $\Delta \phi$.

4.10 Show that a lossless reciprocal matched four-port junction must be a directional coupler.

REFERENCES

[1] P. L. D. Abrie, *The Design of Impedance Matching Networks for Radio Frequency and Microwave Amplifiers*, Artech House, Norwood, MA, 1985.

[2] R. Collin, *Foundations for Microwave Engineering*, McGraw-Hill, New York, 1966.

[3] M. E. Van Valkenberg, *Network Analysis*, Prentice-Hall, Englewood Cliffs, NJ, 1974.

[4] G. L. Matthaei, L. Young, and E. M. T. Jones, *Microwave Filters, Impedance Matching Networks and Coupling Structures*, McGraw-Hill, New York, 1964.

[5] R. J. Wenzel, "The Modern Network Theory Approach to Microwave Filter Design," *IEEE Trans. Electromag. Compt.*, Vol. EMC10, No. 2, pp. 196–209, June 1968.

[6] R. A. Speciale, "Derivation of the Generalized Scattering Parameter Renormalization Transformation," *Proc. IEEE Int. Symp. Circuits Sys.*, pp. 166–169, April 1980.

[7] J. M. Rollett, "Stability and Power-Gain Invariants of Linear Twoports," *IRE Trans. Circuit Theory*, Vol. CT-9, pp. 29–32, March 1962.

[8] R. M. Fano, "Theoretical Limitations on the Broadband Matching of Arbitrary Impedances," *J. Franklin Inst.*, Vol. 249, pp. 57–83, 139–154, Jan./Feb. 1950.

[9] H. A. Haus et al., "Representation of Noise in Linear Two Ports," *Proc. IRE*, Vol. 48, pp. 69–74, January 1960.

[10] H. Fukui, "Design of Microwave GaAs MESFETs for Broad-Band Low-Noise Amplifiers," *IEEE Trans. MTT*, Vol. 27, No. 7, pp. 643–650, July 1979.

[11] U. L. Rhode, A. M. Pavio, and R. A. Pucel, "Accurate Noise Simulation of Microwave Amplifiers Using CAD," *Microwave J.*, Vol. 31, No. 12, pp. 130–141, Dec. 1988.

[12] J. Engberg, "Simultaneous Input Power Match and Noise Optimization Using Feedback," *Proc. Eur. Microwave Conf.*, pp. 385–389, 1974.

[13] T. J. Aprille, "Wideband Amplifier Design Using Major Multiloop Feedback Techniques," *Bell Sys. Tech. J.*, Vol. 54, No. 7, pp. 1253–1275, Sept. 1975.

[14] G. F. Franklin, J. D. Powell, and A. Emami-Naeini, *Feedback Control of Dynamic Systems*, 3rd ed., Addison-Wesley, Reading, MA, 1994.

[15] C. Yuen, M. Day, S. Bandy, and G. Zdasiuk, "A Monolithic Three Stage 2–8 GHz Feedback Amplifier," *IEEE Electron. Device Lett.*, Vol. 7, No. 3, pp. 161–163, March 1986.

[16] S. Ramo, J. R. Whinnery, and T. Van Duzer, *Fields and Waves in Communication Electronics*, Wiley, New York, 1965.

[17] E. L. Ginzton, W. R. Hewlett, J. H. Jasberg, and J. D. Noe, "Distributed Amplification," *Proc. IRE*, Vol. 36, pp. 956–969, Aug. 1948.

[18] Y. Ayasli, R. Mozzi, J. Vorhaus, L. Reynolds, and R. Pucel, "A Monolithic GaAs 1–13 GHz Traveling Wave Amplifier," *IEEE Trans. Microwave Theory Tech.*, Vol. 30, No. 7, pp. 976–981, July 1982.

[19] M. Rodwell, M. Riaziat, K. Weingarten, and D. Bloom, "Internal Microwave Propagation and Distortion Characteristics of Travelling Wave Amplifiers Studied by Electro-Optic Sampling," *IEEE Trans. Microwave Theory Tech.*, Vol. MTT-34, No. 12, Dec. 1986.

[20] J. Beyer, N. Prasad, R. Becker, J. Nordman, and G. Kohenwater, "MESFET Distributed Amplifier Design Guidelines," *IEEE Trans. Microwave Theory Tech.*, Vol. 32, No. 3, pp. 268–275, March 1984.

[21] S. G. Bandy et al., "A 2–20 GHz High-Gain Monolithic HEMT Distributed Amplifier," *IEEE Trans. Microwave Theory Tech.*, Vol. 35, No. 12, pp. 1494–1500, Dec. 1987.

[22] Y. Ayasli et al. "2–20 GHz Traveling Wave Power Amplifier," *IEEE Microwave, Millimeter Wave Circuits Symp. Dig.*, pp. 67–70, June 1983.

[23] M. Riaziat, S. Bandy, L. Y. Ching, and G. Li, "Feedback in Distributed Amplifiers," *IEEE Trans. MTT*, Vol. 38, No. 2, pp. 212–215, Feb. 1990.

[24] R. Majidi-Ahy, M. Riaziat, C. Nishimoto, M. Glenn, S. Silverman, S. Weng, Y. C. Pao, G. Zdasiuk, S. Bandy, and Z. Tan, "5–100 GHz InP Coplanar Wave-

guide MMIC Distributed Amplifier," *IEEE Trans. MTT*, Vol. 38, No. 12, pp. 1986–1993, Dec. 1990.

[25] K. Kurokawa, "Some Basic Characteristics of Broadband Negative Resistance Oscillator Circuits," *Bell Syst. Tech. J.*, Vol. 48, pp. 1938–1955, July 1969.

[26] J. Obregon, A. Khanna, and J. Sautereau, "GaAs MESFET Oscillators," in Soares, Grafeuill, and Obregon (Eds.), *Applications of GaAs MESFETs*, Artech House, Dedham, MA, 1983, Chapter 6.

[27] R. T. Logan, L. Maleki, and M. Shadaram, "Stabilization of Oscillator Phase Using a Fiber Optic Delay Line," *Proc. 45th Ann. Symp. Freq. Cont.*, May 1991.

[28] T. C. Edwards, *Foundations for Microstrip Circuit Design*, Wiley, New York, 1981.

[29] W. M. Louisell, *Coupled Mode and Parametric Electronics*, Wiley, New York, 1960.

[30] C. P. Wen, "Coplanar Waveguide: A Surface Strip Transmission Line Suitable for Nonreciprocal Gyromagnetic Device Applications," *IEEE Trans. Microwave Theory Tech.*, Vol. MTT-17, No. 12, pp. 1087–1090, Dec. 1969.

[31] H. Shigesawa, M. Tsuji, and A. A. Oliner, "Conductor Backed Slotline and Coplanar Waveguide: Dangers and Full Wave Analyses," 1988 *IEEE MTT-S Int. Microwave Symp. Dig.*, pp. 199–202, May 1988.

[32] Y. Fujiki, M. Suzuki, T. Kitazawa, and Y. Hayashi, "Higher Order Modes in Coplanar Type Transmission Lines," *Electron. Commun. Jpn*, Vol. 58-B, No. 2, pp. 74–81, 1975.

[33] Y. C. Shih and T. Itoh, "Analysis of Conductor Backed Coplanar Waveguide," *Electron. Lett.*, Vol. 18, No. 12, pp. 538–540, June 1982.

[34] R. W. Jackson, "Mode Conversion at Discontinuities in Finite Width Conductor Backed Coplanar Waveguide," *IEEE Trans. Microwave Theory Tech.*, Vol. MTT-37, pp. 1582–1589, Oct. 1989.

[35] G. Ghione and C. Naldi, "Analytical Formulas for Coplanar Lines in Hybrid and Monolithic MICs," *Electron. Lett.*, Vol. 20, No. 4, pp. 179–181, Feb. 1984.

[36] J. B. Knorr and K. Kuchler, "Analysis of Coupled Slots and Coplanar Strips on Dielectric Substrate," *IEEE Trans. Microwave Theory Tech.*, Vol. MTT-23, No. 7, pp. 541–548, July 1975.

[37] D. B. Rutledge, D. P. Neikirk, and D. P. Kasilingam, "Integrated Circuit Antennas," in *Infrared and Millimeter Waves*, Vol. 10, Academic, New York, 1983.

[38] R. Collin, *Field Theory of Guided Waves*, 2nd ed., IEEE, New York, 1991.

[39] M. Riaziat, R. Majidi-Ahy, and I. J. Feng, "Propagation Modes and Dispersion Characteristics of Coplanar Waveguides," *IEEE Trans. MTT*, Vol. 38, No. 3, pp. 245–250, March 1990.

[40] M. Riaziat, I. J. Feng, R. Majidi-Ahy, and B. A. Auld, "Single Mode Operation of Coplanar Waveguides," *Electron. Lett.*, Vol. 23, No. 24, Nov. 1987.

[41] K. D. Marx, "Propagation Modes, Equivalent Circuits, and Characteristic Terminations for Multiconductor Transmission Lines with Inhomogeneous Dielectrics," *IEEE Trans. Microwave Theory Tech.*, Vol. MTT-21, No. 7, pp. 450–457, July 1973.

[42] M. Riaziat, I. Zubeck, S. Bandy, and G. Zdasiuk, "Coplanar Waveguides Used in 2–18 GHz Distributed Amplifier," *1986 IEEE MTT-S Int. Microwave Symp. Dig.*, June 1986.

[43] M. Riaziat, E. Par, G. Zdasiuk, S. Bandy, and M. Glenn, "Monolithic Millimeter Wave CPW Circuits," *1989 IEEE MTT-S Int. Microwave Symp. Dig.*, pp. 525–528, June, 1989.

[44] A. Hardy and W. Streifer, "Coupled Mode Theory of Parallel Waveguides," *J. Lightwave Tech.*, Vol. LT-3, No. 5, pp. 1135–1146, Oct. 1985.

[45] R. Majidi-Ahy, K. Weingarten, M. Riaziat, D. Bloom, and B. Auld, "Electro-optic Sampling Measurements of Dispersion Characteristics of Slotline and Coplanar Waveguide Even and Odd Modes," *IEEE MTT-S Int. Microwave Symp. Dig.*, pp. 301–304, May 1988.

[46] B. Bhat and S. K. Koul, *Analysis, Design and Applications of Fin Lines*, Artech House, Norwood, MA, 1987.

[47] C. Montgomery, R. Dicke, and E. Purcel, *Principles of Microwave Circuits*, McGraw-Hill, New York, 1948.

[48] J. A. G. Malherbe, *Microwave Transmission Line Couplers*, Artech House, Norwood, MA, 1988.

[49] B. M. Oliver, "Directional Electromagnetic Couplers," *Proc. IRE*, Vol. 42, No. 11, pp. 1686–1692, Nov. 1954.

[50] M. Kirschning and R. H. Jansen, "Accurate Wide-Range Design Equations for the Frequency Dependent Characteristics of Parallel Coupled Microstrip Lines," *IEEE Trans.*, Vol. MTT-32, No. 1, pp. 83–89, Jan. 1984.

[51] T. G. Bryant and J. A. Weiss, "Parameters of Microstrip Transmission Lines and of Coupled Pairs of Microstrip Lines," *IEEE Trans.*, Vol. MTT-16, No. 12, pp. 1021–1027, Dec. 1968.

[52] S. E. Miller, "Coupled Wave Theory and Waveguide Applications," *Bell Sys. Tech. J.*, Vol. 34, pp. 661–719, May 1954.

[53] P. Ikalainen, G. Matthaei, and M. Monte, "Broadband Dielectric Waveguide 3-dB Couplers Using Asymmetrical Coupled Lines," *IEEE MTT-S Int. Microwave Symp. Digest*, pp. 135–136, June 1986.

[54] J. Lange, "Interdigitated Stripline Quadrature Hybrid," *IEEE Trans.*, Vol. MTT-17, pp. 777–779, Dec. 1969.

[55] A. Presser, "Interdigitated Microstrip Coupler Design," *IEEE Trans.*, Vol. MTT-26, No. 10, pp. 801–805, Oct. 1978.

[56] Y. Tajima and S. Kamihashi, "Multiconductor Couplers," *IEEE Trans.*, Vol. MTT-26, No. 10, pp. 795–801, Oct. 1978.

CHAPTER FIVE

Optical Waveguides and Passive Components

Until very recently, the medium of choice for long-distance transmission of optical signals was air or free space [1]. For high data rate long-distance communications, optical signal transmission through air is considered unreliable due the serious drawback of being adversely affected by bad weather and the possibility of being blocked by moving obstacles. Free space is still a viable medium for limited applications such as satellite-to-satellite communications.

Optical guiding characteristics of transparent dielectrics have been known for over a century [2]. The concept of dielectric waveguides was fully developed for microwaves in the early twentieth century [3]. At optical wavelengths, however, most dielectrics were found to be too lossy for long-distance transmission. In other words, no dielectric material was "transparent" enough to guide light over long distances. First practical implementations of optical fibers were in the form of fiber bundles used for flexible illumination or for viewing objects not directly visible. In order for optical fibers to be able to compete with coaxial cables for signal transmission, they would have to be manufactured with a lower loss than 20 dB/km (i.e., transmit 1% of the optical power through a distance of 1 km [4]). In 1970, this milestone task was accomplished by Corning Glass Works. Since that date dramatic improvements in signal transmission through fiber have been achieved. Presently, silica glass fibers are available with losses lower than 0.2 dB/km (95% transmission in 1 km). Today, no other technique of long-distance signal transmission can compete with the low loss of an optical fiber. In order to put this in perspective, it should be mentioned that for the transmission of a modest 2.0-GHz signal, the expected loss is 10 dB/km for a rectangular waveguide and 90 dB/km for a coaxial cable. The attenuation of microwaves through atmosphere can be very low (0.02 dB/km at 10 GHz). However, for practical antenna sizes, the loss due to beam spreading far exceeds that of optical fibers.

5.1 OPTICAL FIBERS

In non-TEM electromagnetic waveguides including hollow circular guides, one interpretation of the guiding mechanism is a plane wave reflecting back and forth at the conducting walls of the waveguide (Figure 5.1a). The same mechanism can guide the wave in a dielectric waveguide where reflections from the walls occur by *total internal reflection*. The requirement here is that the refractive index of the dielectric be higher than that of the surrounding medium. In other words, the propagation velocity of the wave in the guiding medium needs to be slower than outside. Optical fibers are based on this principle. A general-purpose optical fiber is a circular dielectric waveguide with a surrounding cladding layer of slightly lower dielectric constant. Since total internal reflection generates an evanescent tail in the second medium, the cladding needs to be thick enough to keep the guided wave from interacting with surrounding objects.

There are similarities between circular hollow waveguides and optical fibers, but the differences are also considerable. For example, in a circular waveguide with metallic walls the longest wavelength that can propagate is 3.41 times the radius of the waveguide [6]. The next higher order mode has a wavelength of 2.61 times the radius. Reversing the argument, for single-mode operation, the diameter of the hollow waveguide has to be between 59 and 77% of the wavelength. For optical guiding, this is too small to be practical. Furthermore, reflection from a conducting wall causes excessive optical power loss. For example, a reflection coefficient of 0.999 at the con-

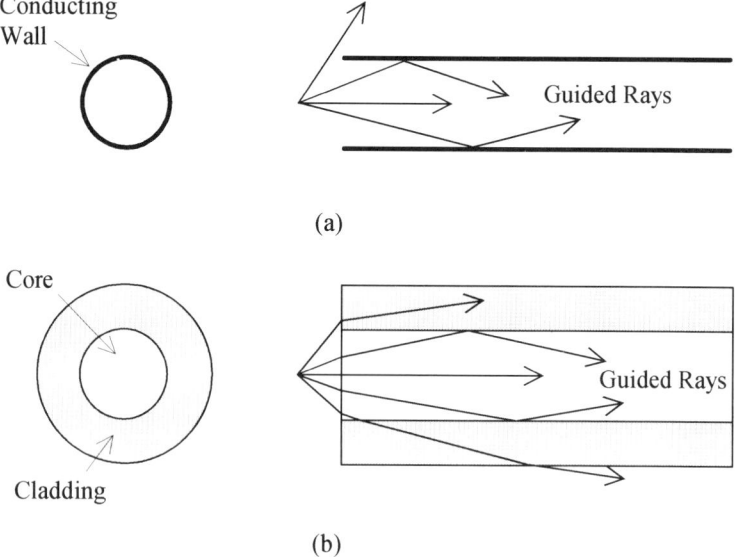

FIGURE 5.1 Wave guiding in multimode waveguides: (*a*) metallic and (*b*) dielectric.

ducting wall means that more than half of the optical power is lost in just 400 reflections. In a dielectric waveguide, on the other hand, the total internal reflection is virtually lossless. Also, a single-mode optical fiber has a core diameter that is close to 10 times the optical wavelength. This leads to practical production dimensions.

The discussion of optical fibers in this chapter is divided into step-index multimode fibers, graded-index fibers, and single-mode fibers. Data transmission limitations are discussed in each case.

5.1.1 Step-Index Fibers

In a step-index fiber the boundary between the core and the cladding is abrupt. Figure 5.1 shows the cross section of a step-index fiber with the path of a typical ray of light (ray optics is valid if the core diameter is much larger than the wavelength of light). Total internal reflection at the boundary between core and cladding can occur only if the angle θ_2 is greater than the critical angle θ_c:

$$\theta_2 > \theta_c = \sin^{-1} \frac{\eta_{clad}}{\eta_{core}}. \tag{5.1}$$

This restriction gives rise to a maximum deviation from normal incidence that the input optical beam may have. Calling this deviation θ_{max}, it is given by

$$\sin \theta_{max} \approx \sqrt{\eta_{core}^2 - \eta_{clad}^2}. \tag{5.2}$$

Here, $\sin \theta_{max}$ is the *numerical aperture*, which is a measure of the light gathering capability of the fiber. Optical fibers are normally *weakly guiding*, meaning that the difference between the refractive indices of the core and the cladding $\delta\eta$ is very small (on the order of 0.5%). Numerical aperture for a communication fiber is in general below 0.15.

An optical ray that enters the fiber along the axis is called an *axial ray*, and one that enters at the maximum deviation angle is called a *marginal ray*. The marginal ray travels a longer path through the fiber. The difference in path lengths can be calculated by geometry, and when converted to a time difference, the result is the *differential time delay*, given by

$$\Delta t = \frac{L(\delta\eta)}{c}, \tag{5.3}$$

where L is the length of the fiber. This is an approximate gage for the *intermodal dispersion* in a step-index fiber. In deriving the above expression, the possibility of having skew rays, that is, those that propagate in a helical path along the fiber, was ignored. Intermodal dispersion strongly limits the speed of long-distance transmission.

Example The differential time delay experienced in a 10-km fiber with a refractive index step of 0.005 is calculated from the equation above to be 0.167 μs. If this time delay is not allowed to be more than half the bit spacing in a digital network, the maximum data rate that can be carried is 3 Mbits/s. ∎

There are two types of fibers that eliminate this effect: graded-index fibers and single-mode fibers.

5.1.2 Graded-Index Fibers

Total internal reflection does not require a sharp boundary. In a mirage over a hot road, for example, light reflection is caused by the gradual variation of the refractive index of air due to temperature near the surface of the road. It will be shown in this section that a similar effect can be used to guide light in a fiber with a reduced differential time delay.

In the step-index fiber discussed above, the refractive index of the fiber as a function of radius has a discontinuity at the core–cladding boundary. This discontinuity is not necessary for confining the optical field. Light can be guided by various index profiles as long as the refractive index decreases with radius. In graded-index fibers, the index profile is chosen to have a lens-like effect on light propagation. A section of graded-index fiber can refocus the light emerging from a single point onto a real image of that point. Along the graded-index fiber, this process is repeated and light is guided by periodic focusing. Guiding light in this fashion inherently reduces the differential time delay. In order to observe this fact more clearly, consider light that travels from a point on the axis of a concave lens to its image point on the other side (Figure 5.2). Despite the available axial and marginal paths, no differential time delay occurs. This is due to the fact that the lens is thinner near the edge and the marginal rays spend less time in the slower lens medium. This compensates for the longer path they take. This effect is essential for the

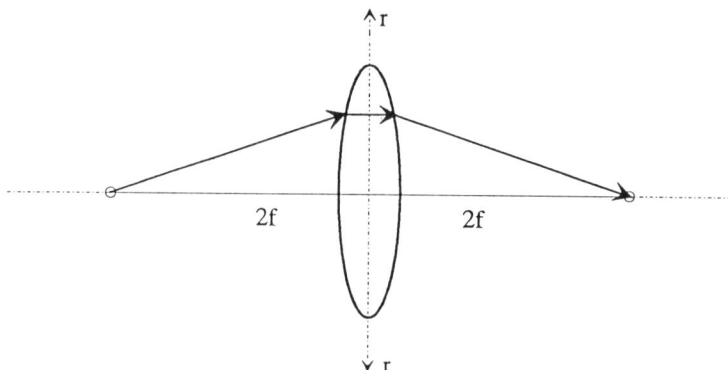

FIGURE 5.2 Real image of a point source formed by a convex lens.

formation of real images and can be traced to Fermat's principle. Fermat's variational principle states that in going from point A to point B, light selects the path among adjacent paths that minimizes travel time. Since every point in a real image is formed by converging rays of adjacent paths, all rays traveling from any point on the object to its corresponding point in the image experience the same time delay.

Consider the thin biconvex lens of focal distance f shown in Figure 5.2. A point on the axis a distance $2f$ from the lens has a real image at the same distance from the lens on the opposite side. The thickness of the lens is $t(r) \ll f$, and its refractive index is η. Any ray connecting the object and image points takes the physical path length $l(r)$ given by[1]

$$l(r) = 2\left[\sqrt{4f^2 + r^2} + \tfrac{1}{2}\eta t(r)\right], \qquad (5.4)$$

where r is the ray's distance from the axis at the lens. Since the on-axis and off-axis rays need to have the same optical path length, the following condition holds for the thickness of the lens:

$$2\sqrt{4f^2 + r^2} + \eta t(r) = 4f + \eta t(0), \qquad (5.5)$$

or

$$\eta t(r) = \eta t(0) + 4f\left[1 - \sqrt{1 + \left(\frac{r}{2f}\right)^2}\right]. \qquad (5.6)$$

This condition on the thickness of the lens is approximately met by thin, symmetrical biconvex spherical lenses. An alternate approach yielding the same optical path length for all rays is to keep the thickness of the lens constant and to change the refractive index as a function of r. The result is a *gradient-index lens*, sometimes referred to as a GRIN rod [7]. The restriction on $\eta(r)$ is found from the above expression, where $\eta t(r)$ is replaced with $t\eta(r)$. The result is

$$\eta(r) = \eta(0) + 4\frac{f}{t}\left[1 - \sqrt{1 + \left(\frac{r}{2f}\right)^2}\right]. \qquad (5.7)$$

When the approximation of $a \ll 2f$ holds, a being the radius of the lens, Eq. (5.7) can be further simplified to give

$$\eta(r) = \eta(0) - \frac{f}{2t}\left(\frac{r}{f}\right)^2. \qquad (5.8)$$

---

[1] The exact expression from geometry is $l(r) = 2\left[\sqrt{[2f - \tfrac{1}{2}t(r)]^2 + r^2} + \tfrac{1}{2}\eta t(r)\right]$. Here, the assumption was made that $t(r) \ll 4f$.

This is the parabolic index profile used in many GRIN lenses.

A graded-index fiber is intended to function in a similar fashion. Marginal rays spend more of their propagation time in the faster outer regions of the fiber, compensating for their longer paths. Consider a graded-index fiber with a core refractive index slightly higher than that of the cladding, that is, $\eta_{clad} = \eta_{core}(1 - \delta\eta)$, where $\delta\eta \ll 1$. In common manufacturing processes, the refractive index profile of the fiber may be approximated by [8, 9]

$$\eta(r) = \eta_{core}\left[1 - 2(\delta\eta)\left(\frac{r}{a}\right)^{\alpha}\right]^{1/2}, \qquad r < a, \tag{5.9}$$

where a is the core radius. The index remains constant in the cladding, where $r > a$. The profile parameter α can range from zero to infinity. Small values of α correspond to very slow variation of the index, while with large α it approaches the profile of a step-index fiber. Intermodal dispersion or differential time delay for the fiber has a sharp minimum for $\alpha \approx 2$ (Figure 5.3). For small $\delta\eta$ this is identical to the parabolic index profile discussed above.

The sharp minimum of the intermodal dispersion curve points to a problem with graded-index fibers, that is, their sensitivity to variations in manufactur-

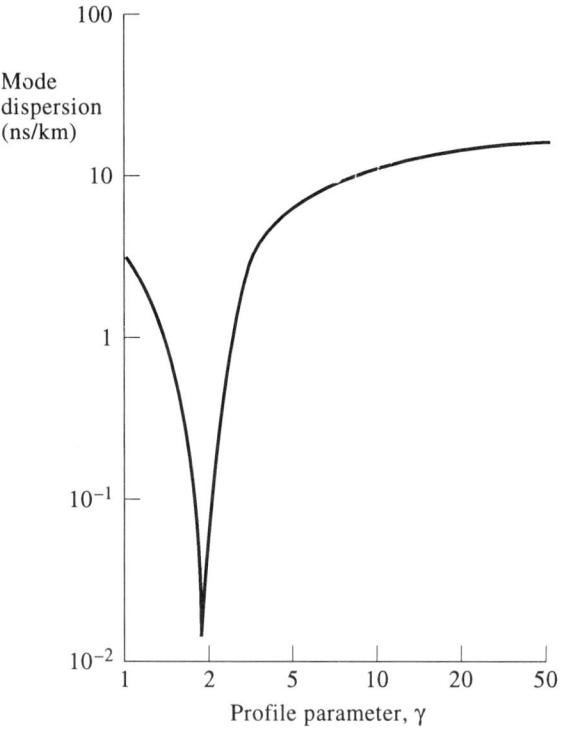

FIGURE 5.3 Differential time delay as a function of profile parameter α [10].

ing parameters. The index profile needs to be accurate and maintained by a tight tolerance throughout the length of the fiber. The index profile also has a wavelength dependence, meaning that a graded-index fiber optimized for a particular wavelength may do a poor compensation of differential time delay at another wavelength [11]. In practice, despite the fact that differential time delay is strongly reduced in graded-index fibers, its total elimination is only possible in single-mode fiber.

5.1.3 Single-Mode Fibers

In the discussion of multimode and graded-index fibers, we restricted our analysis to ray optics for simplicity of visualization. Ray optics gives an accurate picture only in cases where the physical dimensions involved are much larger than the optical wavelength. This condition holds in multimode fibers where the core diameter is at least 50–100 times the guided wavelength. As the core diameter of the fiber is reduced, the picture changes and only "rays" at specific angles to the axis can propagate. This is due to optical interference. Each such angle defines a propagation mode with its own characteristic phase velocity. If the core diameter is reduced further, eventually only a single mode can propagate, and the problem of intermodal dispersion is eliminated.

In strongly guiding electromagnetic circular waveguides such as those with conducting walls, single-mode operation requires that the diameter of the waveguide be smaller than the wavelength of the guided radiation ($D < \lambda/1.3$ for hollow waveguide with conducting walls). In optical fibers, however, since the walls are reflecting only for grazing angles, some of the higher order modes cannot propagate. This makes it possible for fibers with core diameters larger than the optical wavelength to operate in the single-mode regime.

Modes in an optical fiber are classified as four different types. The TE and TM modes are those with either electric or magnetic fields perpendicular to the axis of the fiber. The other two types are the EH and the HE modes with electric and magnetic components in all coordinate axes, but with either dominant transverse E or dominant transverse H, respectively. The EH and HE modes correspond to *skew rays* that do not propagate along a meridional plane of the fiber. These rays follow helical paths. The four sets of exact solutions are usually approximated in weakly guiding fibers by a single set of *linearly polarized (LP) modes*. In an LP$_{lm}$ mode, the index l is a measure of the pitch in the helical propagation. Larger l numbers correspond to larger pitch angles or tighter helixes. The index m is an indicator of the angle of the meridional projection of the propagation direction with the axis. All of the LP modes except LP$_{01}$ have finite cutoff wavelengths. The cutoff condition of the LP$_{lm}$ modes are given by the mth root of the Bessel function [12]:

$$J_{l-1}(V) = 0. \tag{5.10}$$

The number V is the *normalized frequency*, sometimes called *characteristic*

168 OPTICAL WAVEGUIDES AND PASSIVE COMPONENTS

waveguide parameter, but is most often referred to simply as the *V-number*. It is defined as

$$V = 2\pi \frac{a}{\lambda}\left(\eta_{core}^2 - \eta_{clad}^2\right)^{1/2} = 2\pi \frac{a}{\lambda}(\text{NA}), \qquad (5.11)$$

where NA is the numerical aperture and λ is the free-space wavelength. Figure 5.4 shows both the exact modes and the linearly polarized modes

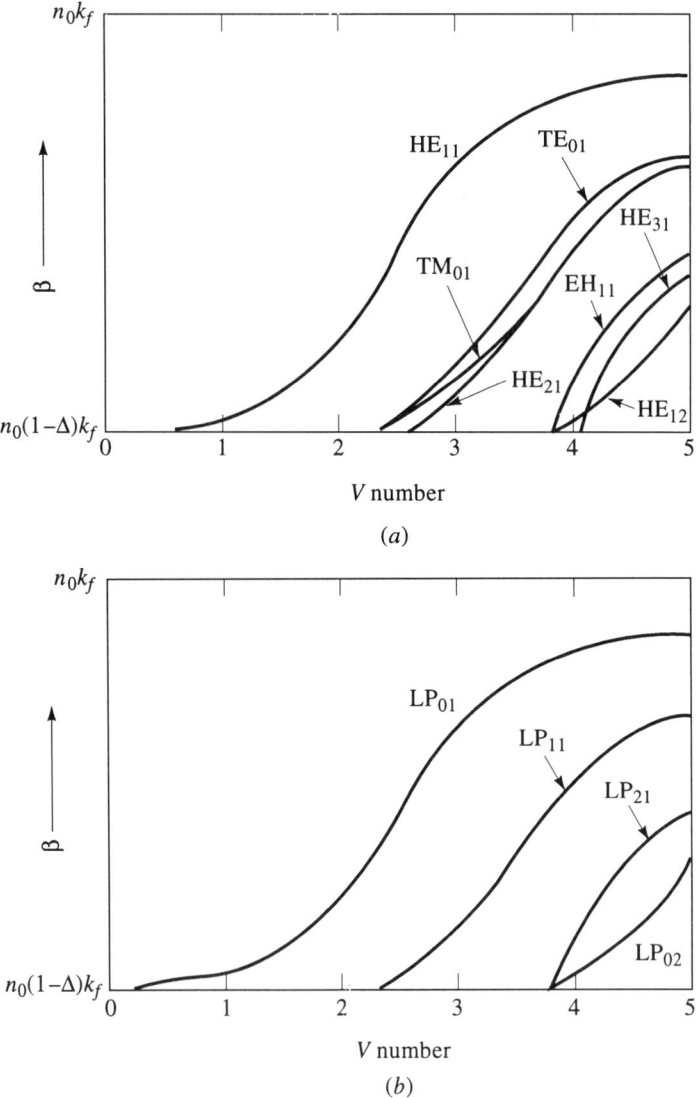

FIGURE 5.4 Propagation constant vs. V-number of (*a*) exact low order modes of optical fiber and (*b*) approximate linearly polarized modes [13].

of an optical fiber as a function of V-number. The fiber is single mode for values of V below the onset of the LP_{11} mode, which occurs at the first root of $J_0(V) = 0$, or $V = 2.405$.

Example A glass fiber with a core refractive index of 1.5 and a core-to-cladding index step of 0.001 is to be used to transmit 1.3 μm radiation. Determine the maximum core diameter for single-mode operation.

Solution Remember that $\left(\eta_{core}^2 - \eta_{clad}^2\right)^{1/2}$ may be written approximately as $\sqrt{2\eta_{core}\delta\eta}$. The rest is simply using Eq. (5.1):

$$V = 2.405 = 2\pi \frac{a}{1.3}\sqrt{2(1.5)(0.001)}.$$

It follows that $a = 9.1$ μm. ∎

5.2 INTEGRATED OPTICAL WAVEGUIDES

In integrated optoelectronics, optical *channel waveguides* are formed on the same substrate with other electronic and optical components. An example of such substrate is InP. In hybrid optoelectronics, passive optical components are processed on a substrate to which other electronic and active optical components are attached. Substrates of choice in the latter case are either various types of low loss glasses on which passive optical components are processed or nonlinear optical materials such as $LiNbO_3$ that can be the host to optical modulators, second-harmonic generators, or other specialized materials. The main requirement for forming an optical waveguide is to surround a low optical loss material with lower refractive index dielectrics. Integrated optical waveguides fall under five basic categories: embedded type, ridge type, buried type, loaded type, and active type, all of which satisfy the above requirement (Figure 5.5).

Embedded-type waveguides are fabricated using either metal diffusion into the substrate, ion exchange at elevated temperatures, ion implantation, or electron beam irradiation. All of these techniques generate a guiding region with a higher refractive index than the host material. Optical waveguides processed in this fashion can be made relatively low loss (less than 0.1 dB/cm) depending on the host. The profile of the guiding region in buried-type waveguides is approximately semi-elliptical. Therefore there could be a considerable deviation between the field patterns of the modes supported by such waveguides and those of optical fibers. This causes a significant coupling loss at transitions to optical fibers. In applications where such coupling losses are to be minimized, a second processing step is added whereby the surface portion of the waveguide is converted back to have a lower refractive index. The resulting buried waveguide is closer to circular in shape.

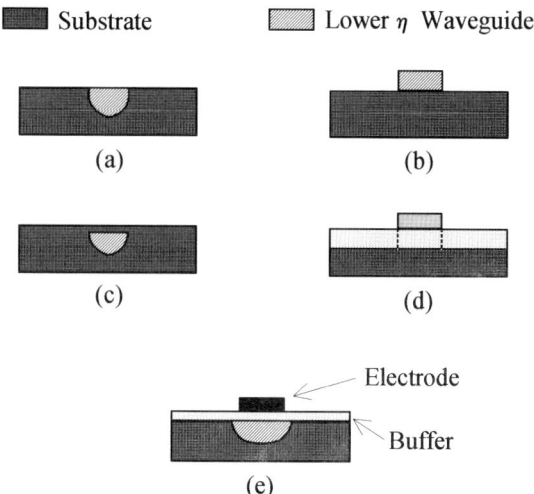

FIGURE 5.5 Five commonly used optical channel waveguide types: (*a*) embedded type, (*b*) ridge type, (*c*) buried type, (*d*) loaded type, and (*e*) active type.

A ridge-type waveguide is formed by either depositing or epitaxially growing a guiding material with a higher refractive index than the substrate. This material is then etched laterally to form the optical waveguide. Such waveguides tend to have a strong lateral confinement that can lead to multimode operation. Also, any roughness of the lateral walls cause significant scattering loss.

Loaded-type waveguides generate a region of increased effective refractive index for the guided mode by loading the guiding film with a dielectric layer. The loading confines the mode in the transverse direction. The refractive index of the added dielectric is lower than that of the guiding film. This is what makes it distinct from the ridge waveguide. The increased effective refractive index of the guiding film is due to the fact that the guided-mode energy partially travels in the added dielectric. The evanescent tail of a similar mode in the unloaded region propagates in air, which has a lower dielectric constant. Loaded-type waveguides are very weakly guiding and are thus limited to straight sections only.

Active waveguides are encountered in gain-guided semiconductor lasers and in electro-optic materials where the needed index variation is induced by an applied voltage. Active guiding is ordinarily done in an active device where guiding is needed to achieve other optical effects.

5.2.1 Propagation Modes of Planar and Channel Waveguides

The cross-sectional geometries of most integrated channel waveguides resemble either a rectangle or a semi-ellipse. Also, the index profiles show

INTEGRATED OPTICAL WAVEGUIDES   171

either gradual variation or step discontinuities at the boundaries [14]. Here, a rectangular shape step-index channel waveguide is chosen for analysis as an example to represent a typical modal behavior for integrated waveguides. The discussion is broken into two parts. First, the planar waveguide is reviewed, for which exact analytical solutions exist. Then the discussion is expanded to an approximate solution of the channel waveguide modes.

The modes of a dielectric slab waveguide without lateral confinement (planar waveguide) and with the same upper and lower boundaries were considered briefly under CPW modes in Chapter 4. The modal behavior is simply derived by assuming a sinusoidal solution in the waveguide dielectric and exponentially decaying solutions in the upper and lower regions. The solutions are matched at the two boundaries. The actual derivation is not presented here, but the results for even TE modes are shown in Figure 4.26. Optical planar waveguides are analyzed somewhat differently from the microwave slab waveguides of Chapter 4. The first deviation comes from the fact that in planar optical waveguides the upper and lower boundaries no longer have the same dielectric constant. The second deviation is the different normalizing conventions used in the two cases.

The two-dimensional waveguide shown in Figure 5.6 consists of three re-

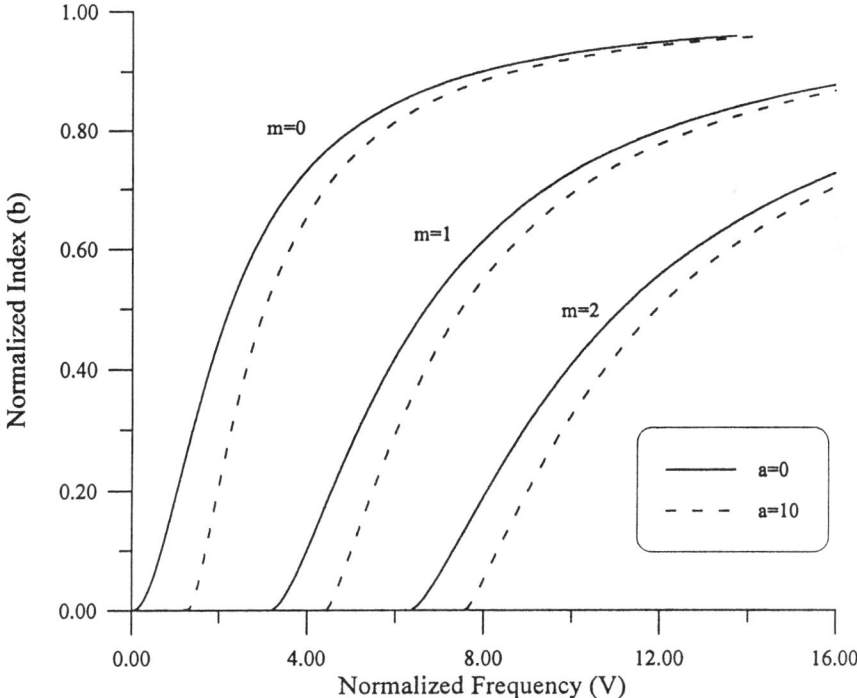

FIGURE 5.6 Geometry of a step-index planar optical waveguide and its normalized modal dispersion curves.

gions: the guiding layer, the substrate, and the cover layer with refractive indices η_g, η_s, and η_c, respectively. In order to account for the asymmetry between the upper and lower refractive indices, an *asymmetry factor* is defined. This factor quantifies the deviation from identical cover and substrate indices. For TE modes the asymmetry factor is defined as

$$a = \frac{\eta_s^2 - \eta_c^2}{\eta_g^2 - \eta_s^2}. \tag{5.12}$$

For TM modes the factor is multiplied by $(\eta_g/\eta_c)^4$. The value of $a = 0$ corresponds to identical substrate and cover refractive indices. The modal dispersion behavior for this case converges to the curves shown in Figure 4.26. For nonzero values of a the dispersion curves are modified slightly such that now instead of one curve per mode there is a family of curves per mode (Figure 5.6).

The convention for normalizing the frequency is similar to the optical fiber. Specifically, the V number is defined as

$$V = 2\pi \frac{h}{\lambda} \sqrt{\eta_g^2 - \eta_s^2}, \tag{5.13}$$

where h is the height of the guiding layer. The effective refractive index of the guided mode η_{eff} is also normalized and both TE and TM modes are represented by a single set of universal dispersion curves.[2] The normalized refractive index b is defined as

$$b = \frac{\eta_{\text{eff}}^2 - \eta_s^2}{\eta_g^2 - \eta_s^2}. \tag{5.14}$$

With these normalizations, the modal dispersion equations are represented by the expression

$$V\sqrt{1-b} = \tan^{-1}\sqrt{\frac{b}{1-b}} + \tan^{-1}\frac{b+a}{1-b} + m\pi, \tag{5.15}$$

where m is the order of the mode. This is the expression plotted in Figure 5.6.

The modal behavior of channel waveguides is more involved due to the extra boundary conditions and is best calculated numerically. There are, however, approximate solutions such as the *effective index method* that can be used for quick calculations of modal dispersion and the number of modes propagating in the waveguide.

---

[2] The curves are approximately valid for TM modes only when the difference between η_g and η_s is very small.

The effective index method (Figure 5.7) deals with the horizontal and vertical light confinement separately. This method is applicable to propagation that is not close to cutoff and yields more accurate results for waveguides with a high aspect ratio. Referring to Figure 5.7, if the vertical confinement is considered first, a V-number and an effective refractive index η_1 can be calculated for the shown planar waveguide using the curves in Figure 5.6. The effective index found in this way is then used in the second planar waveguide with horizontal confinement. The resulting modal propagation constants are the desired values for the channel waveguide. To be more specific, suppose the modes propagating in the channel waveguide have dominant field components E_x and H_y. In the planar waveguide with vertical confinement, the magnetic field vector is parallel to the boundary, and the resulting propagation is in a TM mode. The V-number for this arrangement is $V_1 = 2\pi(h/\lambda)\sqrt{\eta_g^2 - \eta_s^2}$, with the asymmetry factor

$$a = \frac{\eta_g}{\eta_c}^4 \frac{\eta_s^2 - \eta_c^2}{\eta_g^2 - \eta_s^2}.$$

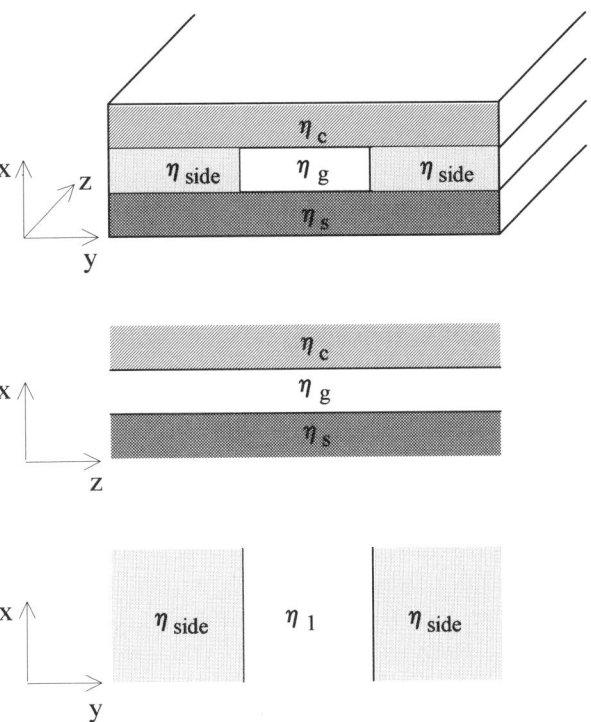

FIGURE 5.7 In effective index approximation vertical and horizontal confinements are treated separately.

From Figure 5.6 or Eq. (5.15) the resulting b value is found, which leads to the effective index η_1. This index is used as the index of the guiding medium (replacing η_g) in the horizontal confinement case. Here, the electric field vector is parallel with the boundary, and the propagation is in a TE mode. The same procedure is followed again to determine the effective refractive index η, which is now the desired value for the channel waveguide.

5.3 INTEGRATED PASSIVE COMPONENTS

Guided optical signals can be processed by integrated passive components such as those that can change the optical path direction (bends and corners), those that can be used for power combining and power dividing (branching waveguides), those that can sample the signal (directional couplers), and those that can do polarization control or wavelength demultiplexing. This section attempts to elucidate some of the physical characteristics of integrated passive components and to explain their relevant design parameters.

5.3.1 Bends and Corners

There are two ways of changing optical propagation direction in integrated channel waveguides: a gradual bend and an abrupt corner. The abrupt corner acts like a mirror and requires either a strongly guided wave or the metallization of one edge of the waveguide. Selective metallization of a waveguide edge is a difficult processing step that is not practical. A 90° abrupt bend without metallization requires a strong enough guiding such that light incident at 45° on the waveguide wall is totally internally reflected. This means that the ratio of cladding to core index should be 0.7 or less. This condition is satisfied only in some ridge-type waveguides. In a vertical 90° bend that couples light out of a channel waveguide, often Bragg reflectors are utilized. In most other cases requiring in-plane bending of light path, a gradual circular bend is normally used. The schematic diagram of a circular gradual bend is shown in Figure 5.8. Any bending of the waveguide is expected to cause some optical loss. This can be seen intuitively in the case of a multimode waveguide. In the ray optics picture, bending of the waveguide causes some of the rays to impinge on the sidewall at less than the critical angle and escape the waveguide. The shorter the radius of curvature, the higher the fraction of the lost rays. The case of a single-mode waveguide is more intricate. But even in this case, the following intuitive way of thinking about the optical loss has been suggested. The propagation mode of the waveguide has evanescent tails extending into the surrounding medium. At the waveguide bend the tail that extends from the outer edge of the waveguide travels a distance longer than the center of the waveguide. The outer portion of this tail is expected to travel faster than the speed of light in that medium. This portion therefore cannot "keep up" with the guided wave and is lost to radiation. This is of course not an accurate explanation, but it does possess intuitive appeal.

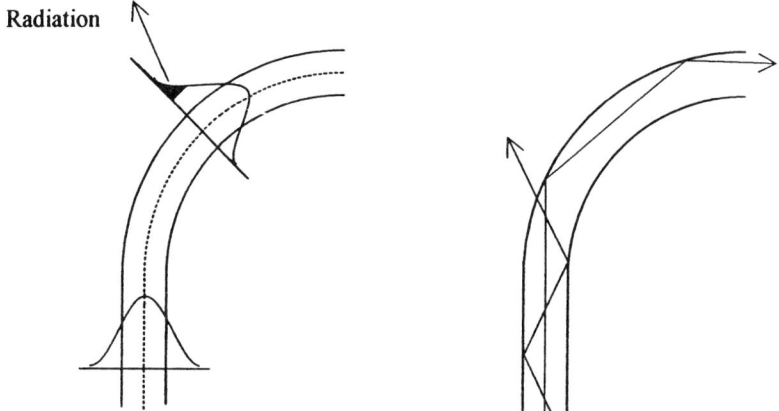

FIGURE 5.8 Optical loss due to waveguide bend.

Analytical expressions exist that characterize propagating modes of a circularly bent waveguide and quantify the associated optical losses [15–17]. We will not derive such expressions here, but simply state that the attenuation coefficient α per unit length of a waveguide of width w and bend radius R in the low-loss approximation of $\alpha R \ll 1$ may be written as

$$\alpha = \frac{\gamma^2 \kappa^2 e^{(2\gamma w)}}{(\eta^2 - \eta_s^2) k^2 \beta (2 + \gamma w)} e^{(-2\gamma^3 R/3\beta^2)}. \tag{5.16}$$

Here, η is the effective index of the guided mode and η_s is the refractive index of the surrounding material, β is the propagation constant along the waveguide, $k = \omega/c$, $\kappa = (\eta^2 k^2 - \beta^2)^{1/2}$, and $\gamma = (\beta^2 - \eta_s^2 k^2)^{1/2}$. Note that for weakly guided modes where η is very close to η_s, even a moderate bending radius causes significant attenuation.

5.3.2 Branching Waveguides

Branching waveguides, otherwise known as power dividers or power combiners, are multiport optical junctions with either one input and a number of outputs or vice versa. Here, we will briefly discuss three-port junctions that are used as equal two-way power dividers or combiners. The schematic diagram of such a device is shown in Figure 5.9. At the branching point, the angle 2θ needs to be as small as possible, forming as sharp a corner as possible in order to avoid radiative losses. For very small branching angles, the effect of mutual coupling between the two output waveguides of the power divider should also be taken into account. This is of concern when power division is to be unequal. Typically, integrated power dividers have full branching angles 2θ of 2°–5°. Connecting three identical waveguides in this arrangement, a tapered section is needed to avoid abrupt width changes (Figure 5.9).

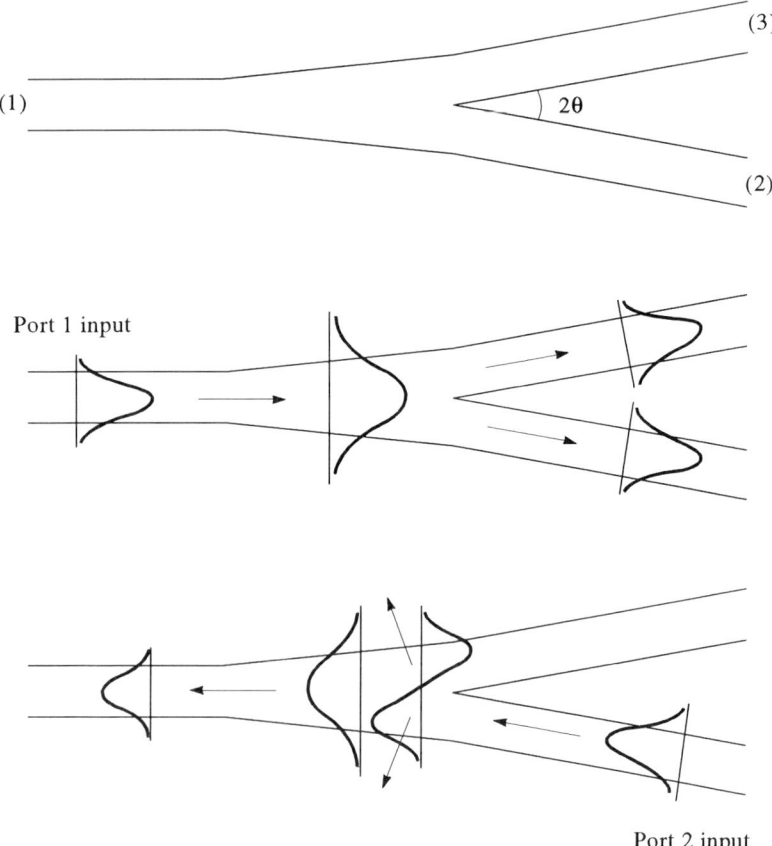

FIGURE 5.9 (a) Three-port optical power divider/combiner using single-mode waveguides. (b) Power division or in-phase combining. (c) Loss of power in uneven combining.

If all three waveguides are single-mode, the tapered section starts as a single-mode waveguide at the narrow end (port 1) but widens to twice the width of a single-mode waveguide. Therefore, at the broad end it can support at least two modes—an odd mode and an even mode—corresponding to in-phase and out-of-phase excitation of two side-by-side single-mode waveguides. The tapered section serves to match the input waveguide mode to the even mode of the output waveguides. The odd mode of the tapered section can only be excited from ports 2 and 3. It cannot propagate to the single-mode end of the taper. Therefore, it escapes the waveguide in weakly guiding structures and is reflected back to ports 2 and 3 in strongly guiding arrangements.

As can be seen in Appendix C, a lossless, reciprocal three-port junction cannot be completely matched. This means that a two-way power divider/combiner cannot, in general, be free from reflections and losses. It is only under certain input conditions that reflections can be minimized. The normally chosen condition for branching waveguides is $S_{11} = 0$. Note also that the branching optical waveguide junction cannot be assumed to be lossless under all excitations, except if the channels are strongly guiding. Referring to Figure 5.9, notice that when an incident wave is split into two outgoing signals, or vice versa, the lowest order even mode is excited in the tapered section. This situation leads to low loss and low reflection. If, on the other hand, an optical wave is incident on port 2 only, its field profile in the tapered section is approximately the superposition of the even and odd modes. The higher order odd mode cannot propagate in any of the waveguides and escapes the junction. Hence this loss mechanism depends on the excitation of the junction. Similarly, if two signals are incident in phase on ports 2 and 3, the odd mode is not excited in the tapered section, and losses in the junction are minimized. In contrast, if the excitations of ports 2 and 3 are out of phase, only the odd mode is excited and most of the power is lost by radiation into the substrate.

5.3.3 Directional Couplers

The construction and analysis of optical directional couplers are similar to the distributed forward coupling case of electrical waveguides explained in Section 4.9. Backward couplers are not practical at optical wavelengths due to the required small dimensions and controlled mode impedances.[3] When two identical channel waveguides are placed side by side such that they have some evanescent field overlap, optical power in one waveguide is coupled to the other. The coupled power increases with distance until all the power in the first waveguide is coupled to the second. The process reverses, and this periodic coupling continues sinusoidally with distance. This behavior, as was discussed before, is a manifestation of the beating phenomenon between the odd and even modes of propagation. The distance required to couple all the power from one waveguide to the other is called the coupling length L_c. Any value of coupling can be achieved by choosing the length of the coupler as a suitable fraction of the coupling length. The required design tool is an expression that gives the coupling length for given physical parameters of two waveguides as a function of the distance between them.

In order to derive such an expression, consider two parallel waveguides a and b carrying signals with amplitudes $A(z)$ and $B(z)$ with propagation constants β_a and β_b, respectively. In the absence of coupling between the waveguides the amplitudes are expected to be constant with distance (neglecting attenuation). In the presence of coupling, the rate of change of each amplitude is proportional to the amplitude of the signal on the other waveguide. The resulting coupled-mode equations are [18]

---

[3]Backward optical coupling is possible if the coupling medium has spatial periodicity along the coupling path.

$$\frac{dA(z)}{dz} = -jKB(z)e^{-j\Delta\beta z},$$
$$\frac{dB(z)}{dz} = -jKA(z)e^{j\Delta\beta z}, \qquad (5.17)$$

where $\Delta\beta = \beta_b - \beta_a$. The proportionality constant K is called the coupling coefficient and is a function of the field overlap between the waveguides:

$$K = \frac{\omega\eta^2}{4P} \int\int (\eta^2(x,y) - \eta_s^2) E_a^* E_b \, dx\, dy, \qquad (5.18)$$

where the integration is over the cross section of either waveguide and η_s is the refractive index of the substrate. One solution to the coupled equations above is the expected periodic power transfer:

$$P_a = F \sin^2\left(\frac{K}{\sqrt{F}}z\right), \qquad P_b = 1 - P_a, \qquad (5.19)$$

where

$$F = 1 + \left(\frac{\Delta\beta}{K}\right)^2. \qquad (5.20)$$

Note that when the propagation constants of the two waveguides are identical, a complete power transfer takes place, with a coupling length given by $L_c = \pi/2K$.

The bandwidth of the directional coupler is another important parameter in its design. In some applications such as wavelength or polarization separators (see the next section) the bandwidth needs to be as narrow as possible. In these applications, the length of the coupler is made longer than one coupling length. The longer the length of the coupler, the narrower the bandwidth. In other applications such as in optical taps or samplers, wavelength insensitivity is desired. The widest bandwidth or the lowest sensitivity to wavelength variation is achieved at the maximum coupling point, that is, when the length of the coupler is equal to one coupling length. One way to design wide-band couplers therefore is to select the coupling ratio by choosing $\Delta\beta$ and keeping the length of the coupler equal to L_c.

5.3.4 Wavelength and Polarization Separators

Most active integrated optoelectronic components used for processing an optical signal are polarization sensitive. For example, electro-optic modulators (with a few exceptions) require linearly polarized optical inputs. On the other hand, the optical input is normally brought to the channel waveguide by a non-polarization maintaining fiber and is therefore elliptically polarized. The selection of the proper polarization or the separation of TE and TM modes is also necessary in general due to the different propagation characteristics of these modes.

The simplest tool for mode suppression in a channel waveguide is a short section of waveguide with metal cladding. Any direct metal contact on a channel waveguide causes the attenuation of all optical propagation, particularly in the lowest order modes. However, this attenuation is much greater for TM modes due to their higher electric fields near the conducting layer. A typical metal-clad section attenuates the lowest order TE mode by 2 dB and the lowest order TM mode by 20 dB. The metal-clad section is therefore a lossy but simple to implement polarizer.

Polarization separation can be performed by devices that are sensitive to the difference between the propagation constants of TE and TM modes. Take, for example, a directional coupler consisting of two dissimilar waveguides. One waveguide may be fabricated on top of the other for maximum degrees of freedom in design parameters. Suppose the waveguides are designed to have the same propagation constant for TE modes ($\beta_{TE}^1 = \beta_{TE}^2$) but not for TM modes ($\beta_{TM}^1 \neq \beta_{TM}^2$). The coupling coefficients K_{TE} and K_{TM} will also be different. If the length of the coupler is selected to be equal to the coupling length for the TE mode ($L = \pi/2K_{TE}$), all of the power in the TE mode is coupled from waveguide 1 to waveguide 2. The TM mode power coupled to waveguide 2 is found from Eq. (4.66) to be

$$P_{TM} = \frac{\sin^2\{(\pi/2)(K_{TM}/K_{TE})[1 + (\Delta\beta/2K_{TM})^2]^{1/2}\}}{1 + (\Delta\beta/2K_{TM})^2}. \quad (5.21)$$

For a large $\Delta\beta$ and the proper value of K_{TM}/K_{TE}, this expression can be made equal to zero. The result is that the TE power couples completely to waveguide 2, while the TM power remains in waveguide 1, and the desired mode splitting is achieved.

PROBLEMS

5.1 For a step-index fiber derive the expression for the differential time delay between axial and marginal rays [Eq. (5.3)]. If we allow optical rays with helical paths through the fiber (skew rays), what maximum differential time delay should we expect?

5.2 For 0.85-μm wavelength transmission determine the maximum core diameter of a step-index fiber that assures single-mode propagation. The refractive index of the core is 1.5. Find the solution for four different values of $\delta\eta$: 0.0005, 0.001, 0.005, and 0.01.

5.3 A channel waveguide is fabricated by diffusion on a substrate with a refractive index of 2.4. The top surface is exposed to air. The width of the waveguide is 10 μm and its depth is 5 μm. The diffusion process generates a refractive index difference of 0.005. Find the effective refractive

180 OPTICAL WAVEGUIDES AND PASSIVE COMPONENTS

index of the waveguide and determine if it is a single-mode guide for 1.06-μm wavelength.

5.4 Determine the coupling coefficients K_{TE} and K_{TM} for a mode splitting directional coupler having a physical length of 5 mm and a $\Delta\beta = (6)10^{-9}$ m^{-1}.

5.5 A branching waveguide is built as an equal power divider matched at port 1 (Figure 5.9). Two optical signals of equal magnitude and phase are incident on ports 2 and 3. Calculate all the reflected and transmitted signals at every port. Assume a lossless junction.

REFERENCES

[1] A. G. Bell, "Apparatus for Signaling and Communicating Called Photophone," U.S. Patent No. 235,199, Dec. 7, 1880.

[2] The guiding of light by a stream of water was demonstrated by J. Tyndall, *Six Lectures on Light*, Longmans, Green, London, 1873.

[3] J.A. Stratton, *Electromagnetic Theory*, McGraw-Hill, New York, 1941.

[4] C. K. Kao and G. A. Hockham, "Dielectric Fibre Surface Waveguides for Optical Frequencies," *Proc. IEEE*, Vol. 113, No. 7, pp. 1151–1158, 1966.

[5] F. Kapron, D. Keck, and R. Maurer, "Radiation Losses in Glass Optical Waveguides," *Appl. Phys. Lett.*, Vol. 17, No. 10. p. 423, 1970.

[6] S. Ramo, J. R. Whinnery, and T. Van Duzer, *Fields and Waves in Communication Electronics*, Wiley, New York, 1965.

[7] SELFOC Product Guide, NSG America, Inc., Somerset, NJ.

[8] D. Gloge and E. A. J. Marcatili, "Multimode Theory of Graded-Core Fibers," *Bell Syst. Tech. J.*, Vol. 52, No. 9, pp. 1563–1578, Nov. 1973.

[9] A. Yariv, *Optical Electronics,* 3rd ed., Holt, Rinehart, and Winston, New York, 1985.

[10] J. Wilson and J. F. B. Hawkes, *Optoelectronics, an Introduction*, Prentice-Hall, Englewood Cliffs, NJ, 1989.

[11] R. Olshansky and D. B. Keck, "Pulse Broadening in Graded Index Optical Fibers," *Appl. Opt.*, Vol. 15, No. 2, pp. 483–491, Feb. 1976.

[12] D. Gloge, "Weakly Guiding Fibers," *Appl. Opt.*, Vol. 10, pp. 2252–2258, 1971.

[13] D. B. Keck, in M. K. Barnoski (Ed.), *Fundamentals of Optical Fiber Communications*, 2nd ed., Academic, New York, 1981.

[14] H. Kogelnik, "Theory of Dielectric Waveguides," in T. Tamir (Ed.), *Integrated Optics*, Springer-Verlag, New York, 1975.

[15] N. Nishihara, M. Haruna, and T. Suhara, *Optical Integrated Circuits*, McGraw-Hill, New York, 1989.

[16] E. A. J. Marcatili, "Bends in Optical Dielectric Guides," *Bell Sys. Tech. J.*, Vol. 48, No. 7, pp. 2103–2132, Sept. 1969.

[17] D. Marcuse, "Bending Losses of Asymmetric Slab Waveguide," *Bell Sys. Tech. J.*, Vol. 50, No. 8, pp. 2551–2563, Oct. 1971.

[18] D. Marcuse, *Light Transmission Optics*, 2nd ed., Van Nostrand, New York, 1982.

CHAPTER SIX

Short-Pulse Generation

The discussion of short pulses can be separated into three categories: (1) short electrical pulses, (2) short microwave pulses, and (3) short optical pulses. The development of short pulses in each category started in response to very different driving forces. Short electrical pulses in the form of high power bursts of electrical energy are normally used in particle accelerators, pulsed microwave tubes, electromagnetic mass accelerators, etc. The need for short microwave pulses has consistently been driven by high resolution radar. Pulsed magnetrons were the initial sources of pulsed microwaves for radar. The need to have not only a very short pulse but enough energy for sufficient range led to the development of chirped radar. Some of the concepts of chirped radar were later applied to optical short pulse generation.

The interest in short optical pulses started with strobe light photography. Very early in the history of photography people realized that it was easier to illuminate the object with a short burst of light rather than to use a very fast shutter. Edgerton of MIT is well known for his pioneering work in this field [1]. Strobe light photography allowed for the first time to "stop the action" and to view in detail events in daily life that occurred too fast for the eye to track (Figure 6.1). Presently, light pulses can be generated that are short enough to "stop the action" in extremely fast events such as microwave propagation [2] and chemical reactions [3, 4]. Pulses of a few femtoseconds duration have been generated [5]. These pulses extend only a few micrometers in space.

Short pulses with high repetition rate can be used in communications. Other applications of short light pulses include time-resolved imaging (Chapter 1) and time-resolved measurements [6].

6.1 OPTICAL PULSE FORMING

A laser cavity is capable of supporting a number of *longitudinal modes*. Each

182 SHORT-PULSE GENERATION

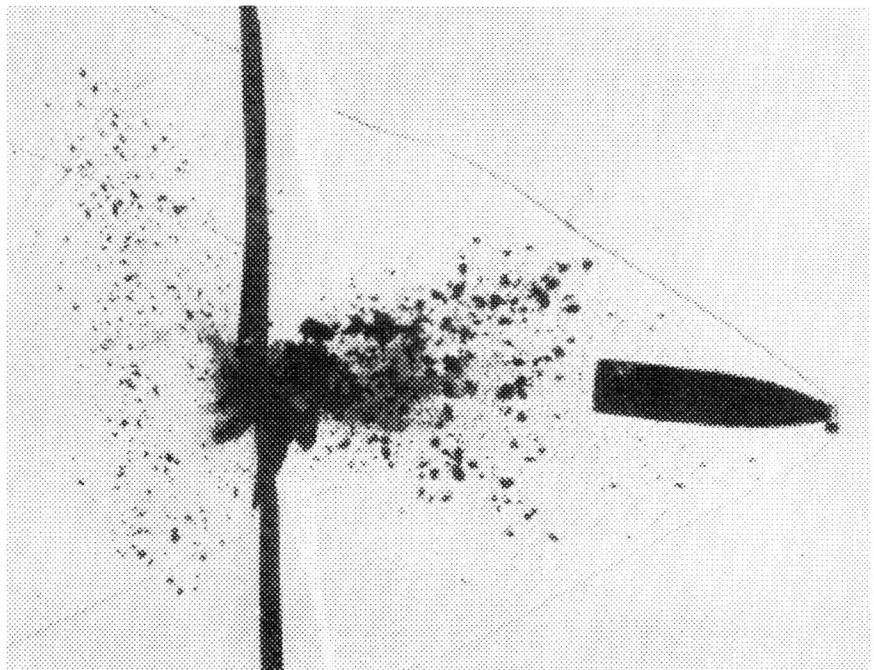

FIGURE 6.1 Bullet after penetrating Plexiglass barrier. (Flash light photography courtesy of McGraw-Hill [1].)

longitudinal mode corresponds to a resonance frequency of the cavity. Resonance conditions were described in Section 3.3. From that discussion it was found that longitudinal mode separation Δf is given by

$$\Delta f = \frac{c}{2\eta L}. \tag{6.1}$$

The mode separation is the inverse of the photon round-trip time in the cavity. Another constraint on the operation frequency of the laser comes from the gain medium. Since the gain process in the laser medium is frequency dependent, there is a bandwidth associated with it. This bandwidth is a function of the medium, the gain process, and the pumping level. Table 6.1 lists typical bandwidths associated with some laser gain media. Laser oscillation occurs at longitudinal cavity mode frequencies where the medium provides sufficient gain. In a short cavity laser, mode separation is very large and the medium can provide gain for only one longitudinal mode. Take, for example, a 10-cm-long He–Ne laser. The mode separation is approximately 1.5 GHz. In this case the gain medium also has a bandwidth close to 1.5 GHz and cannot easily accommodate more than a single longitudinal mode. In cases where the laser cavity is longer or the gain medium has a wider bandwidth, oscillations

TABLE 6.1 Typical Wavelengths and Approximate Bandwidths of Some Laser Gain Media

Gain Medium	Wavelength (μm)	Bandwidth
He–Ne	0.6328	1.5 GHz
Argon ion	0.5145	6 GHz
Nd–YAG	1.064	12 GHz
Ruby	0.6934	60 GHz
RD 6G dye	0.6	10 THz
GaAs	0.8	60 THz
Ti–sapphire	0.7–1.1	150 THz

can build up in a number of longitudinal modes. The output waveforms of such laser cavities depend on phase relationships among the various modes. In a free-running cavity where the phases are not well defined, various frequencies oscillate at random phases that can shift very rapidly. The output power will be effectively free of fluctuations in time but will have a very short coherence length.[1]

If the phases of the various modes do not drift in time, their interference causes periodic variations of the output power. Figure 6.2 shows a single sine wave, the superposition of two sine waves, and the superposition of four sine waves. These figures are representative of a laser in single-mode operation, a laser with two longitudinal modes, and one with four longitudinal modes. In general, if the output consists of the superposition of N longitudinal modes of equal amplitude centered at the frequency ω, one possible output waveform is given by

$$f(t) = \sum_{m=-(N-1)/2}^{(N-1)/2} A e^{j(\omega + m\omega_0)t}. \tag{6.2}$$

Here, $\omega_0 = \pi c/nL$, and the relative phase delays between modes are taken to be zero. The above equation simplifies to

$$f(t) = A e^{j\omega t} \frac{\sin(\tfrac{1}{2}N)\omega_0 t}{\sin \tfrac{1}{2}\omega_0 t}. \tag{6.3}$$

The peak amplitude is N times the amplitude of each mode, and the pulse width shrinks linearly with N. The time between the peak and the first zero is given by

$$\Delta t = \frac{2\pi}{N\omega_0}. \tag{6.4}$$

---

[1] In this discussion it has been assumed that the laser is inhomogeneously broadened.

184 SHORT-PULSE GENERATION

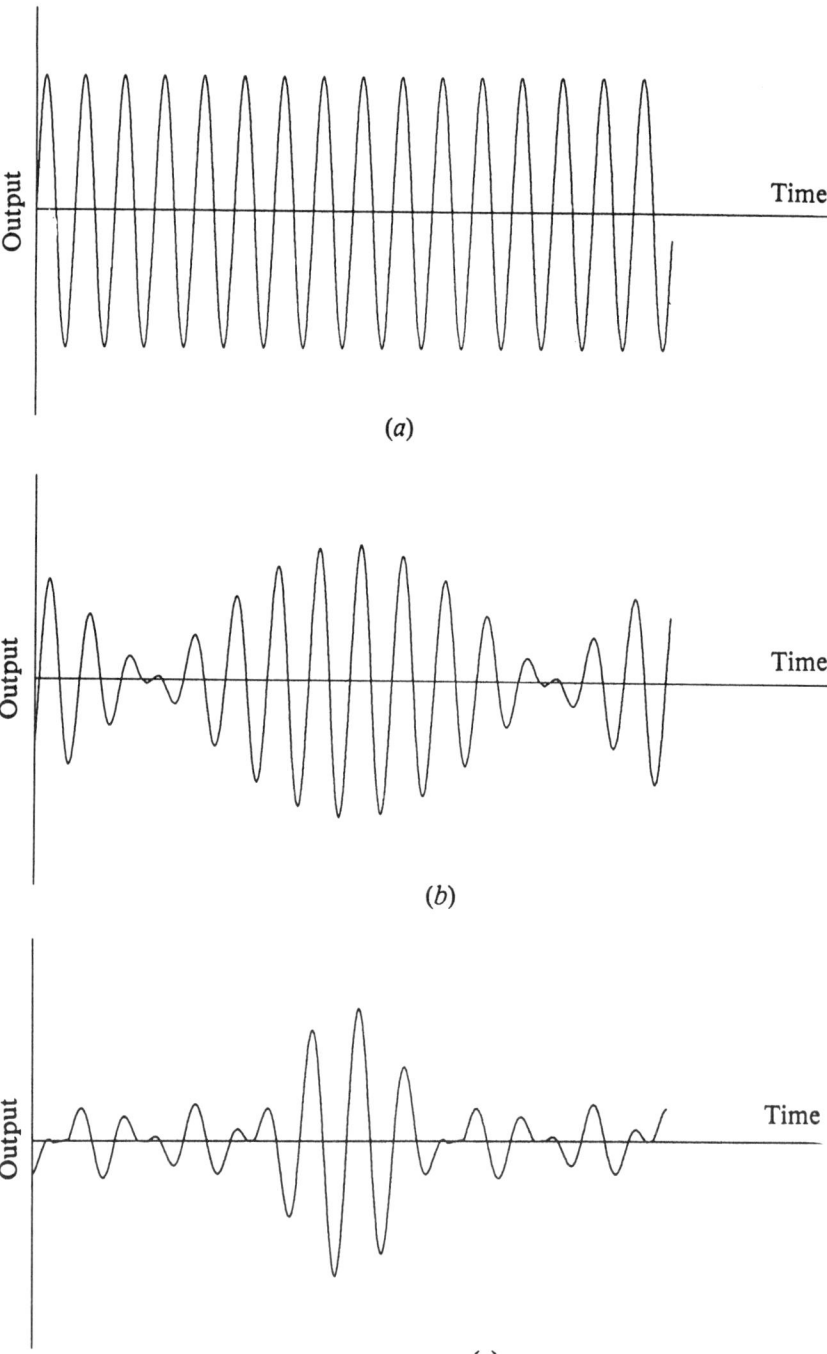

FIGURE 6.2 (a) Single sine wave, (b) superposition of two sine waves, and (c) superposition of four sine waves.

As can be seen, the interference of a large number of longitudinal modes can lead to short output pulses. The requirement is for the longitudinal modes to have fixed relative phases. The method to ensure the satisfaction of this requirement is known as *mode locking*.

6.1.1 Mode Locking

The concept of a physical process that could lock the phases of the various modes in a laser cavity may appear to be perplexing at first. If, however, the end result is examined in the time domain, the picture simplifies considerably. The introduction of mode locking causes the laser to generate periodic short pulses. Turning the argument around, mode locking can be thought of as the means to force the laser cavity to output only periodic bursts of radiation. Modulation of gain or loss in the cavity are two commonly used mode locking techniques. The modulation is done at the fundamental frequency of $c/2\eta L$ or the round-trip time for a single pulse in the cavity. The loss modulation approach is analogous to using a high-speed shutter. It can be thought of as a lossy medium placed in the cavity that prevents the propagation of any signal except for a very brief time in each cycle. This allows the existence of only a single pulse that reflects back and forth in the cavity. Every time this pulse impinges on the output coupler, an output pulse is generated (Figure 6.3).

The same effect can be achieved by modulating the gain in the cavity. In that case the active medium has no gain except for the short period of time when the pump pulse is passing through. This technique, which is known as *synchronous pumping*, is encountered when a pulsed laser is used to pump a gain medium with a fast relaxation time. External cavity semiconductor lasers can be mode locked by gain modulation through pulsing the drive current. These methods fall under the category of *active mode locking*. There are also passive methods for making the propagation of short pulses in the cavity more favorable than CW radiation. The most common method of *passive mode locking* is the placement of a saturable absorber in the cavity. Active and passive mode locking are further analyzed below.

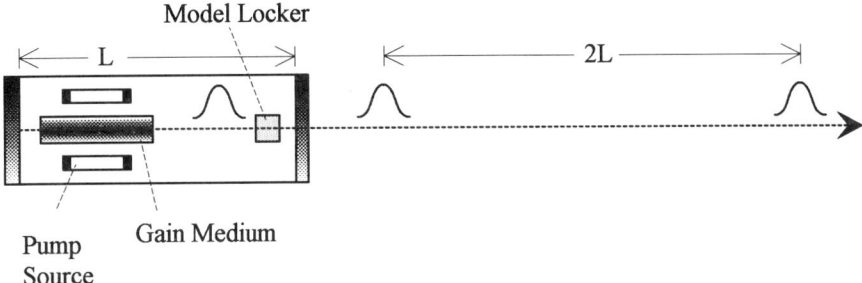

FIGURE 6.3 Mode locking by loss modulation. Separation of pulses in space is equal to cavity round-trip optical path length.

6.1.2 Active Mode Locking

Active mode locking is done by cavity loss modulation at a rate compatible with the round-trip time of a pulse in the laser cavity [7, 8]. Only the pulse that passes through the modulator when it is at its minimum-loss, experiences a net gain in the cavity. The modulator should be at the minimum-loss state again when the same pulse reaches it after one round trip in the cavity. It is desirable for obtaining short pulses that the duration of the minimum-loss state of the modulator be small compared with the round-trip time of the cavity.

An acousto-optic (AO) modulator or mode locker is often used to introduce a periodic loss in the cavity. The AO mode locker achieves this effect through Bragg diffraction from an acoustic standing wave (Figure 6.4). The strain field of the acoustic wave generates periodic local index variations that cause optical diffraction. Bragg diffraction introduces a loss by spatially separating some of the optical energy from the main beam. Twice in each acoustic cycle the acoustic strain field goes to zero, and diffraction loss is minimized. Optical loss in this arrangement is proportional to $\sin^2(\omega_m t)$, where ω_m is the mode locker drive frequency. Since minimum loss occurs twice in each cycle, the drive frequency needs to be $\omega_m = \pi c/2\eta L$, which is half the intermode frequency spacing or a harmonic thereof (L is the cavity length). The electrical oscillator is normally a fixed stable oscillator to which the optical cavity is synchronized by adjusting L.

The electro-optic modulator, described in Chapter 3, can be used in a similar manner. The electro-optic crystal is driven by an external electrical oscillator. The alternating electric field rotates the plane of polarization of the incident beam, which in turn goes through variable attenuation by a fixed analyzer. This is similar to the AO mode locker and the same drive frequency constraint holds.

In order to generalize the picture of mode locking, it is instructive to look at the process in the frequency domain. Consider a CW laser operating in a

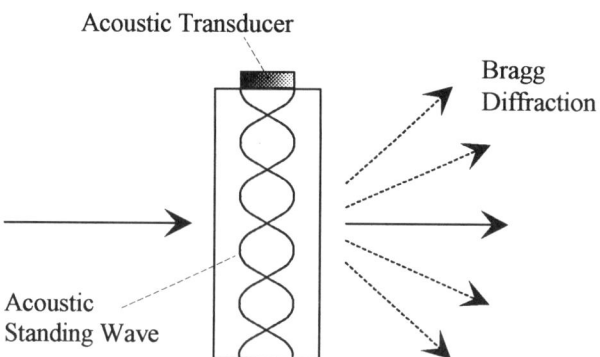

FIGURE 6.4 An AO mode locker introduces periodic loss in cavity by diffraction.

single longitudinal mode. Once an AO mode locker is added to the cavity, the laser mode is amplitude modulated (AM) at the angular frequency $2\omega_m$. This modulation generates upper and lower sidebands separated from the main laser frequency by $2\omega_m$. Since $2\omega_m$ is the intermode frequency spacing of the laser, the result of the modulation is coupling the laser mode to the upper and lower adjacent modes. If the laser is oscillating in multiple modes instead of one, all the modes will be "injection locked" in this way to their neighboring modes. This is the process of mode locking that was described above in the time domain. This type of mode locking is also known as *AM mode locking*. This interpretation of mode locking suggests that frequency modulation (FM) may achieve a similar effect. This is indeed the case, and the electro-optic phase modulator (without the analyzer) can induce mode locking in the laser cavity [9].

The width of the mode-locked laser pulse depends on the number of modes locked together and is a function of the fundamental repetition rate of the laser f_m and the mode locker modulation depth m, among other factors. In general, pulsewidth is reduced as the pulse repetition rate is increased, that is, the cavity is made shorter. This dependence can be seen intuitively as a reduction in the time period when the mode locker is in its minimum-loss state. The exact dependence of the pulsewidth on f_m and modulation depth depends on whether laser transitions are homogeneously or inhomogeneously broadened. An approximate expression derived elsewhere [9] gives the following expression for pulsewidth τ in the case of a homogeneously broadened laser:

$$\tau \approx \frac{(gL)^{1/4}}{2\sqrt{mf_m \Delta \nu}}, \qquad (6.5)$$

where g is the round-trip gain at the center frequency, L is the length of the cavity, and $\Delta \nu$ is the gain bandwidth of the laser medium. Common methods of short-pulse generation by active mode locking include shortening the cavity length and harmonic mode locking. In harmonic mode locking, the modulator operates at a multiple of the fundamental laser cavity rate. Picosecond pulses have been reported using short cavities with high speed mode lockers [10].

Active mode locking is capable of generating optical pulses much shorter than the period of the modulation. However, it falls short of being able to mode lock the entire gain bandwidth of many laser gain media. This task is more closely achieved by passive mode locking.

6.1.3 Passive Mode Locking

Passive mode locking takes advantage of the nonlinear characteristics of *saturable absorbers* to make the conditions of the cavity more favorable to short-pulse propagation. A saturable absorber is a laser dye[2] that absorbs

---
[2] A semiconductor section can also be used as a saturable absorber. This is discussed separately.

the radiation in the emission wavelength range of the laser. In CW mode, it suppresses the lasing. A sufficiently intense pulse, however, can saturate or bleach the absorber, thus reducing the loss for the propagation of such pulses.

The *bleaching* of an absorbing medium due to population inversion or the depletion of the lower level population of an absorbing transition is well known. Any material exhibiting these characteristics can, in principle, be used as a saturable absorber. However, in practice, it is implied that a saturable absorber does not manifest any significant fluorescence at the absorption wavelength. The behavior of a saturable absorber is commonly expressed in terms of the four-level system shown in Figure 6.5. A transition with a very short lifetime $\tau_{fc} \ll \tau_A$ populates level 3 and keeps level 2 population from rising. Photons with energy ν_L are absorbed by transitions either from 1 to 2 or from 3 to 4. A saturable absorber is called either fast or slow depending on the length of the recovery time τ_A compared to the duration of the optical pulse.

Optical transmission T through an absorber of length l can be approximated by

$$T = \frac{I_{\text{out}}}{I_{\text{in}}} = \exp[-\sigma_A N_1 l - \sigma_{\text{ex}} N_3 l], \tag{6.6}$$

where σ is the absorption cross section as indicated in Figure 6.5, and $N_1 + N_3 = N_0$ is the total number density of absorber molecules. Defining the *saturation intensity* I_S^A as

$$I_S^A = \frac{h\nu_L}{\sigma_A \tau_A}, \tag{6.7}$$

for fast absorbers Eq. (6.6) can be written as

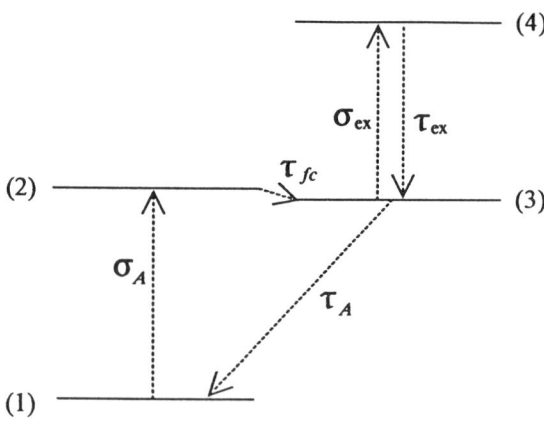

FIGURE 6.5 Energy band diagram of a four-level saturable absorber material.

$$T = \exp\left[(-N_0 l)\left(\frac{\sigma_A}{1 + I_{in}/I_S^A} + \frac{\sigma_{ex}}{1 + I_S^A/I_{in}}\right)\right]. \quad (6.8)$$

Note that the saturation intensity is the intensity that reduces N_1 to half the value of N_0. It is therefore a function of the recovery time τ_A. For slow saturable absorbers where τ_A is large compared to the duration of the laser pulse, recovery time loses its significance and does not enter the transmission equation. In that case it is the integrated energy of the pulse that determines the populations of levels 1 and 3, and Eq. (6.8) can be rewritten as

$$T = \exp\left[(-N_0 l)\left(\frac{\sigma_A}{1 + w_{in}/w_S^A} + \frac{\sigma_{ex}}{1 + w_S^A/w_{in}}\right)\right]. \quad (6.9)$$

A slow saturable absorber segment tends to shorten an optical pulse by removing its leading edge. The energy in the leading edge of the pulse is absorbed to achieve transparency for the rest of the pulse. The slow saturable absorber acts like an optical "shutter" triggered by the pulse itself. The leading edge of the pulse "opens" the shutter. The recovery time of the absorber "closes" the shutter before the next pulse arrives. On the other hand, a fast saturable absorber shortens both the leading edge and the trailing edge of the pulse. Since its recovery time is shorter than the pulsewidth, its action allows only the passage of radiation above a given amplitude threshold.

Saturable absorbers have traditionally been various types of laser dyes. However, sections of semiconductors can also be used as solid saturable absorbers. Consider a semiconductor with a bandgap energy close to or lower than the optical radiation of interest. The leading edge of an optical pulse propagating through this material is absorbed, leading to electron–hole pair generation. As the carrier density increases in the semiconductor, the later arriving portions of the pulse experience less attenuation. With an intense enough pulse, the material reaches transparency before the pulse peak arrives. If the pulse duration is short compared with carrier lifetime in the semiconductor, the trailing edge of the pulse is unattenuated (slow saturable absorber), and vice versa. A semiconductor saturable absorber is a very effective tool for mode locking semiconductor lasers. For use in broadly tunable or femtosecond pulsed lasers, however, the effective bandwidth of such absorbers may not be sufficient. Bandgap grading in semiconductors such as AlGaAs may be used to broaden the bandwidth of such absorbers. A continuous grading of the aluminum mole fraction in the growth of an AlGaAs absorber has been shown to increase its tunability by a factor of greater than 3 [11].

6.1.4 Colliding-Pulse Mode Locking

If two counterpropagating pulses interact coherently in the saturable absorber medium, the result is a higher peak field that leads to the saturation of the absorber at a lower pulse energy density. Simultaneous presence of the two pulses in the absorber medium is equivalent to irradiation with up to 50% higher than twice the energy density of each pulse [12]. The lower intensity requirement for saturation lowers the nonlinearity effects in the cavity. A colliding-pulse laser cavity is normally in the form of a ring oscillator (Figure 6.6). Ordinary Fabry–Perot oscillators can also be used in the colliding-pulse configuration by placing the saturable absorber close to the high reflector of the cavity.

The interaction of the colliding pulses generates a *saturation grating* in the absorber medium. Optical scattering by this grating causes some back reflection from each pulse. This cross-coupling mechanism can be a cause of either pulse broadening or the presence of satellite pulses in the output waveform.

6.1.5 Passive Mode Locking by Kerr-Lens Modulation

Kerr-lens modulation occurs in solid state lasers due to the nonlinearity of the index of refraction: $\eta(I) = \eta_0 + \eta_2 I$. An intense optical pulse induces enough extra phase delay along its axis (highest intensity part of the beam) to cause a self-focusing of the pulse. If an aperture is placed at the proper location in the cavity to accommodate such focusing, a loss mechanism is introduced that favors the propagation of intense pulses. Less intense CW modes that do not undergo the self-focusing will face the extra attenuation due to the aperture (Figure 6.7). In practice, the presence of the aperture is not needed. The self-focusing changes the mode shape in the cavity, and the cavity may be adjusted to present the lowest loss to the mode-locked pulses.

Kerr-lens mode locking (KLM), also referred to as self-mode locking, is analogous in action to a fast saturable absorber, in that the shortest pulses

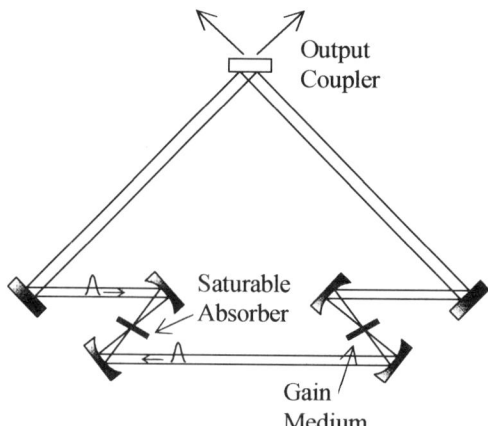

FIGURE 6.6 Essential elements of a colliding-pulse ring laser oscillator.

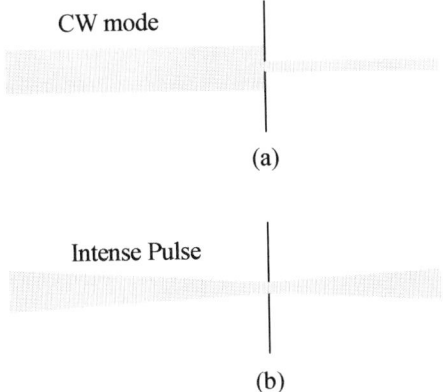

FIGURE 6.7 Kerr-lens mode locking: (a) CW mode loss due to presence of an aperture; (b) self-focusing of intense pulse minimizes loss through aperture.

suffer the least amount of loss in the cavity. Kerr-lens mode locking is not a self-starting mechanism from CW operation. Normal fluctuations in the intensity do not trigger KLM. Methods to start self-mode locking range from mirror vibration by tapping, to additional cavity modulators, synchronous pumping, and weak passive mode locking [11]. To date, Kerr-lens mode locking has generated pulses as short as 11 fs (3.3 μm in length) using a dispersion compensated Ti–sapphire laser [5]. Note that with a center wavelength of approximately 0.8 μm, these pulses correspond to very few cycles of optical radiation.

6.1.6 Dispersion Compensation

Generation of short optical pulses requires high optical bandwidth. In a laser cavity, this may be limited by the bandwidth of the gain element. Presently laser gain materials exist with higher bandwidth than required by the shortest achievable pulses. For example, with a Ti–sapphire laser one should theoretically be able to generate pulses as short as 3 fs (see Table 6.1). Achieving the shortest pulses allowed by the bandwidth is limited mainly by the dispersion of the cavity. Various elements in the laser cavity give rise to dispersive broadening of the pulse (e.g., mirror dispersion and material dispersion).[3] The dispersion in most materials is what is known as normal dispersion, where longer wavelengths have higher group velocities. Consequently, in a short-pulse laser cavity there is a cumulative normal dispersion that needs to be compensated.

Before embarking on the methods of dispersion compensation, it is important to define the dispersion terminology. Consider a propagating signal with a phase function $\phi = (\omega t - \beta z)$. The phase velocity is the velocity of constant

---

[3]Material nonlinearities induce frequency broadening effects such as the Kerr effect chirp in laser dyes. These effects can be combined with dispersion to yield shorter pulses. Refer to the section on pulse compression for more detail.

phase, which is $v_p = \omega/\beta$. The group velocity is the velocity at which a wave packet moves. Group velocity is the first derivative of angular frequency with respect to the propagation constant: $v_g = d\omega/d\beta$. This derivative is evaluated at the center frequency of the packet ω_0. Materials demonstrating *group velocity dispersion* (GVD) have different group velocities for wave packets with different center frequencies. Thus two coinciding wave packets of different center frequencies will develop a relative time delay $\Delta\tau$ after propagating a distance L in the medium:

$$\Delta\tau = L\left[\frac{1}{v_{g2}} - \frac{1}{v_{g1}}\right] = L\left[\left(\frac{d\beta}{d\omega}\right)_{\omega=\omega_2} - \left(\frac{d\beta}{d\omega}\right)_{\omega=\omega_1}\right]. \quad (6.10)$$

If ω_1 and ω_2 are close enough, we can write

$$\left(\frac{d\beta}{d\omega}\right)_{\omega=\omega_2} = \left(\frac{d\beta}{d\omega}\right)_{\omega=\omega_1} + \frac{d^2\beta}{d\omega^2}(\omega_2 - \omega_1). \quad (6.11)$$

The resulting expression written in differential form is

$$\frac{d\tau}{d\omega} = L\left(\frac{d^2\beta}{d\omega^2}\right). \quad (6.12)$$

The second derivative of β with respect to ω is what is often defined as group velocity dispersion (GVD), which can be written in any of the following forms:

$$\frac{d^2\beta}{d\omega^2} = \frac{d}{d\omega}\frac{1}{v_g} = -\frac{1}{v_g^2}\frac{dv_g}{d\omega}. \quad (6.13)$$

It is left as an exercise to show that GVD in a dispersive medium is proportional to the second derivative of the refractive index with respect to wavelength.

Compensation for GVD inside a laser cavity may be done in a number of different ways, including the use of negative-dispersion materials [13], prisms, and diffraction gratings. The topic of dispersion compensation by a diffraction grating pair is treated in the next section under pulse compression. The most common method of dispersion compensation inside a laser cavity is the insertion of a pair of prisms that yield a wavelength-dependent optical path length with controllable dispersion. A possible configuration suggested by Fork et al. [14] is shown in Figure 6.8. This arrangement consists of two identical isosceles prisms cut in such a way that the minimum-deviation angle is the same as Brewster's angle. The optical beam passes close to the apex of each prism. The reference path length L is the distance between the apices. The faces of the prisms are parallel to each other, so the emerging beam is parallel to the incident beam. Normally, a mirror (the high reflector in the cavity) reflects the light back along the same path, thereby doubling the dispersion compensation of the prism pair.

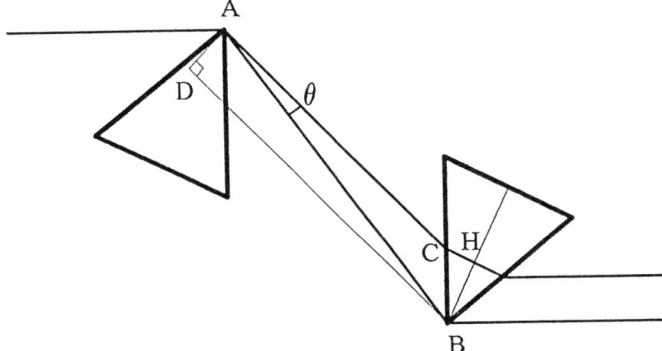

FIGURE 6.8 Prism arrangement used for dispersion compensation.

In Figure 6.8 the difference between the optical path lengths of the two rays shown is the difference between paths AB and ACH. Note that the optical path lengths beyond the prism median, BH, are equal. The rays in that region represent the same plane wave and form parallel lines once they are in the same medium. Similarly, the optical path length ACH is equal to the hypothetical path DB, which could represent the same plane wave. But the path DB is simply $AB \cos(\theta)$. For a round-trip path, going through the prism pair twice, the portion of the optical path length that is a function of θ is given by

$$P(\theta) = 2L \cos(\theta). \tag{6.14}$$

Dispersion through the prism combination is proportional to the second derivative of the variable optical path length P with respect to the wavelength. The second derivative of P can be evaluated indirectly as follows:

$$\frac{dP}{d\lambda} = \frac{dP}{d\theta}\frac{d\theta}{d\eta}\frac{d\eta}{d\lambda} \tag{6.15}$$

$$\frac{d^2P}{d\lambda^2} = \frac{d^2P}{d\theta^2}\left(\frac{d\theta}{d\eta}\frac{d\eta}{d\lambda}\right)^2 + \frac{dP}{d\theta}\left[\frac{d\theta}{d\eta}\frac{d^2\eta}{d\lambda^2} + \frac{d^2\theta}{d\eta^2}\left(\frac{d\eta}{d\lambda}\right)^2\right]. \tag{6.16}$$

In order to evaluate this expression, we need to find the functional dependence of θ on the refractive index of the prism. Figure 6.9 shows the general light path through a prism as well as the light path through a prism at the minimum-deviation angle. From geometry, the following identities always hold among the angles:

$$\psi_1 + \psi_2 = \alpha, \tag{6.17}$$

$$\sin \phi_1 = \eta \sin \psi_1, \tag{6.18}$$

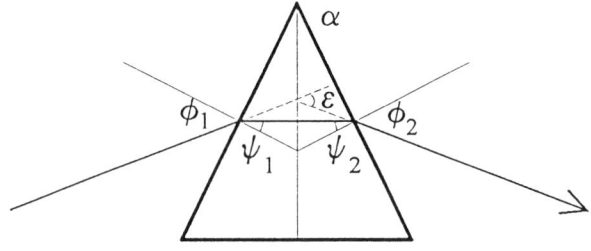

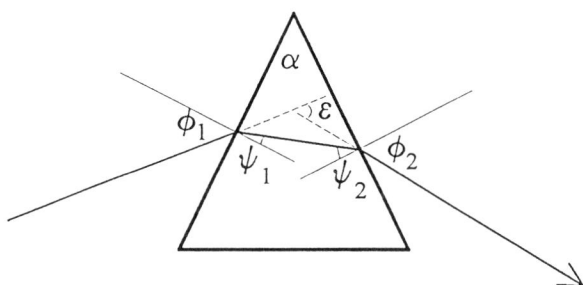

FIGURE 6.9 (*a*) Relevant angles of light path refracted by prism. (*b*) Light path through prism at minimum deviation angle.

$$\sin \phi_2 = \eta \sin \psi_2. \tag{6.19}$$

From the first equation we get $d\psi_1/d\eta = -d\psi_2/d\eta$. Using this relationship and differentiating the last two equations with respect to η, we find

$$\frac{d\phi_2}{d\eta} = \frac{1}{\cos \phi_2} (\sin \psi_2 + \cos \psi_2 \tan \psi_1). \tag{6.20}$$

Differentiating one more time with respect to η yields

$$\frac{d^2\phi_2}{d\eta^2} = \tan \phi_2 \left(\frac{d\phi_2}{d\eta}\right)^2 - \frac{\tan^2 \psi_1}{\eta} \left(\frac{d\phi_2}{d\eta}\right). \tag{6.21}$$

Note that ϕ_2 and θ are defined in the opposite sense to each other, and therefore $d\phi_2/d\eta = -d\theta/d\eta$. If the prism is at the minimum-deviation angle, $\psi_1 = \psi_2$. Furthermore, if the incidence angle is also equal to Brewster's angle, $\tan \phi_2 = \eta$. Substituting these in Eqs. (6.18) and (6.19) and converting from ϕ_2 to θ, we get

$$\frac{d\theta}{d\eta} = -2, \tag{6.22}$$

$$\frac{d^2\theta}{d\eta^2} = -4 + \frac{2}{\eta^3}. \tag{6.23}$$

We have now derived all the terms necessary to evaluate Eq. (6.16). The result is

$$\frac{d^2P}{d\lambda^2} = 4L\left[\frac{d^2\eta}{d\lambda^2} + \left(2\eta - \frac{1}{\eta^3}\right)\left(\frac{d\eta}{d\lambda}\right)^2\right]\sin\theta - 8L\left(\frac{d\eta}{d\lambda}\right)^2\cos\theta. \tag{6.24}$$

Since θ is a small angle, the second term tends to dominate, giving rise to negative dispersion.

Example What is the minimum value of L for a prism pair at minimum deviation and Brewster's angle to yield negative dispersion? The prisms are made of BK7 glass with $\eta = 1.516$, $d\eta/d\lambda = -0.0364\,\mu m^{-1}$, and $d^2\eta/d\lambda^2 = 0.1388\,\mu m^{-2}$ for a 1.0-mm diameter beam of light at 620 μm wavelength.

Solution Since the value of θ is normally very small, $\cos\theta$ may be replaced by unity. The term $L\sin\theta$ is approximately the spread of the beam at the reflecting mirror, which may be taken as twice the beam diameter. Substituting these values in Eq. (6.24), we arrive at

$$\frac{d^2P}{d\lambda^2} = 2.279 - (0.0106)L \quad (mm/\mu m).$$

For values of L greater than 215 mm this prism combination manifests negative dispersion. ∎

The negative dispersion of the prism pair can be adjusted through the choice of L. Another means of controlling the dispersion in this arrangement is to change the position of the second prism along the BC line to allow a longer light path through the prism material (positive dispersion). In this way the negative dispersion of the prism pair can be adjusted through zero. It is this adjustability plus the low loss nature of the prism pair at Brewster's angle that makes it the method of choice for dispersion compensation inside most short-pulse laser cavities.

6.1.7 Pulse Compression

Historically, the need for pulse compression arose in the application of pulsed radar. It was early in the development of pulsed microwave radar that it was realized that there is a trade-off between detection range and range resolution. The detection range R_{max} is given by the radar equation [15]:

$$R_{max} = \left[\frac{P_t G A_e \sigma}{(4\pi)^2 P_{min}}\right]^{1/4}, \tag{6.25}$$

where P_t is the transmitted power, G is the antenna gain, A_e is the antenna effective aperture, σ is the target cross section, and $P_{\min}$ is the minimum detectable signal or the detection noise level. Range resolution is the ability of radar to resolve objects located at different distances from the antenna. It is directly related to time resolution in sequentially received signals. Range resolution is normally higher than angular resolution and is the main mechanism by which multiple objects are distinguished from single objects. High range resolution requires short pulses. But there is always a hardware limit on how short the pulses can get without sacrificing transmitted power. In order to operate beyond this hardware limit, the idea of pulse compression radar was introduced in the 1940s [15].

In its simplest form, pulse compression radar utilizes a chirped transmitted pulse and a dispersive medium to compress the received signal in time. When the frequency of the chirped signal varies linearly with time, the technique is referred to as *linear FM pulse compression*. The medium that compresses the received signal is often a surface acoustic wave (SAW) dispersive delay line. Other devices used for the same purpose include rectangular waveguides near cutoff and charge-coupled devices. Linear FM pulse compression is a special case of the general procedure of radar pulse forming and *matched filtering* used to achieve various goals such as maximizing signal-to-noise ratio [15].

Linear FM pulse compression is also applied to the compression of optical pulses. The implementation of this technique in optics is completely different from radar. Unlike microwave sources, pulsed lasers cannot easily be frequency modulated to produce the desired chirped output. Instead, a nonlinearity effect known as *self-phase modulation* (SPM) in optical fibers is utilized to generate the chirped pulse. If the nonlinearity of the refractive index η is expanded to first order in intensity I, one obtains

$$\eta = \eta_0 + \eta_2 I. \quad (6.26)$$

A positive η_2 means that on the leading edge of an optical pulse, the refractive index is increasing and successive phase fronts are further slowed down. This leads to a red shift proportional to the rate of change of instantaneous intensity at any point along the leading edge. On the trailing edge the effect is reversed and the result is a blue shift. The overall effect on the pulse is the introduction of an approximately linear chirp (proportional to the derivative of the pulse shape) with no appreciable change of the intensity profile. Self-phase modulation is discussed in more detail in the next chapter.

Once the chirped signal is obtained, the red-shifted portion of the pulse needs to be retarded, and the blue-shifted portion needs to be advanced in order to compress the leading edge and the trailing edge toward the peak of the pulse. This task of imparting negative dispersion on the signal is achieved by either a prism pair (discussed earlier) or a *dispersive delay line* formed by a pair of gratings, as shown in Figure 6.10 [16]. Grating pairs are more common in this application mainly due to their broad tunability.

OPTICAL PULSE FORMING 197

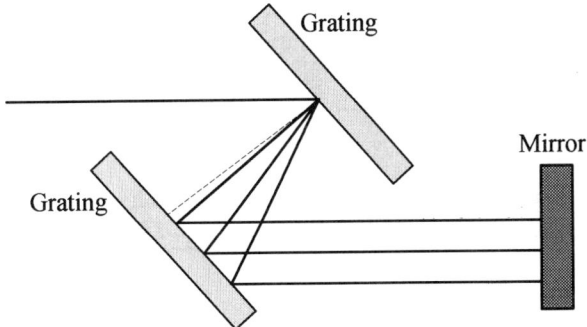

FIGURE 6.10 Grating pair arranged in a dispersive delay line.

Group velocity dispersion for a grating pair is particularly simple and is given by

$$\frac{d\tau(\omega)}{d\omega} = \frac{d^2\phi}{d\omega^2} = \frac{1}{c}\frac{dP(\omega)}{d\omega}. \qquad (6.27)$$

This expression states that the group delay time is determined by the optical path length alone.[4] In order to calculate the optical path length, refer to Figure 6.11. Light is incident on the first grating at the angle γ. The diffracted beam makes an angle θ with the incident beam, and the oblique optical dis-

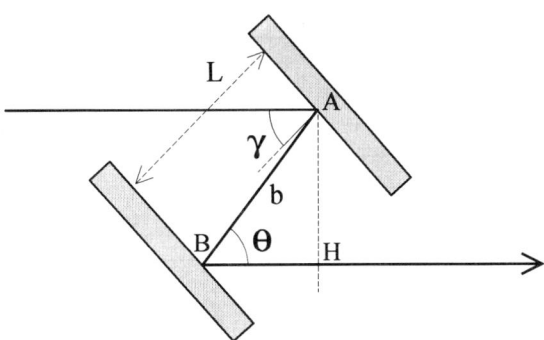

FIGURE 6.11 Optical path and relevant angles for diffraction grating pair. Variable optical path length is ABH.

---

[4]In fact the phase angle as a function of frequency is given by $\phi(\omega) = (\omega/c)P(\omega) + R(\omega)$, where R is a correction term that depends on the groove density of the grating and the angle of incidence. When we take the first derivative of the phase angle with respect to ω to arrive at the delay time τ, we get $\tau = (1/c)P(\omega) + (\omega/c)P'(\omega) + R'(\omega)$. It can be shown that the last two terms cancel, leaving the dependence of the delay time on optical path alone. For more details see ref. [17].

tance between the two gratings is $b = L/\cos(\theta - \gamma)$. Here, L is the distance between the gratings. The relationship between the angles is given by the grating equation:

$$m\lambda = a\left[\sin\gamma + \sin(\theta - \gamma)\right], \tag{6.28}$$

where m is the diffraction order and a is the groove spacing. The angle θ for a first-order diffraction satisfies the following:

$$\sin(\theta - \gamma) = \left(\frac{\lambda}{a}\right) - \sin\gamma. \tag{6.29}$$

The path length of light that varies with wavelength is the path ABH, which is equal to $b(1 + \cos\theta)$. The delay time τ is therefore

$$\tau = \frac{b}{c}(1 + \cos\theta). \tag{6.30}$$

This expression can be differentiated with respect to λ or ω to give the grating pair dispersion. Note that both b and θ are wavelength dependent:

$$\frac{d\tau}{d\lambda} = \frac{b}{ca}\frac{\lambda/a}{1 - (\lambda/a - \sin\gamma)^2}. \tag{6.31}$$

The value of dispersion can be adjusted by the proper choice of b or the grating separation.

Compared to the prism pair, it is easier to generate high values of dispersion with the grating pair. However, the grating pair has higher inherent loss, and the spectral components in higher orders of diffraction are lost by deflecting beyond the physical extent of the opposing grating.

Example Let us find the grating pair separation needed to optimally compress a 20-ps chirped pulse of 1.06 μm radiation with a bandwidth of 12 nm. We assume a grating groove density of 1500 mm^{-1} and an incidence angle of 60°.

Solution The required time delay dispersion $\Delta\tau = (d\tau/d\lambda)\Delta\lambda = 20$ ps. Since $\Delta\lambda = 12$ nm, we need $d\tau/d\lambda = 1.67$ ps/nm. Using Eq. (6.31),

$$\frac{b}{(3 \times 10^{14})(0.67)}\frac{1.59}{1 - (1.59 - 0.87)^2} = 1.67 \times 10^{-9} \text{ s}/\mu\text{m}.$$

The slant distance b between the gratings is found to be 102 mm. ∎

Fiber-grating pulse compressors are capable of achieving high compression ratios (30–90). These compressors are commercially available and are used to generate picosecond to subpicosecond pulses from mode-locked laser outputs.

6.2 ELECTRICAL PULSE FORMING

Electrical pulses are either in the form of a rapid rise and fall of a voltage (or a current) or a short burst of microwave radiation. In the first case, the main frequency component of the pulse is dc. In the second case it is the microwave carrier frequency. The short burst of radiation has its main application in radar and telemetry. The short dc pulse has a variety of applications including impulse radar, time domain spectroscopy, signal sampling, particle accelerator sources, electromagnetic mass accelerators, and power tube drivers, to name a few. In many of these applications large power levels (kilowatts and higher) are required.

6.2.1 Pulse Forming Networks

Since the early days of high voltage power transmission it was well known that closing a power switch is considerably easier than interrupting the current. The technology of opening switches has gone through major advances since those days. But even today, electrical pulses can have much shorter rise times than fall times. Because of this problem, it is customary, in the generation of short high power pulses, to use only closing switches and to rely on energy storage properties of a special circuit to shape the flat portion and the trailing edge of the pulse. A circuit performing this function is called a *pulse forming network* (PFN). The role of the PFN is to feed the stored charge into the circuit at a constant rate such that the voltage or the current remains constant over most of the duration of the pulse. Once the stored charge is consumed, the output voltage drops, which in turn facilitates the opening of the switch. This approach also makes it possible to control more precisely the total energy delivered to the load.

In its simplest construction, a PFN is a section of transmission line open at both ends and connected to a matched load through a closing switch at one end (Figure 6.12). The transmission line is charged to the desired potential. When the switch is closed, the stored charge on the transmission line starts propagating as a nearly constant amplitude pulse, with a duration equal to twice τ, the electrical length of the transmission line. The reason for the pulse duration being equal to 2τ is that the closing of the switch triggers two counterpropagating pulses on the transmission line, each with duration τ. The pulse traveling toward the open end of the line gets reflected and adds to the total duration of the pulse. If the load to which the PFN is connected is not matched, multiple reflections occur. Figures 6.12b, and 6.12c show this ringing behavior for different load impedances.

A PFN can be a combination of transmission line sections in order to

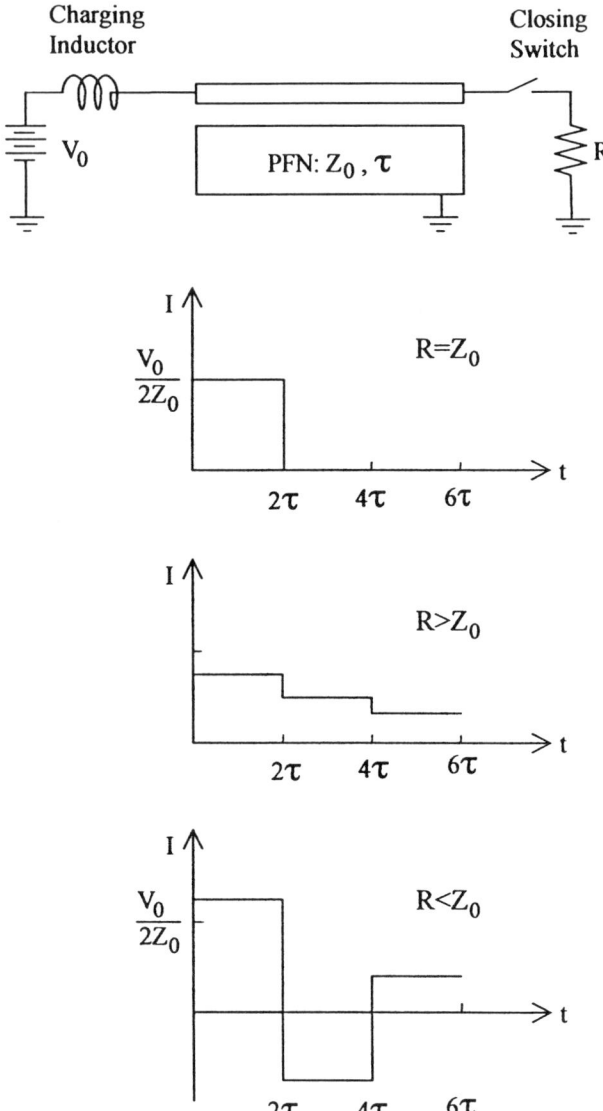

FIGURE 6.12 Pulse forming characteristics of an ideal transmission line segment. Current through load is shown as a function of time after closing switch.

achieve particular effects. As an example, two transmission line sections can be charged in parallel and discharged in series. The resulting network is known as a *Blumlein circuit*, whose schematic diagram is shown in Figure 6.13. Both lines are initially charged to a potential V_0, as shown. The closing of the switch imposes a zero electric field boundary condition at one end of transmission line A. In order to satisfy this boundary condition, a reverse-polarity pulse (amplitude $-V_0$) starts propagating along this line. Once the

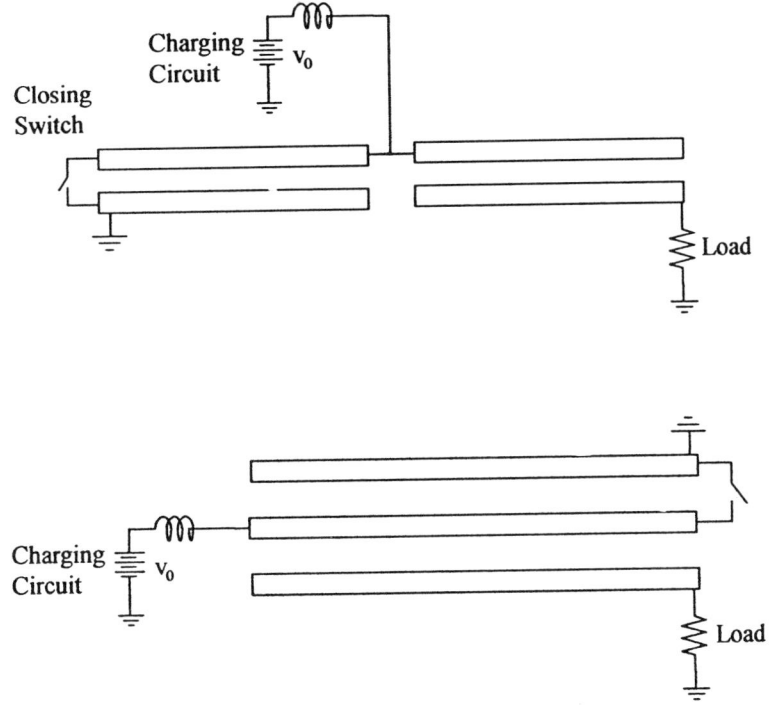

FIGURE 6.13 Blumlein PFN in standard and folded configurations.

reverse polarity reaches transmission line B, the potentials add, and a pulse of amplitude $2V_0$ is delivered to the load.

In a variation of the Blumlein concept, the two transmission lines may be folded on top of each other to form a three-electrode transmission line, as shown in Figure 6.13. In this configuration the voltages on the line may be found by the superposition of odd and even propagation modes. The odd mode has the electric fields between the adjacent electrode pairs in opposite directions. The line voltages are $(0, V_0, 0)$. The even mode has the electric fields in the same direction, causing the potentials $(0, -V_0, -2V_0)$. The boundary condition imposed by closing the switch is satisfied by a superposition of the two modes, resulting in voltages $(0, 0, -2V_0)$.

Using transmission lines is an intuitive and simple method of fabricating PFNs. It is possible, however, to try to synthesize the function of the PFN using lumped circuit elements. The following approximate method is an example. The Fourier series expansion of a flat current pulse of amplitude I_m and finite duration may be written as

$$I(t) = I_m \sum_{m=0}^{\infty} \frac{1}{2m+1} \sin(2m+1)\omega t, \tag{6.32}$$

where the fundamental frequency ω has an oscillation period equal to twice the pulse duration T. The flat current pulse can be approximately reproduced by synthesizing a circuit that has $\omega, 3\omega, 5\omega, \ldots$ as a finite number of natural oscillation frequencies. This may be done by combining series LC branches in a network, as shown in Figure 6.14. In this topology, the summation is truncated after five terms. The well-known problems associated with truncated series are overshoot at pulse edges and amplitude ripple over the duration of the pulse. In order to avoid these problems, specific rise times and fall times may be built into the design. In fact the example shown in Figure 6.14 has been optimized for low ripple. It is known as a Guilleman type C network. For a more detailed discussion of PFN design see reference [18] or [19].

6.2.2 Microwave Radiation Bursts

As was mentioned before, the first major application of radiation bursts was pulsed radar. Radar pulses are generated by fast turn-on and turn-off of microwave power tubes such as magnetrons. The dc power pulses discussed in the previous section are used for this purpose. In order to generate microwave pulses as short as a few cycles, there is growing interest in direct synthesis of radiation bursts. Even though the principles behind this idea are not new, it is the availability of high speed, high power solid state switches that makes compact and efficient sources of radiation bursts feasible. A number of techniques have been developed for the generation of a few cycles of microwave signal [20–23]. One such technique, known as the *frozen wave generator*, stores charge on a transmission line and sets it into propagation by a number of closing switches triggered by a short optical pulse. A variation of this technique referred to as the *injection wave generator* uses toggle switches instead of closing switches. These techniques are further discussed below.

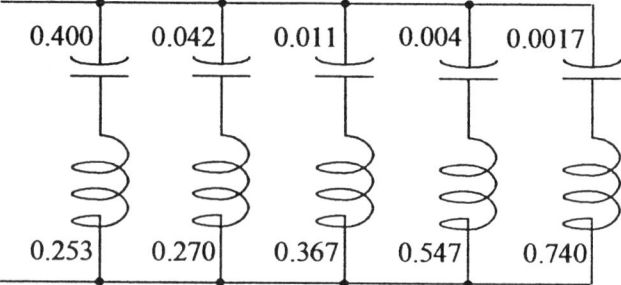

FIGURE 6.14 Guilleman type C PFN. For scaling to proper impedance level Z and pulse duration T, inductance values should be multiplied by TZ and capacitance values by T/Z.

Injection and Frozen Wave Generators

A *frozen wave generator* (FWG) is a circuit designed to generate a few cycles of an electromagnetic signal of a given frequency. It consists of a transmission line whose length is an integral multiple of $\frac{1}{4}\lambda$ at the frequency of interest (λ is the guided wavelength). The transmission line is broken into a number of segments connected together by switches that can be closed simultaneously by an external, usually optical, signal [24]. Each segment is charged to a predetermined potential. When the switches are closed, the potential distribution on the transmission line starts propagating, giving rise to the output signal (Figure 6.15).

In order not to distort the signal, the closing time of the switches should be small compared to the period of the signal to be generated. At the same time the switches should stay closed for the duration of the microwave pulse so that the entire waveform can pass through. The fast rise time is ensured by an ultrashort optical pulse used for closing the switch. High speed photoconductive switches can easily generate subpicosecond rise times in response to short optical pulses. In a linear photoconductive switch (non-avalanche), the resistance of the closed switch is controlled by the power in the optical pulse. If the switch is to be kept at a low resistance state by an optical pulse,

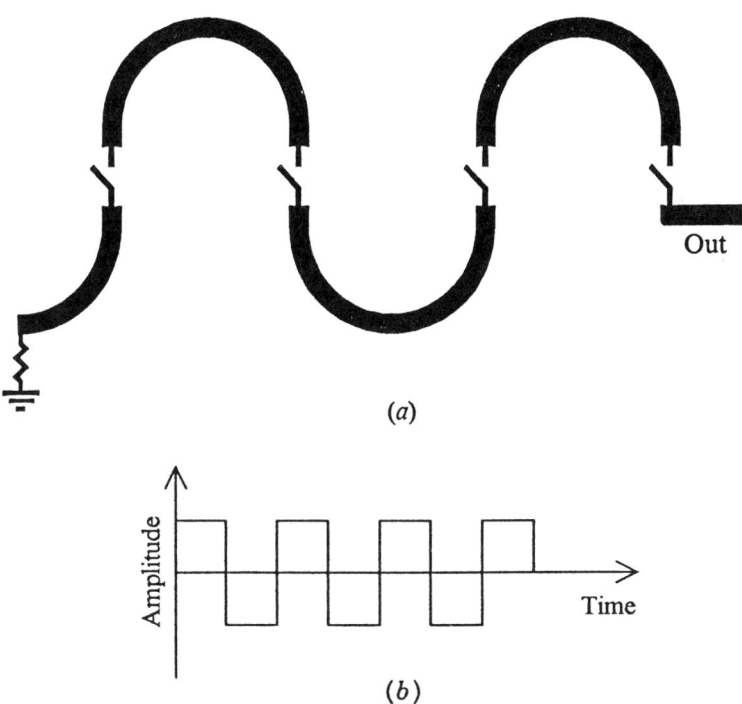

FIGURE 6.15 Frozen wave generator for generating four cycles of radiation. Shown waveform is its desired output.

the pulse should have the same duration as the microwave output. This is not practical in view of the prohibitive optical power requirement. One solution is to use a photoconductive switch on a material such as silicon that provides a fast closure time and a long opening tail due to its long carrier lifetimes. However, due to the decaying tail of the current in the switch, the output waveform will also be a decaying pulse. Another potential switch for this purpose is a photoconductive switch on GaAs or InP in an operation mode known as "lock-on" [25]. In this mode the switch will stay closed after a short triggering pulse and will not open until the applied electric field is reduced. The difficulty with lock-on switches is keeping them conducting after each reversal of the electric field that occurs in FWGs.

Toggle switches can be used in a similar scheme for microwave pulse generation. Toggle switches are those with short closing and opening times. Since in this case the switches will not stay closed long enough for the pulse to propagate through, they have to be placed in parallel with the transmission line rather than in series with it (Figure 6.16). This type of generator is known as an *injection wave generator* (IWG). Positive and negative pulses are injected simultaneously at different points along the line. Injected pulses travel in both directions along the transmission line. Each injected pulse generates half a cycle of radiation. The generator of Figure 6.16 will ideally generate a signal consisting of four microwave cycles. Two of the cycles are generated by the backward-going portion of the pulses that reflect from the shorted end of the transmission line. In practice, the four cycles will not terminate abruptly, and extra ringings continue due to multiple reflections. Also, due to transmission line losses, the four cycles will have a decaying amplitude in time. These effects are more pronounced at higher frequencies. In order to compensate for such problems and to be able to generate an output waveform of the desired shape, various circuit parameters need to be optimized. Optimization routines available for high speed circuit design generally work in the frequency domain. It is therefore useful and instructive to analyze and optimize the pulse generators in the frequency domain.

Frequency Domain Analysis
The optical pulse that triggers the switches is normally much shorter in duration than the microwave signal generated. Therefore the microwave signal can be approximated as the impulse response of the generator circuit. In order to facilitate the design procedure, it is convenient to model the circuit as an all-electrical circuit rather than one with an optical input and an electrical output. Figure 6.17 shows an all-electrical equivalent circuit of an IWG. Since the application of the optical pulse leads to the injection of four different pulses at different points along the transmission line, it can be represented by an electrical signal that controls a set of four voltage-controlled current sources that drive the circuit. The parasitics associated with the switch as well as the transmission lines leading to the injection point are also included in the model.

ELECTRICAL PULSE FORMING 205

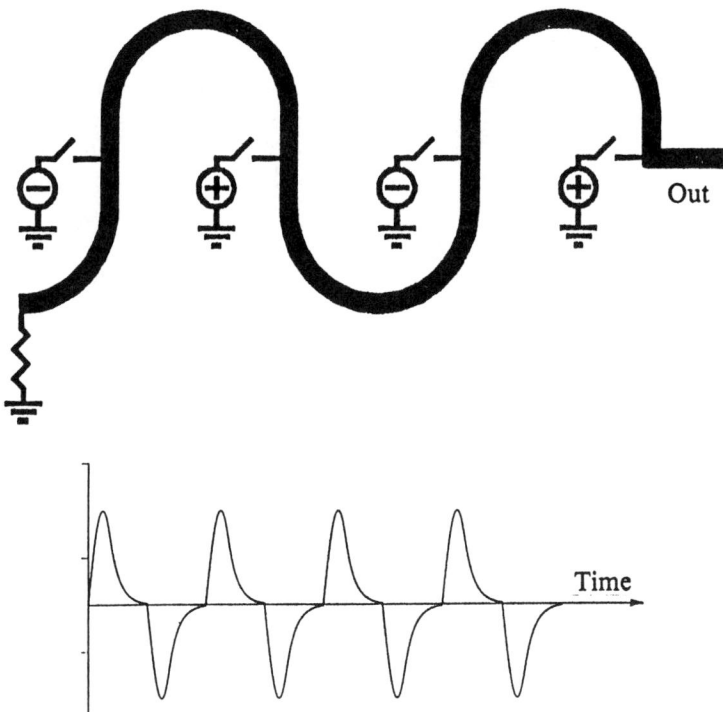

FIGURE 6.16 Injection wave generator with four switches.

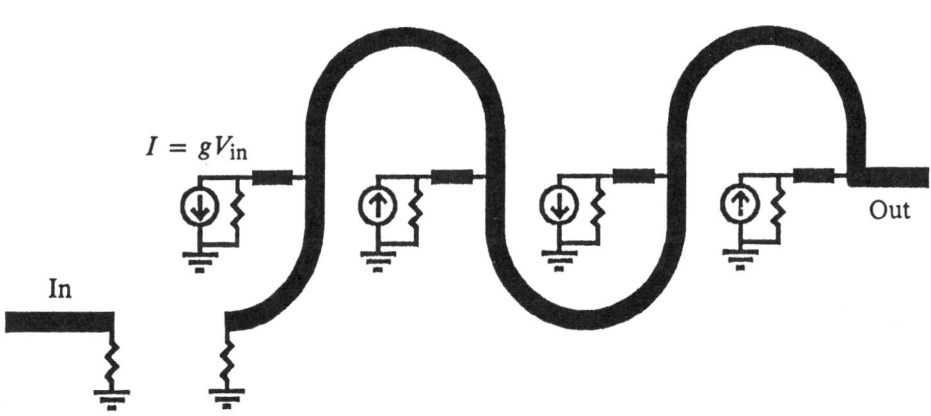

FIGURE 6.17 All-electrical equivalent circuit of an IWG.

Consider the case where the desired output is four cycles of radiation with a constant (nondecaying) amplitude and a sharp cutoff in time. The Fourier transform of four cycles of radiation at any frequency f_0 is shown in Figure 6.18.[5] The horizontal axis is normalized with respect to f_0. The Fourier transform can be regarded as the frequency response of a filter that when driven by a pulse at the input will generate the four cycles of radiation as desired.

Since the output waveform has a constant amplitude over the four cycles and a sharp cutoff in time (square pulse envelope), the passband of the filter shown in the figure is a sinc function convolved with the transform of the sinusoid. The width of the main passband is 30%. Magnitudes and phases

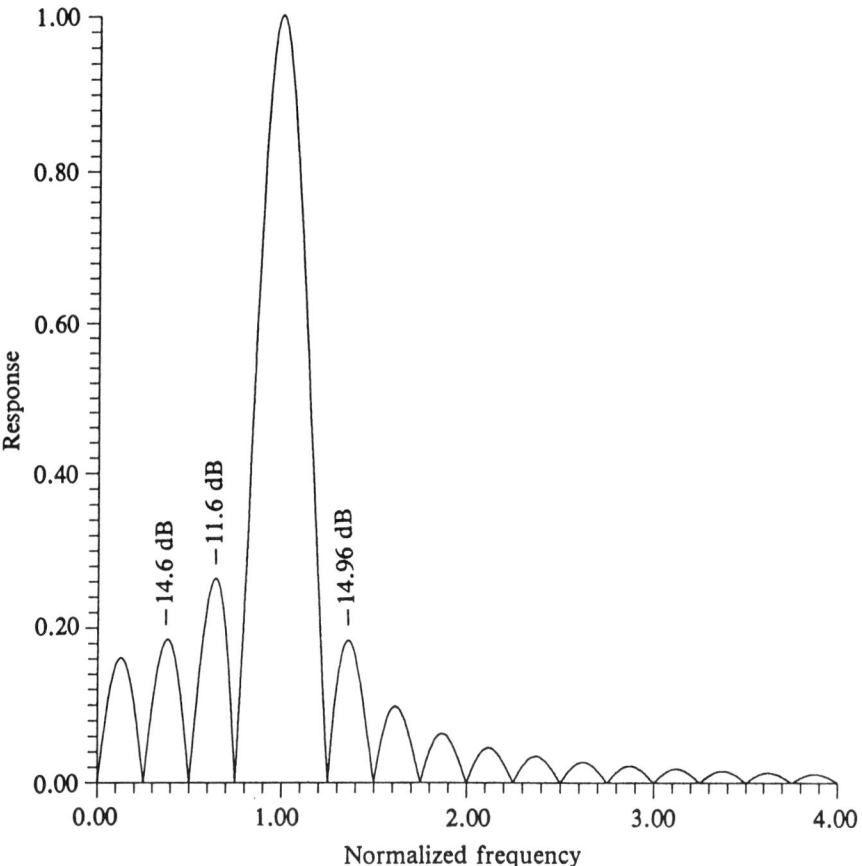

FIGURE 6.18 Spectrum associated with four cycles of radiation. Horizontal axis is normalized to frequency of the four cycles.

---

[5]The case of four cycles has been chosen as an example. In general, the spectrum of n cycles of radiation has $n - 1$ sidebands between 0 and f_0.

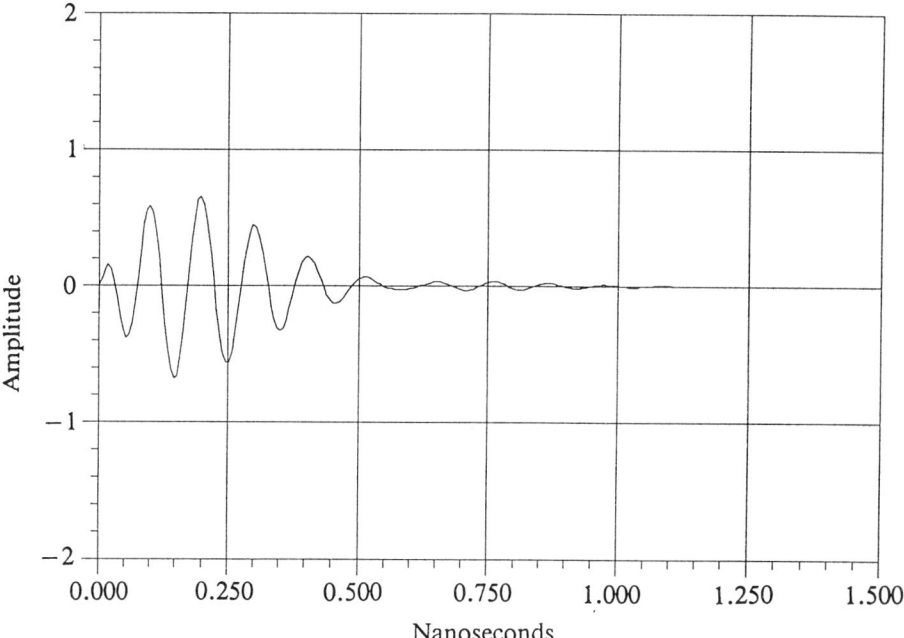

FIGURE 6.19 Example of time waveform associated with only a single passband in frequency domain.

associated with the "sidebands" are indicated in the figure. It is important to note that a filter with a single passband at the frequency of interest will not generate a sharp time waveform. The waveform generated by such a filter is shown in Figure 6.19 and has a profile with a slow rise and fall.

It is possible to approximate the response shown in Figure 6.18 using one band-pass filter for each sideband in a parallel configuration. This approach is undesirable because of its wasteful and complex nature. As will be seen in the following example, constructive and destructive interference of portions of the signal can perform a similar filtering function.

Example Consider a circuit (Figure 6.20) that splits the input eight ways, delays each one by an appropriate time, and combines them to generate the output. Ignoring for a moment the question of realizability of such a filter, let us find the output signal in both the time and frequency domains.

Solution In the time domain the output is

$$V(t) = \sum_{n=0}^{7} (-1)^n \delta(t - n\tau),$$

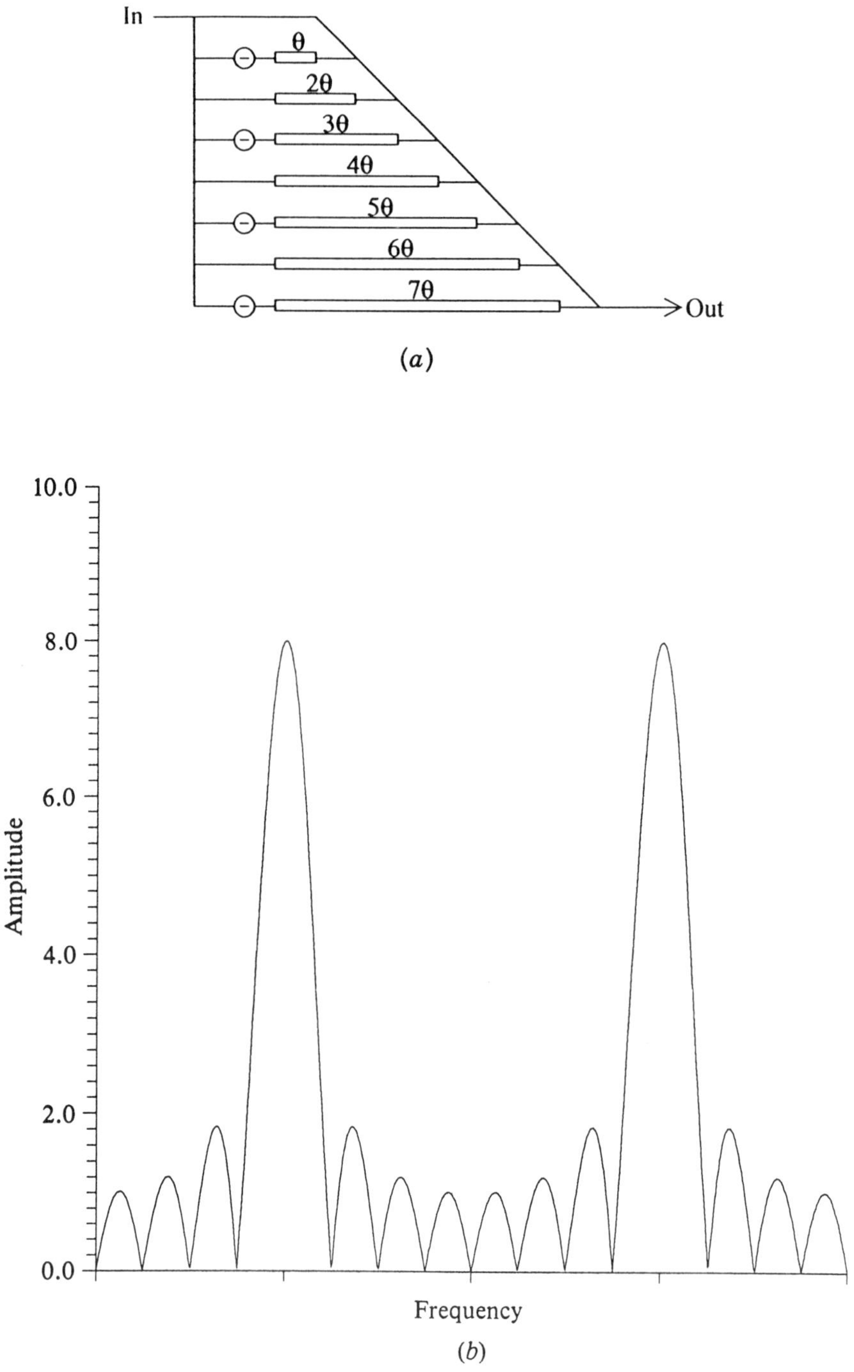

208 SHORT-PULSE GENERATION

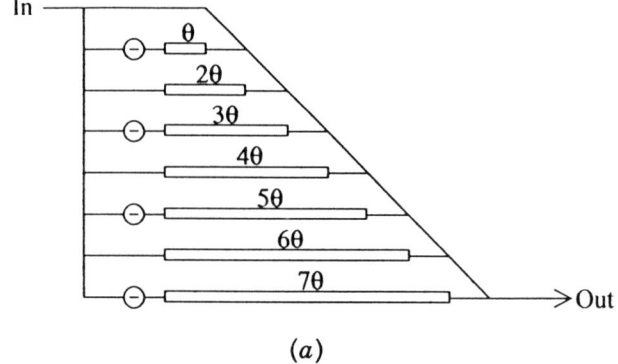

FIGURE 6.20 Hypothetical eight-section filter and its frequency response; θ is time delay in each section.

where $\delta(0) = 1$ and τ is the smallest delay time. The shape of the output is of course a series of positive and negative pulses that, once passed through a low pass filter, will give four cycles of radiation. In the frequency domain, the shape of the output is given in Figure 6.20. The fact that the filter has a repetitive response gives rise to the sharpness of the pulses. Only the first passband and its associated sidebands are needed to generate four cycles of radiation. ∎

The IWG model shown in Figure 6.17 is a variation of the approach just discussed. In the case of the IWG, all the necessary delays are generated by a single transmission line. Also, instead of splitting the signal eight ways, it is split four ways. The other four components are supplied by reflection from one termination of the transmission line.

The ideal situation represented by the summation expression above can only be achieved in an IWG if at the injection points there is no reflection in any direction. But since the microwave junction at any of these points is a three-port, this cannot be attained. With the impedance levels indicated in the figure, the calculated impulse response of this generator is shown in Figure 6.21. The extra ringing is due to multiple reflections on various line segments. Also, signal attenuation on the transmission lines gives rise to the decaying output.

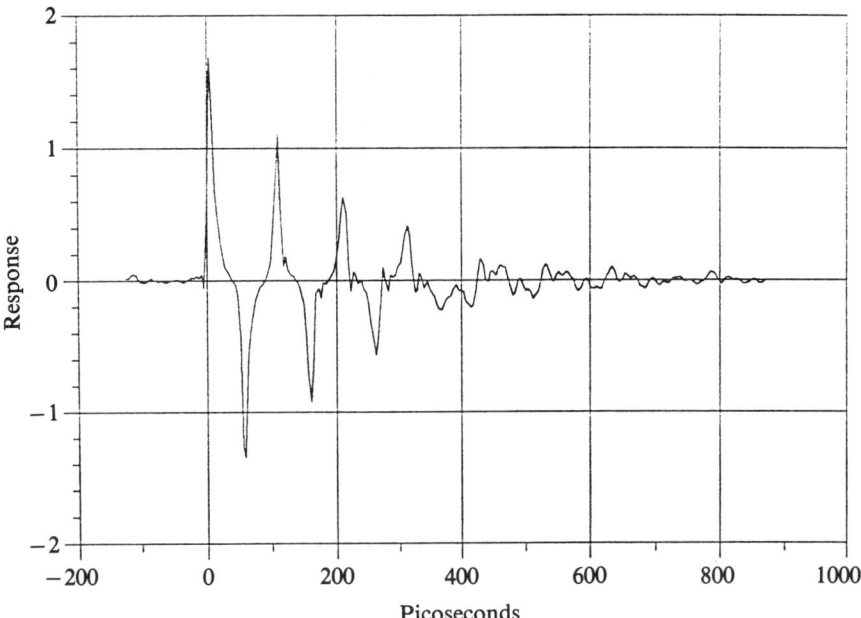

FIGURE 6.21 Time response of unoptimized IWG with four switches. Model includes transmission line loss, junction discontinuities, and switch parasitics.

Lumped-Element Design

The pulse generator circuit once presented in its all-electrical equivalent circuit can be optimized as a multiple-passband filter. This filter can be designed using either distributed or lumped elements. Standard designs of both frozen and injection wave generators generally utilize distributed elements. The lumped-element approach has the advantage of a considerable reduction in size at lower frequencies [26].

The first step in the circuit design consists of replacing the transmission line sections in a conventional IWG by combinations of series inductances and shunt capacitances (Figure 6.22a). At low frequencies ($\omega^2 LC \ll 1$) this combination behaves as a section of transmission line. In an infinite LC network the phase shift per section is $\phi = \omega\sqrt{LC}$. At high frequencies the response deviates from that of a transmission line. In particular, this network is a low pass filter and has a cutoff frequency given by

$$f_c = \frac{1}{2\pi\sqrt{LC}}. \tag{6.33}$$

Unlike the transmission line, the lumped-element approach has a low pass filter response built in. This keeps the output from having the appearance of positive and negative spikes and approximates a sinusoid more closely. Next, the whole circuit may be optimized to give the desired central passband and sidebands. The resulting time response is given in Figure 6.23. Frequency response deviations from this picture cause distortions of the output waveform. The advantage of optimizing in the frequency domain rather than in the time domain is that frequency domain optimization is generally available in circuit simulation softwares while time domain optimization may not be.

6.2.3 Swept-Beam Generators

Instead of a pulsed optical beam it is possible to utilize either a bunched electron beam or a dc electron beam that can be swept across the switch in a very short time. The speed at which the beam is swept can exceed the speed of light. It is therefore possible to generate electrical pulses of a few picoseconds duration using a swept electron beam and a semiconductor switch with a width on the order of 1 mm. This concept was used in a class of devices known as electron-bombarded semiconductor (EBS) devices in the 1970s [27]. Unlike the optical excitation that generates a single electron–hole pair for every absorbed photon, an electron in a typical 20-kV beam is capable of generating a few thousand electron–hole pairs in the semiconductor. So the current generated in the switch is three orders of magnitude higher than the current in the electron beam.

In practice, the scanning of an electron beam in a straight line can be done at speeds up to a few gigahertz with conventional deflection techniques. In a reasonable size tube, the linear path that can be covered by the beam is

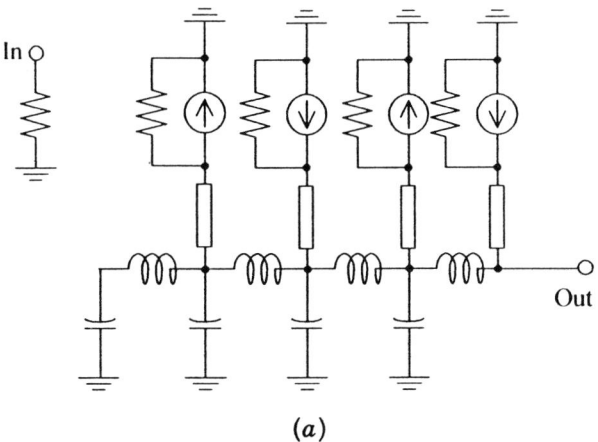

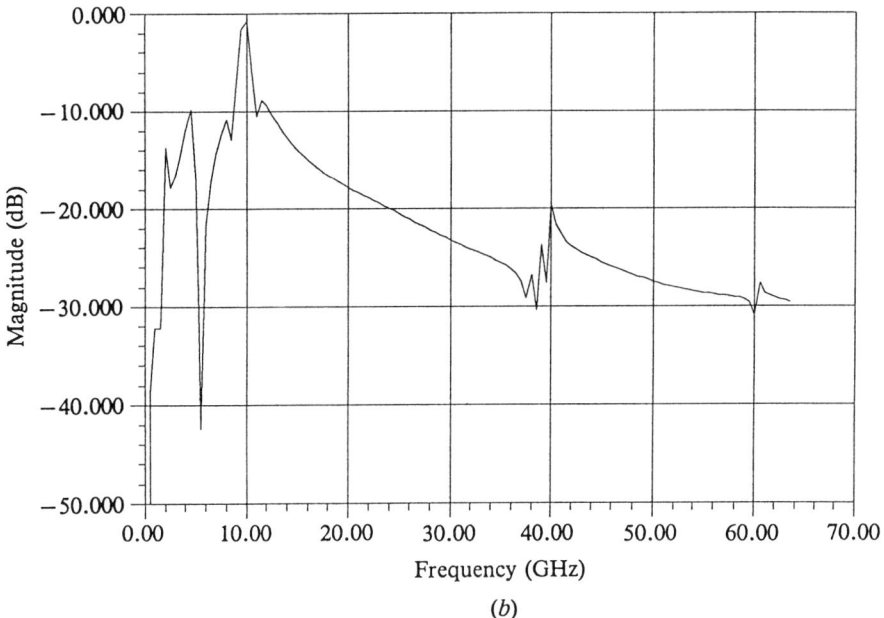

FIGURE 6.22 Lumped-element IWG model and its associated frequency response.

about 10 cm. This corresponds to an average linear beam scan of 2×10^{10} cm/s for a beam swept at 1.0 GHz. Maximum scan rate at the center of the screen exceeds the speed of light. This beam is capable of sweeping across a 100-μm size device in a fraction of a picosecond. The swept beam causes a short electrical current to flow through the device. The duration of this current is limited by carrier lifetime in the device. A number $2N$ of such devices biased consecutively at opposite polarities can generate N cycles of radiation with each sweep of the electron beam.

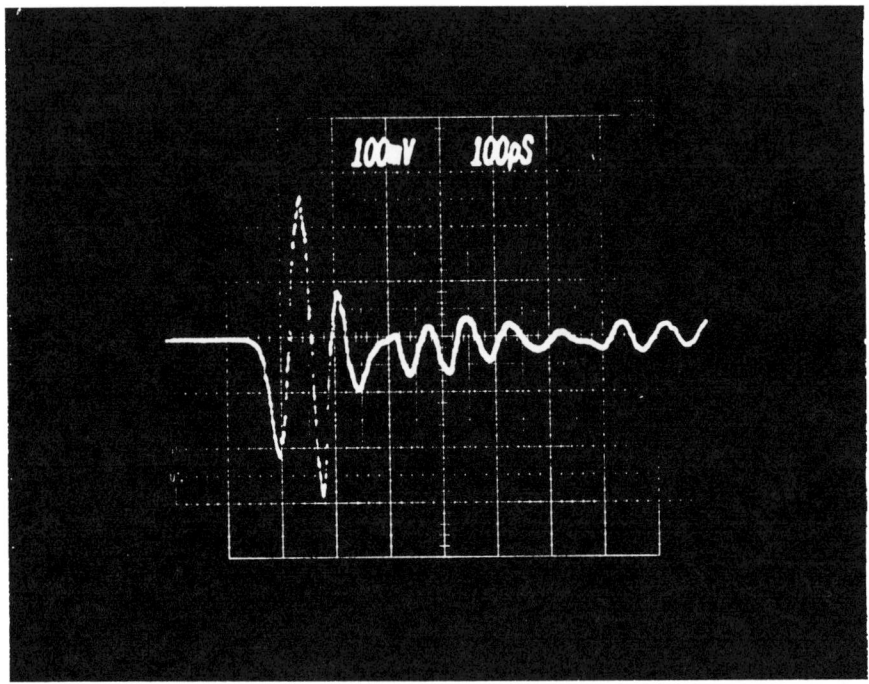

FIGURE 6.23 Observed time response of lumped-element IWG.

In the commutator arrangement shown in Figure 6.24 all switches feed the same point along the transmission line. The necessary time delay between the arrival of consecutive pulses is performed by the electron beam. This concept can be extended to CW power generation if the path of the electron beam is chosen to be circular instead of linear. If the beam is swept at the frequency f, an output frequency of Nf is generated by using $2N$ switches in the circular path of the beam. Neighboring switches are biased at opposite voltages, and carriers are swept through each device in a time much shorter than the half period of radiation it generates. As an example, a circular scan at 1.0 GHz rate can produce a CW output at 35 GHz if 70 switches are triggered in every scan. For high power applications this arrangement has the advantage that each switch is active only one-seventieth of the time, thereby reducing its duty cycle by that factor.

Combining the output pulses of the switches can be done in a variety of ways. One approach is for the pulses to be combined using microstrip lines leading to the center conductor of a coaxial transmission line (Figure 6.24). The coaxial line then couples into a rectangular waveguide through a loop antenna. Each microstrip section includes a matching network that presents an open circuit to the coaxial line when the switch is open.

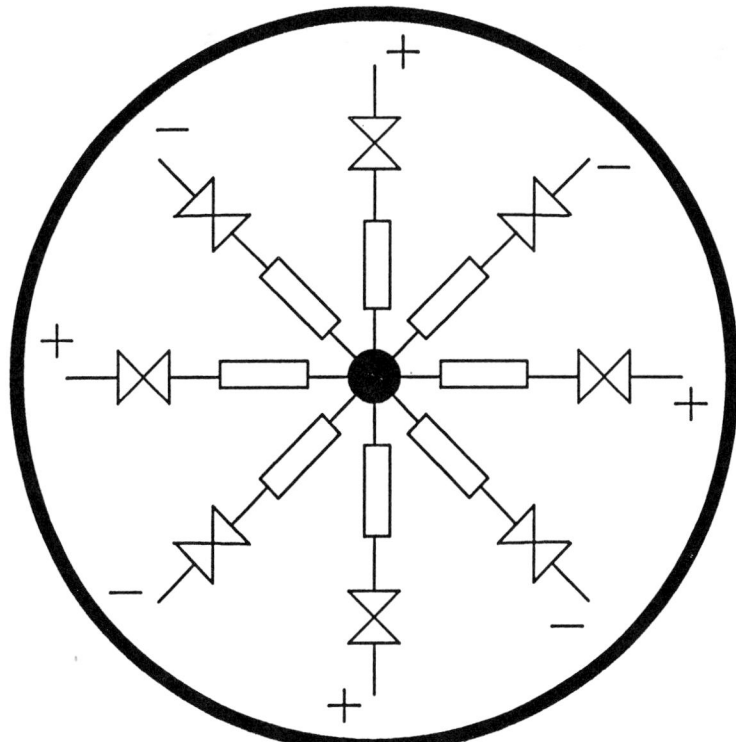

FIGURE 6.24 Commutator arrangement of switches to generate radiation pulses of variable duration. Switches are triggered by a swept electron beam.

In Figure 6.25 you will find simulations of a 5-GHz signal generated using 10 push–pull switches. The output is the result of one sweep of the electron beam over the 10 switches. In the first simulation the microstrip section is very short (5° at 5 GHz). The result is very desirable, but the arrangement cannot physically be realized. The second simulation shows the case with an optimized long section of microstrip. The slow buildup and decay are characteristics of this approach. If the desired output is narrow-band CW and not a short pulse, this effect does not pose any problem.

6.2.4 Shock Wave Generators

Consider a nonlinear transmission line characterized by an amplitude-dependent phase velocity

$$v_p = v_{p0} + \delta v_p(A). \tag{6.34}$$

214 SHORT-PULSE GENERATION

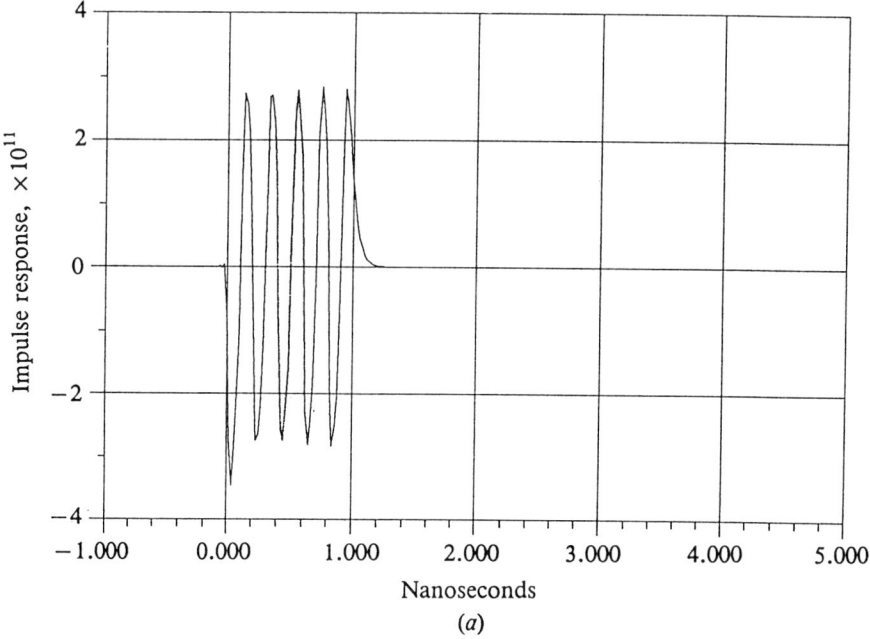

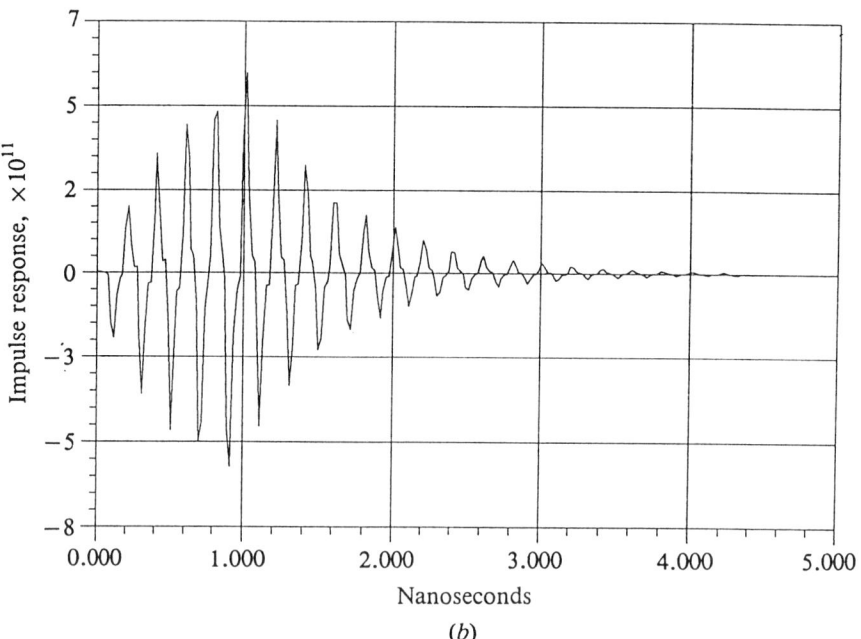

FIGURE 6.25 A 5-GHz signal generated by 10 switches combined with (*a*) short and (*b*) long microstrip sections.

A first-order expansion of v_p with respect to amplitude A gives

$$v_p = v_{p0} + \alpha A. \tag{6.35}$$

Here, α is the first-order nonlinearity coefficient and may have a positive or a negative value. A pulse propagating along such a nonlinear transmission line can develop a sharpening slope either on its leading edge or its trailing edge depending on the sign of α. This process, if continued, can lead to a discontinuous wave or a shock wave. In the case of electrical signals, the sharpening edge eventually alters the characteristics of the transmission line in such a way that a mathematical discontinuity does not form. Nonlinear transmission lines have been used to obtain subpicosecond fall–time electrical pulses with a few volts of amplitude [28]. High speed data sampling is one major application for such pulses.

An electromagnetic pulse with a profile $A(t)$ propagates along a linear transmission line as $A(x - v_{p0}t)$, conserving its profile in time. In a nonlinear transmission line, on the other hand, the higher amplitude portions of the pulse travel faster (or slower), thereby changing the pulse shape. The function $A(x - v_p t)$ will evolve into a different shape if v_p is a function of A. For mild nonlinearities (small α), it is possible to calculate the evolution of the pulse shape by taking $\delta v_p(A)$ to be a small correction to v_p that can be treated as a function of x. Converting $\delta v_p(A)$ to a function of x requires the assumption of a pulse shape $A(x)$ that is taken to be the initial pulse shape. This analysis yields only an incremental modification to the pulse shape corresponding to a small propagation distance. Numerical calculations can apply this correction repeatedly to simulate the evolution of the pulse shape.

Consider a Gaussian pulse of the form $A(x) = A_0 e^{-x^2}$. Its nonlinear propagation is characterized by a position-dependent phase velocity:

$$v_p = v_{p0} + \alpha A_0 e^{-x^2}. \tag{6.36}$$

After a time t the pulse evolves as

$$A(x - v_p t) = A(x - v_{p0}t - \alpha A_0 e^{-x^2} t), \tag{6.37}$$

which can easily be calculated numerically. Figure 6.26 shows, in three steps, the evolution of a sharper trailing edge starting with a Gaussian pulse along a nonlinear transmission line. The same figure shows the corresponding broadening of the spectrum of the pulse.

Among the various techniques available for fabricating nonlinear transmission lines, one that has been studied extensively [29, 30] and may be used in an integrated circuit form is the Schottky contact transmission line. The capacitance per unit length of this transmission line is dependent on the signal amplitude. This nonlinearity is caused by the voltage-dependent capacitance of a Schottky contact. It is possible to either fabricate the entire transmis-

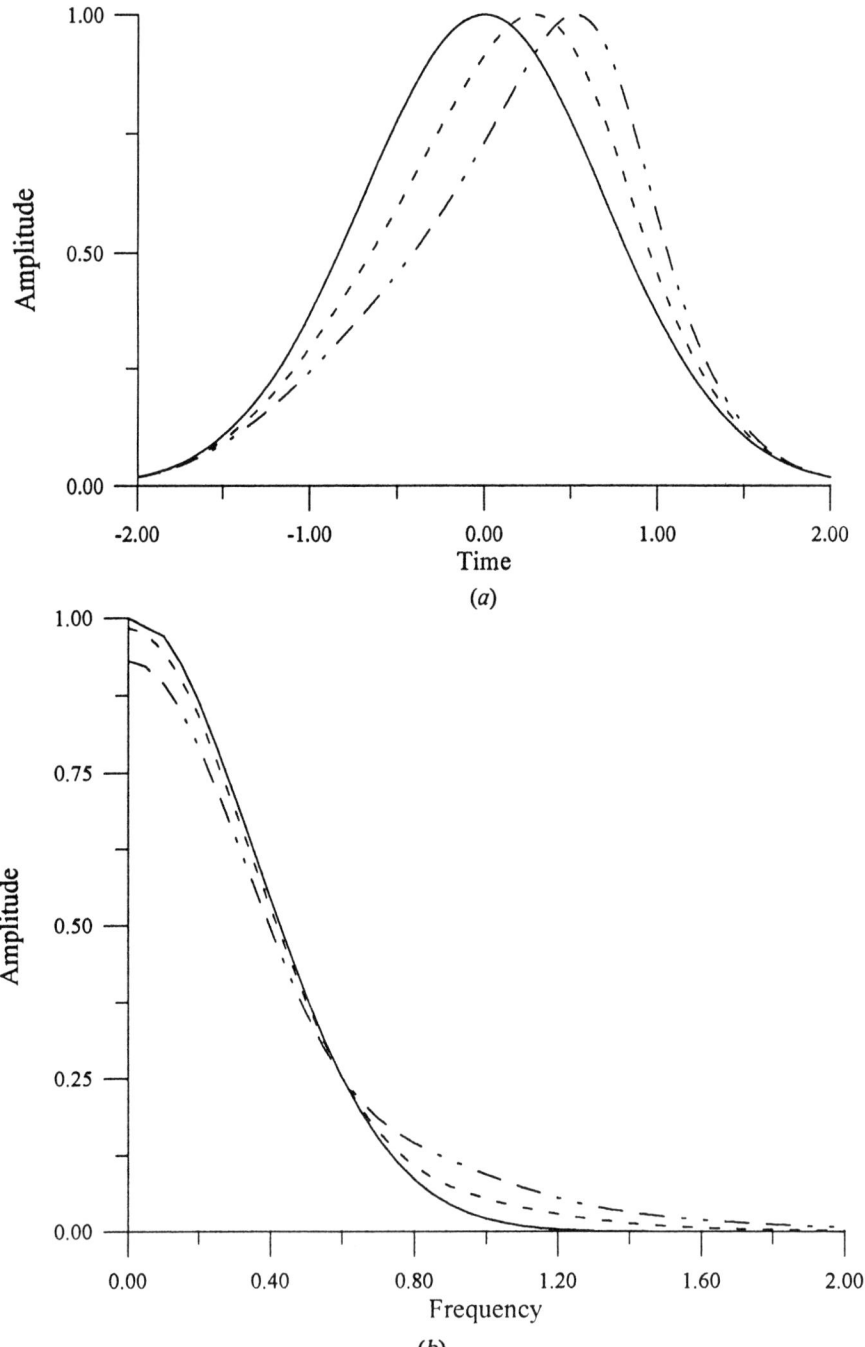

FIGURE 6.26 Propagation of a Gaussian pulse along a nonlinear transmission line: (*a*) evolution of the pulse in time; (*b*) broadening of the spectrum.

sion line on doped semiconductor or periodically load the line with varactor diodes. For simplicity of the discussion we assume the first case, where the distributed capacitance of the transmission line is entirely due to the Schottky contact. Recall from the discussion in Chapter 2 that the capacitance of a Schottky contact as a function of the applied voltage is

$$C(V) = \frac{C(0)}{\sqrt{1 - V/V_{bi}}}. \tag{6.38}$$

If the nonlinearity is assumed to be weak, the amplitude-dependent phase velocity may be written as

$$v_p = \frac{1}{\sqrt{LC}} = \frac{1}{\sqrt{LC(0)}}(1 - V/V_{bi})^{1/4} \approx v_{p0}\left(1 - \frac{V}{4V_{bi}}\right). \tag{6.39}$$

For this case therefore, $\alpha = -1/4V_{bi}$.

Sharp electrical steps generated by shock wave generators or subpicosecond electrical pulses generated by optical switches cannot propagate very far on an ordinary quasi-TEM transmission line due to dispersion. The required bandwidth for such pulses easily extend to a few hundred gigahertz. It was seen in Chapter 4 that in order for quasi-TEM transmission lines to have low dispersion through the entire millimeter-wave range, it is required that their dimensions be exceedingly small. This in turn leads to high conductive losses as well as tight tolerances that severely limit the usefulness of these transmission lines. Even though losses can in certain cases be used to lower the dispersion of the line, this is in general not a practical solution. These problems have led some researchers to use free-space propagation for ultrabroadband electrical pulses. Broadband radiators such as exponentially tapered antennas have been used to launch subpicosecond pulses in air. In a particular absorption spectroscopy application, the pulses go through a test medium, after which they are collected by another antenna. Electro-optic sampling is used to detect the received pulse shape. The Fourier transform of the received pulse gives the absorption spectrum of the test medium [30].

PROBLEMS

6.1 In a dispersive optical medium, express the first three derivatives of the propagation constant with respect to frequency in terms of the derivatives of refractive index with respect to wavelength. Identify phase velocity, group velocity, and GVD in the resulting expressions.

6.2 Derive Eq. (6.31) for $d\tau/d\lambda$ due to a grating pair. Convert this expression to $d^2\phi/d\omega^2$ and compare it with Eq. (6.24) for a prism pair. Use the geometry of Figure 6.11.

6.3 Design a fiber grating pulse compressor to compress 50 ps optical pulses at 1.06 μm wavelength to 2 ps. Assume 125 nJ transform limited pulses with Gaussian profiles. For the fiber $n_2 = 0.001$ W^{-1}. Optical attenuation is 0.5 dB/m in the fiber, and the gratings have 1000 grooves per millimeter. You may want to refer to the nonlinearity discussion in the next chapter.

6.4 A frozen wave generator is open circuited at one end and has transmission line lengths of 5, 9, 8, and 7 mm in that order between the switches (Figure 6.15). Assuming an effective dielectric constant of 6.0 for the transmission lines, sketch the expected output waveform with ideal switches.

REFERENCES

[1] H. E. Edgerton, *Electronic Flash, Strobe*, McGraw-Hill, New York, 1970.

[2] M. Rodwell, M. Riaziat, K. Weingarten, and D. Bloom, "Internal Microwave Propagation and Distortion Characteristics of Travelling Wave Amplifiers Studied by Electro-Optic Sampling," *IEEE Trans. Microwave Theory Tech.*, Vol. 34, No. 12, pp. 1356–1362, Dec. 1986.

[3] A. H. Zewail, "The Birth of Molecules," *Sci. Am.*, pp. 76–82, Dec. 1990.

[4] I. W. M. Smith, "Exposing Molecular Motions," *Nature*, Vol. 343, No. 6260, pp. 691–692, Feb. 1990.

[5] M. Asaki, C. Huang, D. Garvey, J. Zhou, H. Kapteyn, and M. Murnane, "Generation of 11-fs Pulses from a Self Mode-Locked Ti–Sapphire Laser," *Opt. Lett.*, Vol. 18, No. 12, pp. 977–989, June 1993.

[6] R. S. Miranda et al., "Use of Time-Resolved IR Reflection and Transmission as a Probe of Carrier Dynamics in Semiconductors," *Opt. Lett.*, Vol. 16, No. 23, pp. 1859–1861, Dec. 1991.

[7] H. A. Haus, "A Theory of Forced Mode Locking," *IEEE J. Quant. Electron.*, Vol. 11, No. 7, pp. 323–330, July 1975.

[8] A. Yariv, *Optical Electronics*, 3rd ed., Holt, Rinehart and Winston, New York, 1985.

[9] D. J. Kuizenga and A. E. Siegman, "AM and FM Mode Locking of the Homogeneous Laser," *IEEE J. Quant. Electron.*, Vol. 6, No. 11, pp. 694–708, Nov. 1970.

[10] U. Keller et al., "High-Frequency Acousto-Optic Mode Locker for Picosecond Pulse Generation," *Opt. Lett.*, Vol. 15, No. 1, pp. 45–47, Jan. 1990.

[11] G. Jacobovitz-Veselka, U. Keller, and M. Asom, "Broadband Fast Semiconductor Saturable Absorber," *Opt. Lett.*, Vol. 17, No. 24, pp. 1791–1793, Dec. 1992.

[12] J. C. Diels, "Femtosecond Dye Lasers," in F. Durante and L. Hillman (Eds.), *Dye Laser Principles*, Academic, New York, 1990.

[13] M. A. Dugway and J. W. Hansen, *Appl. Phys. Lett.*, Vol. 14, p. 14, 1969.

[14] P. L. Fork, O. E. Martinez, and J. P. Gordon, "Negative Dispersion Using Pairs of Prisms," *Opt. Lett.*, Vol. 9, No. 5, pp. 150–152, May 1984.

[15] M. I. Skolnik, *Introduction to Radar Systems*, 2nd ed., McGraw-Hill, New York, 1980.

[16] W. J. Tomlinson, R. H. Stolen, and C. V. Shank, "Compression of Optical Pulses Chirped by Self-Phase Modulation in Fibers," *J. Opt. Soc. Am. B*, Vol. 1, No. 2, pp. 139–149, April 1984.

[17] E. B. Treacy, "Optical Pulse Compression with Diffraction Gratings," *IEEE J. Quant. Electron.*, Vol. 5, No. 9, pp. 454–458, Sept. 1969.

[18] G. Glasoe and J. Lebacqz, *Pulse Generators*, Radiation Laboratory Series, McGraw-Hill, New York, 1948.

[19] W. Sarjeant and R. Dollinger, *High-Power Electronics*, TAB Books, Blue Ridge Summit, PA, 1989.

[20] D. Auston, "Picosecond Optoelectronic Switching and Gating in Silicon," *Appl. Phys. Lett.*, Vol. 26, No. 3, pp. 101–103, Feb. 1975.

[21] M. Frankel et al., "High Voltage Picosecond Photoconductor Switch Based on Low Temperature Grown GaAs," *IEEE Trans. Electron. Devices*, Vol. 37, No. 12, pp. 2493–2498, Dec. 1990.

[22] C. Lee, "Picosecond Optics and Microwave Technology," *IEEE Trans. Microwave Theory Tech.*, Vol. 38, No. 5, pp. 596–607, May 1990.

[23] W. Nunnally, "High Power Microwave Generation Using Optically Activated Semiconductor Switches," *IEEE Trans. Electron. Devices*, Vol. 17, No. 12, pp. 2439–2448, Dec. 1990.

[24] C. H. Lee, "Optical Control of Semiconductor Closing and Opening Switches," *IEEE Trans. Electron. Devices*, Vol. 37, No. 12, pp. 2426–2438, Dec. 1990.

[25] G. Loubriel, F. Zutavern, H. Hjalmarson, and M. O'Malley, "Closing Photoconductive Semiconductor Switches," *Proc. 7th IEEE Int. Pulsed Power Conf.*, Monterey, CA, June 1989.

[26] M. L. Riaziat and C. K. Nishimoto, "Compact Optically Triggered Microwave Pulse Generator," *Microwave Opt. Technol. Lett.*, Vol. 5, No. 5, pp. 211–215, May 1992.

[27] D. J. Bates, R. I. Knight, S. Spinella, and A. Silzars, "Electron Bombarded Semiconductor Devices," in *Adv. Electron. Electron Phys.*, Vol. 44, pp. 221–281, 1977.

[28] D. W. Van Der Weide, J. S. Bostak, B. A. Auld, and D. M. Bloom, "All-Electronic Generation of 880 Femtosecond, 3.5 Volt Shock-Waves and Their Application to a 3 Terahertz Free-Space Signal Generation System," *Appl. Phys. Lett.*, Vol. 62, pp. 22–24, Jan. 1993.

[29] M. Rodwell, "Picosecond Electrical Wavefront Generation and Picosecond Optoelectronic Instrumentation," Ph.D. Dissertation, Stanford University, Stanford, CA, Dec. 1988.

[30] M. Rodwell et al., "Active Nonlinear Wave Propagation Devices in Ultrafast Electronics and Optoelectronics," *Proc. IEEE*, Vol. 82, No. 7, pp. 1037–1058, July 1994.

CHAPTER SEVEN

High Speed and Long-Distance Communications

Communication systems perform the task of information transfer between two points using the modulation of a carrier wave. The information is normally in the form of an electronic signal with which the modulated parameter has a one-to-one correspondence. The carrier wave is generated and modulated in the *transmit module*. The *propagation medium* carries the wave to the destination point, where the *receive module* detects the signals and converts them to the desired form. The modulation rate is in general much lower than the carrier frequency.

Limitations on the performance of communication systems arise mainly from insufficient bandwidth or excessive noise level. Bandwidth is the highest modulation frequency that the system can accommodate. The maximum rate of information transfer through the system is directly proportional to the bandwidth. To turn the argument around, the bandwidth requirement is set by the information content of the signal. For example, a single telephone conversation may require only 3 kHz of bandwidth, while a broadcast television channel requires 6 MHz. Both noise and bandwidth are discussed in more detail in this chapter.

The age of electronic communications started with the telegraph in the early nineteenth century. Initially Morse code was capable of transmitting one alphabet character, or the equivalent of 6 bits, per second. By 1860, "high speed" telegraph printers were available that handled 15 bits per second. Time division multiplexing was invented for the telegraph in the 1870s. The telephone came into practical use in the late nineteenth century. Initially, single wires and twisted pairs were used as telephone lines. Later, coaxial cables were employed that could carry multiple conversations through frequency division multiplexing. The next major expansion in data transmission capacity and availability came with microwave line-of-sight links. Microwave transmission through communication satellites revolutionized the industry by allowing coverage everywhere in the world. Today, there is a trend toward replacing a major fraction of all data, voice, and video transmission links with fiber optics. Fiber-optic links presently carry only a small fraction of the information rate that their vast bandwidth allows; nevertheless a single fiber bundle of 0.5 in. diameter can easily carry the information rate equivalent to 100 microwave links.

7.1 COMMUNICATIONS LINK

A communications link consists of a transmitter, a transmission medium, and a receiver, as mentioned above. Even though a lot more detail goes into any one of these modules, system analysis is normally done in the three blocks mentioned. The block diagram of a typical link is shown in Figure 7.1. The

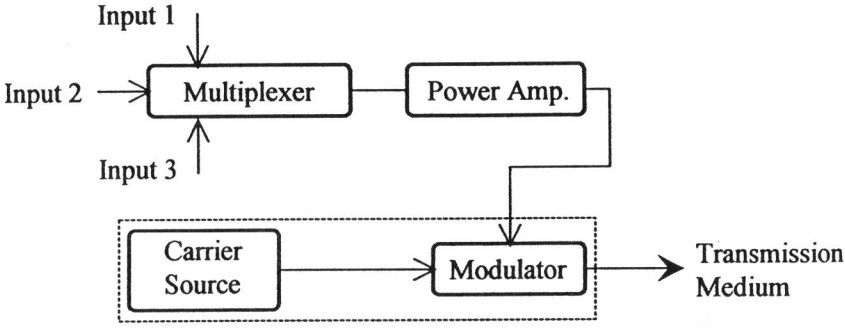

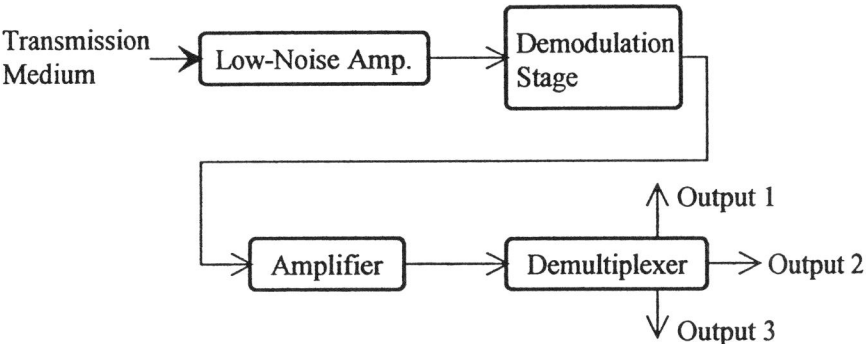

FIGURE 7.1 Various components of a communications link: (*a*) transmitter section; (*b*) receiver section.

transmitter multiplexes the inputs and uses the resulting signal to modulate a carrier wave. The modulated carrier, after traveling through the transmission medium, is demodulated and demultiplexed by the receiver. Various stages of amplification exist in both the transmitter and the receiver.

Information carrying capability of a communications link is characterized by (1) the rate at which it can carry the information and (2) the maximum distance between the transmitter and the receiver. The first characteristic is proportional to system bandwidth, while the second is a function of system noise and power. It will be seen that some aspects of system design will be based on a trade-off between bandwidth and distance. It is therefore sometimes desirable to use the bandwidth–distance product as a figure of merit for the characterization of a communications link. System trade-offs will be discussed in more detail.

7.1.1 Noise and Distortion

Both noise and signal distortion may be perceived as noise in the communications link. However, one is uncorrelated noise such as thermal or shot noise (Appendix B), while the other may be considered a correlated noise that is caused by nonlinearities in the system. Here, by separating the discussion of distortion from noise, we adopt the common assumption that noise is always uncorrelated. First, consider thermal noise and its effect on the communication link capacity. At the receiver end of the link, the carrier-to-noise ratio C/N is defined as the predetection signal-to-noise ratio of the wide-band carrier (including the relevant sidebands). The minimum detectable C/N is established by the receiver noise characteristics. Increasing the information capacity of the link broadens the information bandwidth of the carrier, and thermal noise power increases proportional to this bandwidth. For a given transmitter power the increasing bandwidth eventually reduces the C/N below the minimum detectable level. In this way, the information capacity of a communications channel is proportional to its carrier-to-noise ratio.

Since thermal noise power is proportional to bandwidth, carrier to noise ratio may be written as $C/(BN_0)$, where N_0 is the noise density and B is the bandwidth. Noise density is simply kT_e per hertz, where T_e is the *effective temperature*. The effective temperature takes into account the noise arriving at the receiver plus the thermal noise generated in the preamplifier section of the receiver. The carrier-to-noise density ratio of a communications link using free-space propagation is commonly calculated by a *link budget* analysis using the following relationship:

$$\frac{C}{N_0} = \frac{P_t G_t L G_r}{kT_e}, \qquad (7.1)$$

where P_t is the transmitted power, G_t and G_r are the transmit and receive antenna gains, respectively, and L is the sum of all losses including spherical propagation loss. Its value is less than unity. The product $P_t G_t$ is often referred to as the *effective isotropic radiated power* (EIRP). If the above equation is written in a logarithmic form, the following link budget equation results:

$$\frac{C}{N_0} = \text{EIRP (dBW)} - L_F \text{ (dB)} - L_a \text{ (dB)} + \frac{G_r}{T_e} \text{ (dB/K)} - 10 \log k, \quad (7.2)$$

where $(10 \log k = -228.6 \text{ dBW/K-Hz})$ is the Boltzmann constant. Losses due to spherical propagation L_p and absorption losses L_a have been separated. Link budget analysis identifies gains and losses associated with various system parameters leading to a carrier-to-noise ratio. It is possible to calculate the required C/N_0 for achieving the allowed (worst case) error rate in the received signal. Any higher value of C/N_0 is called the *link margin*. Link margins typically range from 0 to 5 dB and depend on the accuracy and dependability of the link budget calculations for the particular system. Communications links employing established technologies operate at a lower link margin.

Optical links may operate in a regime where the main source of noise is shot noise rather than thermal noise. The primary sources of shot noise in the system are the laser source and the photodetector dark current in the receiver [1]. This is further analyzed in the discussion of detection schemes in this chapter.

Distortion

A device or a system is said to be nonlinear if its transfer function is dependent on the input amplitude. This results in waveform distortion. Nonlinearity in various components of the communication link is represented mathematically by expanding the output of each component in a power series. Considering a voltage input V_{in} and a voltage output V_{out}, the expansion has the form

$$V_{\text{out}} = A_1 V_{\text{in}} + A_2 V_{\text{in}}^2 + A_3 V_{\text{in}}^3 + \cdots. \quad (7.3)$$

For a linear system the only nonzero coefficient is A_1. The first term in the expansion is referred to as the fundamental, denoted by V_1. The other terms are called the harmonics, where the nth harmonic is defined as $V_n = A_n V_{\text{in}}^n$. The nth *harmonic distortion* is defined as V_n/V_1, and the *total harmonic distortion* (THD) is given by

$$\text{THD} = \frac{\sqrt{V_2^2 + V_3^2 + \cdots}}{V_1}. \quad (7.4)$$

Nonlinearity of the system carrying a frequency multiplexed signal causes a correlated noise known as *intermodulation distortion*. In order to understand this, consider two signals V_a and V_b separated by a small frequency difference, being simultaneously transmitted through a nonlinear component of a communication link. The output signal can be written as

$$V_{\text{out}} = \begin{cases} A_1(V_a + V_b) + A_2(V_a + V_b)^2 + A_3(V_a + V_b)^3 + \cdots, \\ A_1(V_a + V_b) + A_2(V_a^2 + V_b^2) + A_3(V_a^3 + V_b^3) + \cdots \\ + 2A_2 V_a V_b + 3A_3(V_a V_b^2 + V_a^2 V_b) + \cdots . \end{cases} \quad (7.5)$$

Products of sinusoidal signals generate sum and difference frequency components. In the above expansion, quadratic terms generate frequency components that are either an octave higher than V_a and V_b, or very close to dc. In either case they can be easily eliminated by filtering. The third-order cross-product terms, otherwise known as *third-order intermodulation* products, on the other hand give rise to frequency components very close to the original signals. These components may coincide with other signals of interest in the communication band and cannot be easily filtered (Figure 7.2).

Intermodulation distortion can occur in nonlinear amplification, nonlinear transmission, or nonlinear detection. As expected, most of the contribution to signal distortion comes from the transmission side of the link where signal levels are high.

7.1.2 Modulation and Demodulation

Modulation and demodulation are shown in Figure 7.1 as single blocks in the communication link. The choice of modulation and demodulation techniques affect the signal-to-noise ratio of the communication link considerably.

In the early days of radio, amplitude modulation was used together with

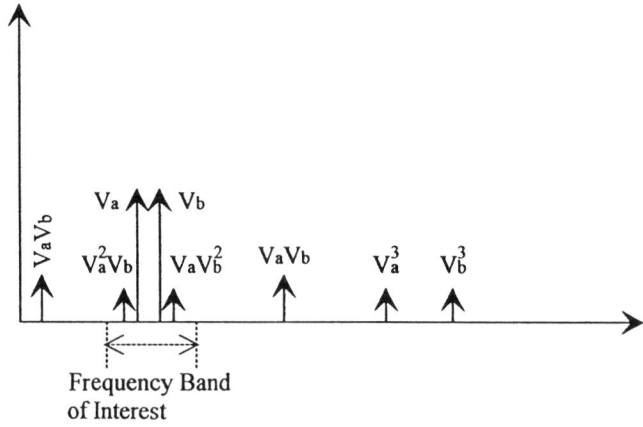

FIGURE 7.2 Frequency components generated by signals V_a and V_b in the presence of distortion. Third-order intermodulation products generally fall close to or within the frequency band of interest.

direct detection at the receiver. This combination required very little sophistication in the instrumentation used. For the same reason, this is also what is being used commonly in fiber-optic communications today. In this section, direct detection is compared with a "coherent" communication scheme with emphasis on signal-to-noise ratio.

Direct Detection

Consider a photodetector whose current output is proportional to the optical signal power: $I_s = \alpha P_s$. If the detector converts every absorbed photon to a *single* charge carrier, the proportionality constant α is found to be $\alpha = \eta_q e / h\nu$, where η_q is the quantum efficiency of the detector (fraction of photons absorbed) and $h\nu$ is the photon energy. Therefore,

$$I_s = \frac{\eta_q e}{h\nu} P_s. \tag{7.6}$$

This is the process of *direct detection* or *quantum counting*. It is instructive to compare the signal current of Eq. (7.4) with the noise current that is generated by the detector. Let us assume that we operate in the shot-noise-limited regime. The noise current for operation bandwidth B is given by

$$I_n^2 = 2eBI = \frac{2\eta_q e^2 B}{h\nu} P_s. \tag{7.7}$$

The ratio of signal power to noise power is the ratio of the currents squared:

$$\frac{S}{N} = \frac{\eta_q}{2Bh\nu} P_s. \tag{7.8}$$

Other noise sources such as thermal and detector dark current were ignored. Since these types of noise are independent of signal level, Eq. (7.6) is valid only when there is enough signal power to make the shot noise dominant. This is called *shot-noise-limited* operation.

Example Compare signal power to thermal noise power for a 40% quantum efficiency detector operating at 500 nm wavelength and terminated in an 800-Ω resistor. Take the input optical power to be 0.5 nW and the detection bandwidth to be 1.0 MHz:

Solution The signal power across the resistor is

$$I_s^2 R = \left(\frac{0.4(1.6 \times 10^{-19})}{(6.6 \times 10^{-34})(6.0 \times 10^{14})} (0.5 \times 10^{-9}) \right)^2 \times 800 = 5.2 \times 10^{-18} \text{ W}.$$

At room temperature, thermal noise power is given by

$$kTB = (1.38 \times 10^{-23})(300 \times 10^6) = 4.1 \times 10^{-15} \text{ W}.$$

As can be seen, signal power is lower than noise power by approximately a factor of 1000, or 30 dB. ∎

Another source of noise in direct detection is background light. In fiber-optic systems this light is effectively blocked, but it may pose a problem in applications such as space communications. Optical filtering, spatial filtering, and heterodyne detection can reduce noise from background light. Optical filtering is limited in practice by the relatively wide bandwidth of available optical filters. The effectiveness of spatial filtering, which limits the acceptance angle of ambient light, depends on the application and the acceptance angle needed for the signal. As will be seen, heterodyne detection is a very effective tool in eliminating the noise associated with background light.

Heterodyne Detection

Figure 7.3 shows the block diagram of a heterodyne receiver. The sinusoidal signal impinging on the detector $E_s = A_s \cos(\omega_s t)$ is combined with a fixed local oscillator (LO) signal $E_0 = A_0 \cos(\omega_0 t)$. Signal power P_s and local oscillator power P_0 are related to the amplitudes through the proportionality constant k: $P_s = kA_s^2$ and $P_0 = kA_0^2$. Current in the detector is proportional to the square of the sum of the amplitudes by the same proportionality constant,

$$I = k\alpha \left[A_0 \cos(\omega_0 t) + A_s \cos(\omega_s t) \right]^2 . \tag{7.9}$$

Cross products resulting from the squaring process give rise to sums and differences of the frequencies involved. Sum frequencies are too high to be detected. The remaining terms are the difference frequencies and the dc terms:

$$I = k\alpha \left[\tfrac{1}{2}(A_0^2 + A_s^2) + A_0 A_s \cos(\omega_0 - \omega_s)t \right] . \tag{7.10}$$

The difference frequency is known as the intermediate frequency (IF). In heterodyne detection, IF signal current is proportional to the geometric mean of LO power and signal power:

$$I_s = k\alpha A_0 A_s = \alpha \sqrt{P_0 P_s} . \tag{7.11}$$

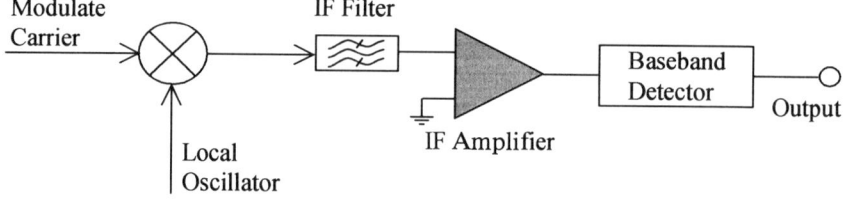

FIGURE 7.3 Block diagram of a heterodyne detection system.

This gives the possibility of "amplifying" the IF signal by increasing LO power. There is also increased detection dynamic range, since a reduction of signal power by a factor of n reduces the IF current only by $\sqrt{n}$.

The shot noise, mainly generated by the dc portion of the detected signal, is given by

$$I_n^2 = ek\alpha(A_0^2 + A_s^2)B = e\alpha(P_0 + P_s)B. \quad (7.12)$$

This should be added to contributions from other sources of noise, unless this term is dominant. Unlike in direct detection, the shot noise term is not dependent only on signal strength and can always be made dominant by having sufficient LO power. Signal to noise ratio at the IF level is given by

$$\frac{S}{N} = \frac{\alpha P_0 P_s}{e(P_0 + P_s)B}. \quad (7.13)$$

In most cases, LO power is much larger than signal power. Under these conditions, the dominant noise is the shot noise due to the LO, and Eq. (7.13) is reduced to

$$\frac{S}{N} = \frac{\alpha P_s}{eB} = \frac{\eta_q P_s}{Bh\nu}. \quad (7.14)$$

This is twice the optimum (shot-noise-limited) signal-to-noise ratio of direct detection.[1]

It should also be pointed out that the noise contribution of background light is considerably reduced in heterodyne detection due to the sharp filtering available at the IF signal level. Any ambient light that contributes to a background current in heterodyne detection has a frequency that is identical to ω_s plus or minus the IF frequency. This is a severe spectral restriction on background light.

Example Repeat the previous example replacing direct detection with heterodyne detection with an LO power of 1.0 mW.

Solution The value of the input optical power P_s in direct detection is replaced by $\sqrt{2P_s P_0}$ in heterodyne detection. This increases the signal power at the resistor by a factor of $\sqrt{2P_0/P_s} = 4.5 \times 10^3$. Signal power with heterodyne detection is therefore 2.34×10^{-14} W, which, for the same detection bandwidth, is 7.5 dB above the thermal noise. ∎

The signal-to-noise ratio of heterodyne detection can be improved further by employing a balanced detection scheme. Balanced detection is shown in

---

[1] The simple argument presented here should be viewed only qualitatively. Quantitative comparison of detection methods should be made after specifying modulation schemes and the filtering steps utilized.

Figure 7.4 for both optical and electrical signals. The core element of various balanced detection schemes is a 3-dB coupler (power divider) with controled phase differences between outputs (Chapters 4 and 5). Microwave and RF balanced mixing is done by splitting the signal into out-of-phase components and the LO into in-phase components. The IF currents resulting from the mixing in each branch are 180° out of phase, while currents generated by LO noise are in phase. The differential combination of the IF signals results in the suppression of LO noise contribution. Another advantage of balanced detection is that it provides isolation between the LO and RF without the need for filters.[2]

In optics, power splitting is normally done by either a fiber-optic coupler or a power splitter cube. The same device is also used for combining the optical signal with the LO beam. Balanced detection is necessary to make use of all of the optical power. The power splitter generates a 90° phase shift between its two outputs (Chapter 5). The IF current generated by mixing two optical signals in a photodetector is given by Eq. (7.10). Adding specific phase angles ϕ_0 and ϕ_s to the LO and the input signal, we get

$$I_{IF} = (k\alpha)A_0 A_s \cos(\omega_0 t + \phi_0 - \omega_s t - \phi_s). \tag{7.15}$$

In balanced detection, one of the photodetectors receives the input signal with no phase shift and the LO signal with $\frac{1}{2}\pi$ phase shift. The other photodetector receives the opposite phase relationship. The two photodetector currents are therefore

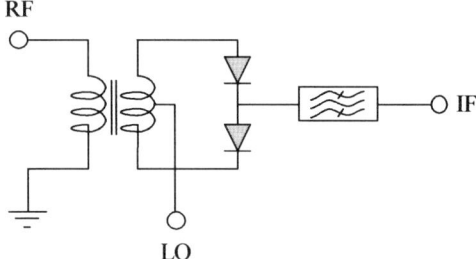

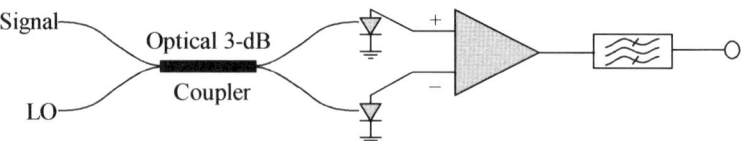

FIGURE 7.4 Electrical and optical implementations of balanced detection.

[2] In rectangular waveguides both the in-phase and the out-of-phase splitting are done by a single "magic T" junction.

$$I_1 = (k\alpha)A_0 A_s \cos(\omega_0 t + \tfrac{1}{2}\pi - \omega_s t),$$
$$I_2 = (k\alpha)A_0 A_s \cos(\omega_0 t - \omega_s t - \tfrac{1}{2}\pi),$$
(7.16)

which are 180° out of phase. Besides these signals, there is a portion of the photocurrent that is mainly due to the detection of the LO power. This current, which contains the noise due to LO intensity fluctuations, is nominally the same in both photodetectors. If the two photocurrents are subtracted or differentially amplified, the IF currents add in phase, while the LO intensity noise is suppressed [2]. Another advantage of heterodyne detection over direct detection is its high degree of frequency selectivity. The sharp filtering available at the IF level allows closely spaced carrier frequencies to be multiplexed for higher information carrying capability.

7.1.3 Multiplexing and Transmission Hierarchies

Multiplexing is what allows multiple communication channels to share the same link. Frequency division, time division, and wavelength division multiplexing utilize different technologies to achieve this goal [3, 4]. In *frequency division multiplexing*, which is heavily used in microwave and satellite communications, multiple modulated carriers at different frequencies are transmitted together and separated after reception, by filtering. Each modulated carrier in turn can use *time division multiplexing* whereby multiple bit streams of digital data are combined onto a single stream with a higher clock rate. At the receive end, the bits are distributed sequentially to the corresponding channels. *Wavelength division multiplexing* is used with fiber-optic communications and uses multiple optical wavelengths as carriers. It is theoretically identical to frequency division multiplexing, but its implementation is different. Presently, wavelength division multiplexing technology does not allow the carrier wavelengths to be as closely spaced as allowed by their information bandwidths. The optical fiber remains highly underutilized in this respect. All three multiplexing technologies can be combined for maximum information transfer by fiber optics. The process of combining multiple microwave subcarriers onto a modulated optical carrier is known as *subcarrier multiplexing* [5].

When multiplexing and demultiplexing are done numerous times at various stages of a communication network, they lead to transmission hierarchies. In particular, the switching and multiplexing of voice channels has led to the creation of extensive transmission hierarchies adopted internationally by telephone companies. These hierarchies are formed by sequential multiplexing of groups of channels into higher capacity trunks. At the receiving end, the reverse is performed by demultiplexing and switching the lower channels in succession. The transmission rates of the North American digital telephone hierarchy is shown in Table 7.1

This architecture gave satisfactory performance for voice communications but is proving insufficient for data and mixed-signal communications. The

TABLE 7.1 Transmission Rates of North American Digital Telephone Hierarchy

Name	Data Rate (Mbits/s)	Equivalent Voice Circuits
Single	0.056	1
DS1	1.5	24
DS2	6.3	96
DS3	45	672
DS3C	90	1344
DS4	274	4032

main drawback is the fact that the user has a fixed capacity line at his disposal. Data and mixed-signal communications are characterized by dramatically varying data rates in time. The ISDN (Integrated Services Digital Network) standards were developed for delivering digital voice and data to the subscriber. Before the implementation of ISDN in North America, the interest shifted to a family of standards known as broadband ISDN (B-ISDN), which have been adopted to address both the high bandwidth requirements and the flexibility needed for data, video, and voice communications. B-ISDN consists of multilevel standards that include SONET (Synchronous Optical Network, known in Europe as the Synchronous Digital Hierarchy, or SDH) and ATM (Asynchronous Transfer Mode). ATM is the switching format that interfaces with the various user services on one side and with the physical transmission format (SONET) on the other. The SONET transmission format is shown in Table 7.2. The STS-1 signal level in SONET has the same capacity as the present DS3 line except that it carries more overhead information to facilitate better monitoring and switching capabilities. Comparing the network capability needed for transferring computer data files with that needed for voice and video communications illustrates the challenge that modern networks have to meet. Data transfer is best performed in large packets of information transmitted infrequently, while a telephone conversation requires small packets of information with a continuous connection. ATM is a compromise solution that transmits information in small (53-byte) packets but at rates that are varied on demand. Each packet contains all the necessary source, destination, and control information needed for its transmission. These packets are combined by the network for physical transmission. The

TABLE 7.2 Sample SONET Transmission Rates

Name	Data Rate (Mbits/s)	Equivalent Voice Circuits
STS-1	51.84	672
STS-3	155.52	2,016
STS-48	2488.32	32,256
STS-96	4976.64	64,512
STS-192	9953.28	129,024

small size of the packets simplifies the task of information flow optimization by the network.

7.2 SATELLITE COMMUNICATIONS

The first serious satellite communication efforts started with the use of the moon as a passive reflector for a communication link between Washington, DC, and Hawaii in 1956. High flying balloons used as reflectors relayed the first trans-Atlantic messages in 1960. These initial attempts suffered from an immense loss of signal power due to the passive nature of the reflector.[3] To gain an appreciation for the magnitude of this loss, consider a satellite with equal distances R to the transmitter and the receiver. The received signal can be calculated from a modified form of the radar equation that quantifies the loss due to the three-dimensional spreading of the waves in space:

$$P_{\text{received}} = \frac{P_t G_t A_e \sigma}{(4\pi)^2 R^4}. \tag{7.17}$$

In this equation P_t is the transmitted power, G_t is the gain of the transmit antenna, A_e is the effective aperture of the receive antenna, and σ is the cross section of the reflector.

Example Let us calculate the signal attenuation between a transmitter and a receiver separated by 600 km and communicating through a passive reflecting satellite with a cross section $\sigma = 8\,\text{m}^2$ at an altitude of 300 km. Take $G = 100$ and $A_e = 80\,\text{m}^2$.

Solution If the satellite is spaced halfway between the two stations, the distance of each station to the satellite is $R = 300\sqrt{2} = 424$ km. Inserting this and the other parameters into the equation above, we find the ratio of the received power to the transmitted power to be 199 dB. ∎

The revolution in active communication satellites happened during the first half of the 1960s. Experiments with active satellites made rapid progress during those five years and led to the launch of the first commercial communication satellite, INTELSAT I, in 1965. Presently, over 300 communication satellites are in operation, and large constellation projects such as Milstar and Iridium are forming satellite networks accessible from the entire globe. The Iridium personal communications system was originally scheduled to employ 77 satellites in its network.

---

[3] In an early reflector balloon link across the United States, the received power was 180 dB below the transmitted power!

7.2.1 Satellite Orbits

A satellite in an elliptical orbit of semimajor axis A around Earth has a period given by [6]

$$T = 2\pi \sqrt{\frac{A^3}{\mu}}, \qquad (7.18)$$

where μ is the gravitational constant times the mass of Earth and is equal to $3.99 \times 10^5 \text{ km}^3/\text{s}^2$. Of particular interest is a circular orbit $A = R$, with a period equal to that of Earth's rotation or the *sidereal day*. The sidereal day takes into account the rotation of Earth around the sun, which causes one day and night period per year. Therefore the length of the sidereal day is

$$24(1 - 1/365.2) = 23 : 56 : 03. \qquad (7.19)$$

This rotation period requires a rotation radius of 42,178 km. Subtracting the Earth radius (6357 km) from this number, we arrive at the needed altitude of approximately 35,800 km. This is called the *geostationary* orbit, which is the highest Earth orbit used for communication satellites. Advantages of this orbit are numerous: A satellite in this orbit does not change its position in the sky, so an Earth station can be in contact with a single geostationary satellite at all times. Furthermore, there is no significant doppler shift, and broad coverage of up to one-third of Earth's surface is possible. Some disadvantages of this orbit are difficult launching, high power needed both for uplink and downlink, and enhanced problems with transit time and echo (round-trip time for signal to this orbit is 240 ms).

The distance to geostationary orbit is approximately three times Earth's diameter. This is called a high altitude orbit. It is considerably less expensive to work with satellites at medium or low altitude orbits. Medium altitude satellites are in orbits ranging from 9000 to 20,000 km in altitude. Rotation period for these satellites range from 5 to 12 h. They stay above the horizon between 2 and 4 h at a time for any Earth station. Satellites deployed in low Earth orbit range in altitude from 200 to 600 km. Low altitude satellites generally have circular orbits and rotation periods of 1–2 h. They remain above the horizon for as short as 15 min per orbit.

Polar regions and latitude locations on Earth encounter special problems with communication satellites. Geostationary satellites flying above the equator (equatorial orbit) are very low on the horizon, thereby compounding the problem of atmospheric absorption. Geostationary satellites with inclined or polar orbits stay above the horizon for only half of the period. A solution to communication with high latitude locations is the so-called *Molniya* orbit. This is a highly elliptical orbit with an apogee of about 40,000 km and a perigee of 1000 km. If the apogee is placed over the north pole, the satellite spends 11 h of its 12-h orbit period over the Northern Hemisphere.

Information Capacity

The concept of "channel capacity" is defined by Shannon's theorem as the information transmission rate for a given link below which it is theoretically possible to have an arbitrarily small error rate in the detected signal. Channel capacity, C, is a function of the signal-to-noise ratio. For white Gaussian noise it is given by [7]:

$$C = B \log_2 \left(1 + \frac{S}{N}\right) \text{ bits/s} \qquad (7.20)$$

where S/N is the signal-to-noise ratio.

In practice, the information carrying capacity of a communications satellite is below the theoretical channel capacity and is determined primarily by (1) its multiple-access technology, (2) its operation frequencies and the corresponding bandwidths, and (3) its on-board signal processing capability.

Multiple access utilizes the wide-area coverage of the satellite to simultaneously work with multiple uplinks and downlinks separated by location on Earth. Space division multiple access (SDMA) is hardwired inside the satellite to utilize multiple antenna beams each with a dedicated amplifier circuit. Space division multiple access has limited configuration flexibility. Frequency division multiple access (FDMA) is similar to frequency division multiplexing except that the frequencies are combined at the satellite. Each transmitting station is assigned a different frequency, but the amplification is done on the combined frequencies in the satellite. As might be expected, intermodulation products need to be controlled in this multiple-access scheme. Time division multiple access (TDMA) avoids the intermodulation noise problem by assigning sequential time slots to various stations. This is analogous to time division multiplexing. Finally, spread spectrum or code division multiple access (CDMA) changes the frequency of the carrier as a function of time according to a specified pseudorandom code. The shifting of the frequency causes each transmission to occupy the full-bandwidth spectrum of the satellite. The receiving station for each channel has to use the same code used by the transmitting station to extract the information by cross correlation. Code division multiple access offers the advantage of built-in encryption and security. Many satellite systems use a combination of the multiple-access technologies mentioned here.

The operation bandwidth associated with each antenna beam determines the number of multiplexed channels or the maximum information transmission rate allowed by that beam. Table 7.3 lists three microwave bands assigned to satellite communications in the United States. The assigned bandwidth per beam is 500 MHz in all cases.

7.2.2 Current Trends in Satellite Communications

A major evolving application of satellite communications is the expansion of direct-to-satellite voice/data links for use by individuals or small organiza-

TABLE 7.3 Most Popular Satellite Bands and Their Frequency Allocations (GHz)

Band	Uplink	Downlink
C	5.9–6.4	3.7–4.2
X	7.9–8.4	7.25–7.75
Ku	14–14.5	11.7–12.2

tions. Many low Earth orbit (LEO) or medium Earth orbit (MEO) satellite networks are being planned for these markets. Networks such as Motorolla's Iridium project plan to provide personal communications coverage for the entire globe. Other competing systems such as Globalstar, Aries, and Odyssey have been designed with similar goals in mind. Satellites in higher altitude orbits are beginning to place a higher emphasis on direct to user digital broadcasting and data transmission. For these applications, the transmission time delay associated with higher orbits does not pose any problem.

7.3 FIBER-OPTIC COMMUNICATIONS

The concept of optical communications and exchanging light messages has a long history. In modern times, A. G. Bell filed a patent on the so-called "photophone" in 1880 and demonstrated its operation. The main drawback of this and most other optical communication techniques was reliance on the atmosphere as the transmission medium. The turbulence and unpredictability of the atmosphere made commercial long-distance optical communications impractical. Recently the progress made in fiber optics has completely changed the picture. Optical fiber has emerged as the lowest loss medium for long-distance signal transmission (Chapter 5). The technology of transmission and reception of optical signals using fibers is making rapid progress, leading to a continuous refinement of this communication modality and its increased practical use.

7.3.1 Transmission Rate and Distance

Early fiber-optic communication links used multimode step-index fibers. One dominant problem with such fibers was the intermodal dispersion, that is, different modes propagating at different phase velocities. In Chapter 5, an approximate expression was given for the differential time delay caused by this effect. Imagine a digital signal being transmitted by this fiber at the bit rate or frequency of f per second. It is easily seen that in order for the data bits to remain distinct the differential time delay needs to be kept lower than half the bit spacing: $\Delta t \leq 1/2f$. Combining this with Eq. (5.3), one obtains

$$\frac{L(\delta \eta)}{c} \leq \frac{1}{2f}. \tag{7.21}$$

This leads to a restriction on the bandwidth–distance product:

$$Lf \leq \frac{c}{2(\delta\eta)}. \qquad (7.22)$$

Early interoffice links in the 1970s that used multimode fiber could transmit data at the approximate rate of 35 Mbits/s through a distance less than 10 km. Multimode step-index fibers are losing their popularity in long-distance data transfer; nevertheless, the above discussion was an example to demonstrate the concept of the bandwidth–distance product limitation. This limitation represents the trade-off between data rate and transmission distance in the presence of dispersion.

In the presence of loss and the absence of dispersion, a similar trade-off exists, but by a different mechanism. Simply stated, signal attenuation limits the transmission distance according to the detectable signal-to-noise ratio. Increased bandwidth reduces this ratio by raising the noise floor. This leads to a bandwidth–distance limitation independent of dispersion.

Chromatic Dispersion

Chromatic dispersion is encountered when propagation velocity in fiber has a wavelength dependence. In general, in single-mode fibers, the propagation constant β varies slowly with wavelength. For a pulse with a carrier frequency ω_0 the propagation constant can be expanded in terms of a power series:

$$\beta(\omega) = \beta_0 + \beta_1(\omega - \omega_0) + \tfrac{1}{2}\beta_2(\omega - \omega_0)^2 + \cdots. \qquad (7.23)$$

Since group velocity is defined as $v_g = d\omega/d\beta$, the term β_1 is simply the inverse of group velocity with which the pulse envelope moves, while β_2 is responsible for dispersive pulse broadening. In other words,

$$\beta_2 = \frac{d}{d\omega}\frac{1}{v_g} = \frac{d\beta_1}{d\omega}. \qquad (7.24)$$

The variation of group velocity with wavelength is also referred to as group velocity dispersion (GVD) (see Section 6.1). The *dispersion parameter D* of the fiber is defined in a slightly different way:

$$D = \frac{d(1/v_g)}{d\lambda} = -\frac{2\pi c}{\lambda^2}\beta_2. \qquad (7.25)$$

If longer wavelengths propagate faster ($D < 0$), the dispersion is called normal; otherwise it is referred to as anomalous.

Dispersion in single-mode optical fibers is a combination of material and waveguide dispersion. Material dispersion is caused by the dependence of the refractive index on wavelength. It is proportional to the second derivative of the refractive index with respect to wavelength. Waveguide dispersion, on the other hand, is caused by dependence of the propagation constant on the diameter of the waveguide normalized to the wavelength. Pulse broadening due to the combined dispersion effects of a typical silica fiber is shown in Figure 7.5 [8].

Example Let us determine the pulse broadening that is expected when a 1.0-ps pulse of 1000 nm radiation is propagating in a fiber with dispersion $D = 15$ ps/km-nm.

Solution The bandwidth of a transform-limited 1.0-ps pulse is expected to be approximately 440 GHz. In wavelengths, this bandwidth corresponds to 1.5×10^{-3} nm. The dispersion is therefore

$$(15)(1.5 \times 10^{-3}) = 0.0225 \text{ ps/km}.$$

This pulse will be dispersed by 2.25 ps after 100 km of propagation. ∎

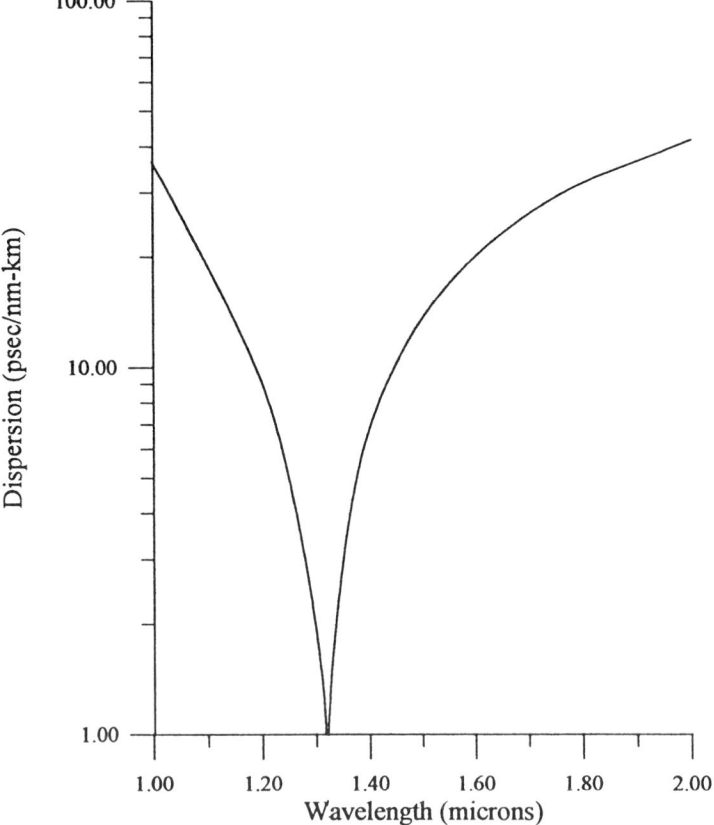

FIGURE 7.5 Dispersion as a function of wavelength for silica fiber with a V number of approximately 2.

In digital modulation, if we place a dispersion limit of one-half bit spacing on the transmission medium, that is,

$$DL\Delta\lambda \leq \frac{1}{2f}, \qquad (7.26)$$

it follows that

$$Lf \leq \frac{1}{2D\,\Delta\lambda}. \qquad (7.27)$$

For $\lambda = 1.55$ nm and $D = 15$ ps/nm·km, this frequency–distance limit is 6.67 (Gb/s)·km for $\Delta\lambda = 5$ nm. Note that signal amplification in the fiber does not alter this limit.

Nonlinearity

Nonlinearity effects are those that depend on the intensity of the optical signal. Of particular interest is the effect known as *self-phase modulation* (SPM), which leads to spectral broadening of the signal propagating in a fiber. Self-phase modulation is caused by the dependence of the refractive index on optical intensity. The refractive index η expanded to first order with respect to intensity I is

$$\eta = \eta_0 + \eta_2 I. \qquad (7.28)$$

Consider an optical pulse broad enough to be considered quasi-monochromatic with angular frequency ω_0. The propagation constant β of the pulse is given by

$$\beta = \omega_0 \frac{\eta}{c} = \beta_0 + \omega_0 \frac{\eta_2 I}{c}, \qquad (7.29)$$

where I is the intensity profile of the pulse. The phase angle ϕ becomes

$$\phi = \omega_0 t - \beta z = \omega_0 t - \beta_0 z - \Delta, \qquad (7.30)$$

where Δ is given by

$$\Delta = \omega_0 \frac{\eta_2 I}{c} z. \qquad (7.31)$$

The instantaneous frequency of the pulse is the time derivative of the phase angle:

$$\omega = \omega_0 - \frac{d\Delta}{dt} = \omega_0 - \frac{\omega_0 \eta_2}{c}\frac{dI}{dt}z, \qquad (7.32)$$

where z is the distance traveled. The time derivative of the intensity profile may be calculated using time T of the reference frame of the pulse. If the intensity profile in this time frame is a Gaussian given by

$$I(T) = I_0 e^{-(T/\tau)}, \qquad (7.33)$$

the instantaneous frequency becomes

$$\omega = \omega_0 + \frac{\omega_0 \eta_2}{c} \frac{2T}{\tau^2} I(T) z. \qquad (7.34)$$

Note that $T = 0$ corresponds to the peak of the pulse, meaning that for a positive η_2, the leading edge of the pulse experiences a red shift and the trailing edge experiences a blue shift. The amount of shift is proportional to the intensity of the pulse and inversely proportional to the square of the pulsewidth. Figure 7.6 shows SPM in the time domain and the corresponding broadening of the spectrum.

7.3.2 Soliton Transmission

In order to avoid the limitation that GVD places on the bandwidth–distance product, it is possible to utilize the nonlinearity of the fiber to counterbalance the effect of GVD. As a result of GVD alone, a pulse traveling in the fiber develops a chirp. The SPM nonlinearity alone can impart a similar chirp on the pulse. In spectral regions where these two effects have opposite signs, solitons may be formed. Solitons in optical fibers are pulses that retain their original shape by the continuous processes of SPM and dispersive pulse compression [9].

In the absence of attenuation, the amplitude $A(z, T)$ of a pulse propagating in a single-mode fiber changes slowly with distance z due to dispersion and nonlinearity. These effects are normally quantified as follows [10]:

$$j \frac{\partial A}{\partial z} = \frac{1}{2} \beta_2 \frac{\partial^2 A}{\partial T^2} - \gamma A |A|^2, \qquad (7.35)$$

where $T = t - z/v_g$ is time in the reference frame of the pulse moving with the group velocity v_g. This equation is known as the nonlinear Schrödinger equation. The constant β_2 is the GVD parameter and γ is the nonlinearity parameter, given by

$$\gamma = \frac{\eta_2 \omega_0}{A_{\text{eff}} c}. \qquad (7.36)$$

In the above equation A_{eff} is the effective core area of the fiber.

It can be seen by inspection that a purely sinusoidal signal of amplitude A_0 (independent of T) and phase $\gamma A_0^2 z$ is a solution of Eq. (7.35). This means that a CW signal propagates along a lossless fiber undisturbed except for

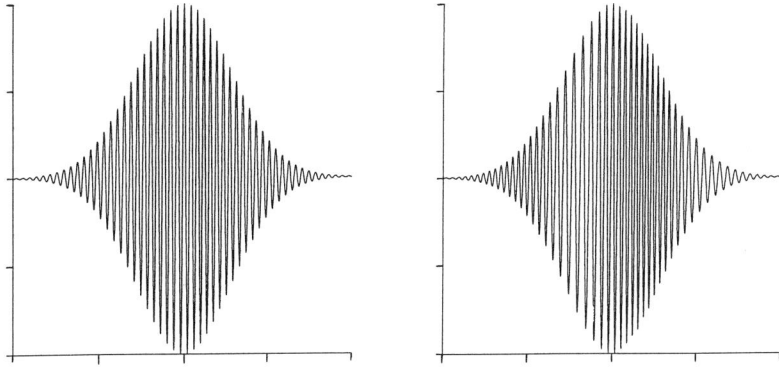

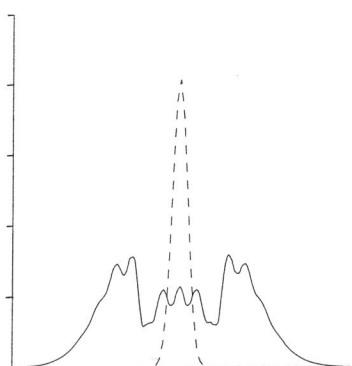

FIGURE 7.6 SPM in time domain and frequency domain.

developing an amplitude-dependent phase. This solution may or may not be stable under small perturbations to the amplitude. In fact, if the sign of β_2 is negative (anomalous GVD), the CW solution suffers from what is known as modulation or self-pulsing instability. Soliton formation occurs in the same regime.

Soliton solutions of the nonlinear Schrödinger equation are those that have either nonvarying or periodic amplitudes. It can be shown that pulses with hyperbolic secant shapes satisfy this condition given restrictions on their initial peak power P_0. Specifically,

$$A(0, \tau) = \sqrt{P_0}\, \text{sech}(\tau) \tag{7.37}$$

initiates a soliton propagation given the restriction that

$$P_0 = N^2 \frac{|\beta_2|}{\gamma T_0^2}, \tag{7.38}$$

where T_0 is the pulsewidth and $\tau = T/T_0$. Here, N is an integer called the *soliton order*. A soliton with $N = 1$ is called a *fundamental soliton*. It propagates with a constant amplitude. The peak power required to launch an Nth-order soliton is N^2 times the peak power for a fundamental soliton. Higher order solitons lose their original shape and acquire multiple peaks both in time and in frequency, but the process reverses, and the original shape is reached again. The minimum distance needed for a higher order soliton to regain its original shape is called the *soliton period* z_0, given by

$$z_0 = \frac{\pi}{2} \frac{T_0^2}{|\beta_2|}. \tag{7.39}$$

Example Determine the required power for a 10-ps fundamental soliton at 1550 nm wavelength in a fiber with $D = 27$ ps/nm·km, $\eta_2 = 10^{-3} \mu m^2/W$, and an effective core area of 100 μm^2.

Solution Here, β_2 can be found from the dispersion of the fiber using Eq. (7.25):

$$|\beta_2| = \frac{D\lambda^2}{2\pi c} = (3.44)10^{-26} s^2/m.$$

Next, using Eq. (7.36) we find γ to be 4.1×10^{-5} (W-m)$^{-1}$. Finally, combining these values in Eq. (7.38), we arrive at a P_0 of 8.4 W. ∎

The above discussion on solitons was based on the assumption that the fiber is lossless. In long-distance communications where fiber loss is significant, the soliton continuously loses strength, and its peak power falls below the required level. Fortunately solitons tend to be very tolerant of variations in peak power and pulse shape. Pulses whose peak powers fall in the range corresponding to $0.5 < N < 1.5$ tend to adjust their widths and shapes such that they will evolve into fundamental solitons. For long-distance communications this means that solitons may be periodically amplified by doped fiber amplifiers. In laboratory experiments soliton transmission has been demonstrated through many thousands of kilometers of fiber, and the record is improving annually [11].

7.3.3 Trends in Fiber-Optic Communications

Intensity modulation and direct detection that currently dominate fiber-optic communication systems have inherent shortcomings. They include less than optimum signal-to-noise ratio and low dynamic range [12]. Despite these deficiencies the low loss, low cost, and tremendous bandwidth of optical fibers have created enough performance advantages over other communication modalities to make them widely accepted in a relatively short time. Presently, a lot of emphasis is placed on system development and definition of standards that will form the foundation of the new information industry with the emphasis on integrated services. Meanwhile, more advanced technologies such as coherent communications [13] and soliton networks are evolving rapidly. Stable distributed Bragg reflector lasers, broadband modulators, optical fiber amplifiers, fast and stable soliton sources, high efficiency balanced detectors, and other components necessary for efficient fiber-optic communications are currently available for exploratory use. These technologies will inevitably find their way into widespread use in the near future. It will however take a much longer time for the practical information carrying capability of the optical fiber to come close to its theoretical limit.

PROBLEMS

7.1 Calculate the link margin for a microwave transmitter and receiver separated by 20 km, a transmitter power of 5 kW, antenna gains of 15 dB, and an effective temperature of 350 K. Assume an atmospheric absorption of 0.1 dB/km.

7.2 In an amplifier with nonlinearities defined by Eq. (7.3), show that the gain at a single frequency saturates with increased input power. The *1-dB compressed power* is defined as the output power for which the gain is 1 dB below its small signal value. Find the 1-dB compressed power in terms of parameters of Eq. (7.3) for an amplifier with 50 Ω input and output impedances.

7.3 A heterodyne detector consists of two detection stages. At the first stage, a 10-GHz signal is down converted to 50 MHz with an IF filter bandwidth of 1.0 MHz. At the second stage, the IF is further down converted to 500 kHz with an output band-pass filter of 20 kHz bandwidth. The second IF is low pass detected at a bandwidth of 20 kHz. At each down conversion we get $P_{IF} = 0.4\sqrt{P_{RF}P_{LO}}$. Find the minimum LO powers needed for a 10-dB signal-to-noise ratio at the output. The input power is 1.0 nW, and impedances are 50 Ω throughout. Discuss possible trade-offs.

7.4 Calculate the geometric shape of the Molniya orbit based on the description given in Section 7.2.1.

7.5 Find the theoretical channel capacity of a link with a bandwidth of 3 MHz and a signal-to-noise ratio of 30 dB. Discuss how it may be possible to achieve this capacity in practice.

REFERENCES

[1] W. K. Pratt, *Laser Communication Systems*, Wiley, New York, 1969.
[2] G. Abbas, V. Chan, and T. Yee, "A Dual-Detector Optical Heterodyne Receiver

for Local Oscillator Noise Suppression," *IEEE J. Lightwave Tech.*, Vol. 3, No. 5, pp. 1110–1122, Oct. 1985.

[3] B. Sklar, *Digital Communications, Fundamentals and Applications*, Prentice-Hall, Englewood Cliffs, NJ, 1988.

[4] H. Taub and D. Schilling, *Principles of Communication Systems*, 2nd ed., McGraw-Hill, New York, 1986.

[5] R. Olshansky, "Microwave Subcarrier Multiplexing: New Approach to Wideband Lightwave Systems," *IEEE Circ. Devices*, pp. 8–14, Nov. 1988.

[6] W. L. Pritchard, "Satellite Communication—An Overview of the Problems and Programs," *Proc. IEEE*, Vol. 65, pp. 294–307, Mar. 1977.

[7] C. E. Shannon, "A Mathematical Theory of Communications," *Bell System Tech. J.,* Vol. 27, pp. 379–623, 1948.

[8] W. Gambling, H. Matrumara, and C. Ragdale, "Zero Total Dispersion in Graded Index Single Mode Fibers," *Electron. Lett.*, Vol. 15, p. 474, 1979.

[9] L. F. Mollenauer, R. H. Stolen, and J. P. Gordon, "Experimental Observation of Picosecond Pulse Narrowing and Solitons in Optical Fibers," *Phys. Rev. Lett.*, Vol. 45, No. 13, pp. 1095–1098, Sept. 1980.

[10] G. P. Agrawal, *Nonlinear Fiber Optics*, Academic, San Diego, 1989.

[11] M. Nakazawa, "Ultra Long Distance Soliton Transmission," 1992 *CLEO Dig.*, p. 314, May 1992.

[12] Y. Yamamoto and T. Kimura, "Coherent Optical Fiber Transmission Systems," *IEEE J. Quant. Elect.*, Vol. 17, No. 6, pp. 919–934, June 1981.

[13] P. R. Herczfeld, A. S. Daryoush, A. Rosen, P. Stabile, and V. M. Contarino, "Optically Controlled Microwave Devices and Circuits," *RCA Rev.*, Vol. 46, pp. 528–551, Dec. 1985.

CHAPTER EIGHT

Measurement Techniques

Accurate measurement of the performance of high speed devices is a challenging task whose technology tends to evolve in parallel with that of the devices to be characterized. Refinement of measurement technology allows higher precision characterization, which in turn leads to better optimized, higher performance devices. Those devices in turn generate new challenges for measurement systems, and the cycle continues. This chapter deals mainly with electrical measurements made on microwave and millimeter-wave devices and circuits. Measurement of short optical pulses is also briefly discussed. Issues in high speed electrical measurements are divided into sensor issues and system issues. Sensor issues deal with the interface between the device under test (DUT) and the signal processing part of the system. System issues in high speed testing are usually related to calibration, sources of error, and accuracy of the measurement system.

8.1 TEST FIXTURES AND SURFACE CONTACTING PROBES

Test ports of most microwave measurement equipments are either rectangular waveguides or coaxial cables. Yet it is often the case that the DUT has terminals that are not compatible with either type. Particularly, semiconductor devices tend to have terminals in the form of small contact pads. High speed integrated circuits may have compact planar transmission lines at input and output ports. It is the role of the test fixture to form the transition between the DUT and the test port of the measurement system. Test fixtures also often provide the means of dc power delivery to the device. The schematic diagram of a microstrip coaxial test fixture is shown in Figure 8.1.

Most high speed test fixtures require that individual devices be mounted in the fixture and electrical connections be made with bond wires. This is a labor-intensive process. Advances in high speed monolithic circuits generated the need to perform tests on the processed wafer and to avoid the cost of placing

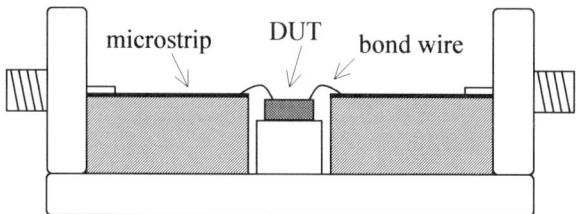

FIGURE 8.1 Microstrip test fixture with coaxial ports.

unacceptable devices in fixtures. Also for the characterization and modeling of active devices for consequent circuit insertion, it is advantageous to have the measurement port as close to the device as possible.[1]

On-wafer measurement at dc and low frequencies have been common practice for many decades. Needle probes make contact to the appropriate device terminals on the wafer, and assorted dc and low frequency tests are performed. Such measurements are used to either monitor various processing steps or screen devices before fixturing. The inductance associated with the needle makes these probes unusable at higher frequencies. In the mid-1980s, the first microwave probes were introduced that brought the test ports for high frequency measurements directly to the contact points on the wafer [1]. The idea was to fabricate a controlled impedance transmission line extending to the contact point of the probe. This was achieved using a CPW as the transmission line. An impedance preserving taper connected the large dimensions of the coaxial cable to the small sizes and spacings of the contact points (Figure 8.2).

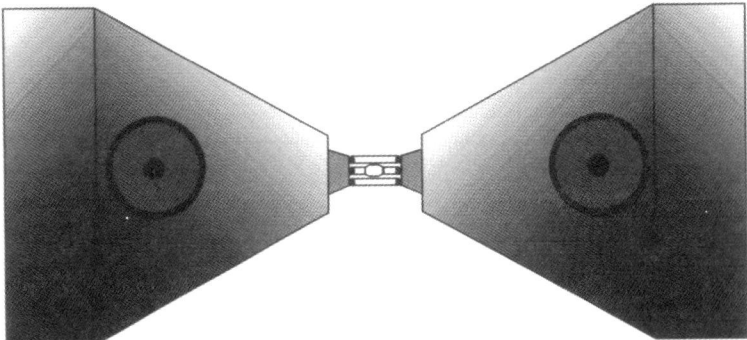

FIGURE 8.2 A CPW probe can be used for microwave measurements of unpackaged devices and devices on wafer. Here the DUT has input and output ports particularly designed for contact probing.

---

[1] It is possible with fixtured devices to move the test ports close to the device by either fixture deimbedding or direct calibration using microstrip standards. This process is simplified by wafer probes that allow calibration at probe tips without any assembly requirement.

It is important for a surface contacting probe to have low and reproducible contact resistance, low loss, and rugged probe tips. Devices to be tested by such probes need to have input and output ports that accommodate the footprint of the probe. Currently, contact probes are commercially available for testing at frequencies as high as 60 GHz and beyond. One problem that limits the usability of surface probes at higher frequencies is the round-trip signal loss from the coaxial or rectangular port to the probe tip and back. The presence of this loss reduces the dynamic range of the measurement. Attempts have been made to build frequency upconversion and down conversion modules into the probe itself in order to minimize the source-to-device travel distance of the high frequency signal. Encouraging experiments have been reported using such probes [2], but presently they are not available as commercial products.

8.2 NONCONTACT PROBING

The accuracy of measurements with contacting probes degrade at higher frequencies, due to the following factors: (1) uncertainty in the positioning of the probe especially during calibration, (2) increased insertion loss of the probe, and (3) poor electrical contact due to imperfect positioning or degraded probe tips. At frequencies close to or beyond 100 GHz, as well as for measuring time waveforms with rise and fall times in the picosecond range and below, it is not practical to use contacting probes. An alternative approach is afforded by short optical pulses that can be used for sampling and detection of electrical signals. Two techniques of optical sampling are discussed in this section: electro-optic sampling and photoconductive sampling.

8.2.1 Electro-Optic Sampling

Some compound semiconductors such as GaAs and InP that are used for the fabrication of high speed circuits exhibit a linear electro-optic effect. This means that the presence of an electrical signal at any point on a circuit built on these materials affects the refractive index of the substrate. This phenomenon can be utilized to monitor the propagation of high speed electrical signals [3].

A change in the refractive index of the substrate produced by a variable electric field modulates the phase of an optical beam that passes through the field region. This phase modulation is converted to a polarization change and subsequently to an amplitude or intensity modulation using crossed polarizers. The schematic diagram of this arrangement for the typical GaAs IC substrate crystal orientation, (100), is shown in Figure 8.3. If the applied voltage is a result of a microwave signal on a transmission line, the laser beam noninvasively measures this signal by sensing the amount of intensity modulation. It is assumed that the optical energy is sub-bandgap, that is, there is no significant absorption by the material.

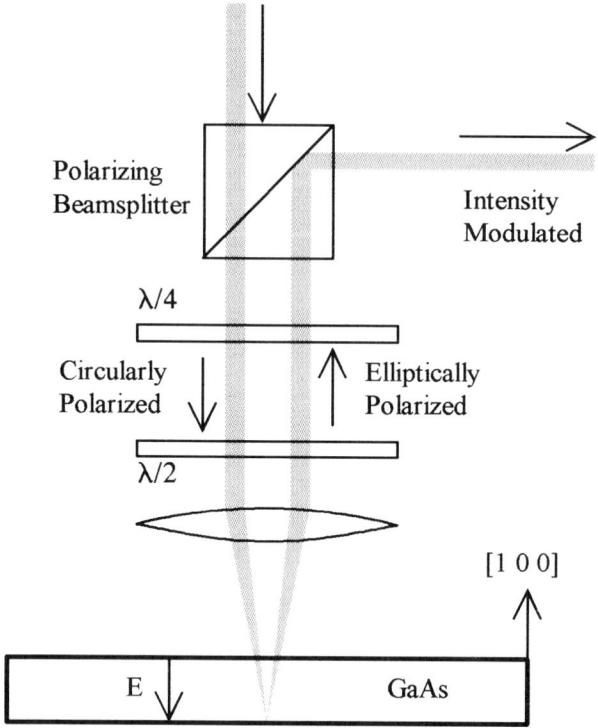

FIGURE 8.3 Experimental arrangement for converting refractive index perturbation to intensity modulation of optical probe beam.

For testing MMICs, the optical probe beam is focused through the GaAs substrate in one of the two geometries shown in Figure 8.4. For the probing of coplanar waveguide circuits, the probe beam enters from the backside of the substrate and gets reflected by the center conductor. For MMICs using microstrip, frontside probing is used where the beam passes through the conductor's fringing field and is reflected back by the ground plane.

The electro-optic tensor (Section 3.5) of GaAs has only three nonzero elements, given by [4]

$$r_{41} = r_{52} = r_{63} = 1.4 \text{ pm/V}, \qquad (8.1)$$

where the axes are defined along the cleave directions of the crystal, with the x axis being along the crystal [100] axis. The principal axes of the crystal share the x axis as defined above. The other axes are rotated by 45° about the x axis. To find the relationship between the intensity modulation and the electric field on the circuit, consider an optical beam along the [100] or x axis. If the electric field due to the circuit is also applied in the x direction, a change in the refractive index η_4 or η_{23} will result. Due to the orientation of

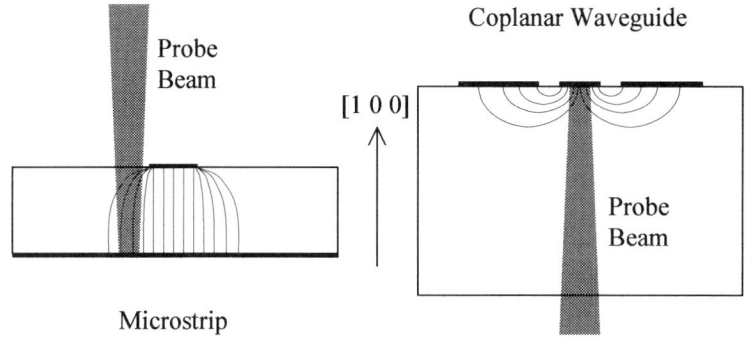

FIGURE 8.4 Frontside and backside electro-optic sampling.

the principal axes, this translates to an equal and opposite refractive index change for polarizations along the y and z axes:

$$\Delta \eta_z = -\Delta \eta_y = \tfrac{1}{2} \eta_0^3 r_{41} E_x . \tag{8.2}$$

After propagating through a distance L in the crystal, the phase difference between the two polarizations will be

$$\Delta \phi = \frac{2\pi \eta_0^3 r_{41}}{\lambda} \int_{x=0}^{L} E_x \, dx . \tag{8.3}$$

Notice that the integral of the electric field along the x axis translates to the potential on the signal conductor in both microstrip and CPW configurations. Equation (8.3) is therefore simplified to

$$\Delta \phi(t) = \frac{2\pi \eta_0^3 r_{41}}{\lambda} V(t) . \tag{8.4}$$

Once this phase modulation is converted to an intensity modulation, the output power is

$$P(t) = P_0 \{ 1 + \sin[\Delta \phi(t)] \} . \tag{8.5}$$

Since $\Delta \phi$ is very small in all practical cases, the first-order expansion of the sine function may be used. The resulting intensity modulation is

$$P(t) = P_0 \left(1 + \frac{2\pi \eta_0^3 r_{41}}{\lambda} V(t) \right) . \tag{8.6}$$

A high speed signal on the circuit under test generates a high speed intensity modulation that in turn requires high speed detection systems. In order to avoid this limitation and to obtain measurement bandwidths exceeding 100 GHz, pulsed lasers and sampling techniques are used. A train of short optical pulses can sample the microwave voltage waveform and in effect slow down the signal to allow the use of readily available slow detectors (Figure 8.5). Sampling with laser pulses is analogous to the function of a strobe light, which can produce the effect of "stopping" or slowing down repetitive motions under investigation. If the sampling rate is an exact subharmonic of the repetitive signal, only a single point in the time waveform is detected. If the sampling rate is changed slightly, the waveform appears to slowly evolve in time. For example, a pulse train with a repetition rate of 80 MHz can sample a 19-GHz sine wave and replicate its shape at the frequency of $19,000 - 237(80) = 40$ MHz.

Another way to analyze the sampling process is to treat it as harmonic mixing. A train of Gaussian laser pulses of width τ and repetition frequency f_0 has a Fourier transform that consists of harmonics of f_0 extending to $f = 0.44/\tau$ (half amplitude). The sampling of an electrical signal at frequency f_e corresponds to the mixing of the optical and the electrical signals due to the product term in Eq. (8.6) above. The result of the mixing is the generation of sum and difference frequencies of f_e and nf_0 for all n. The lowest frequency generated this way is $|f_e - Nf_0|$, where Nf_0 is the harmonic frequency closest to f_e.

In typical electro-optic samplers, optical pulses with durations of hundreds of femtoseconds to a few picoseconds are used. Repetition rates are normally in the megahertz range.

Limitations of Electro-Optic Sampling

The bandwidth of the sampling system is primarily determined by the optical pulsewidth. If the timing jitter between the laser and the microwave excitation source is comparable to or exceeds the optical pulsewidth, it could signifi-

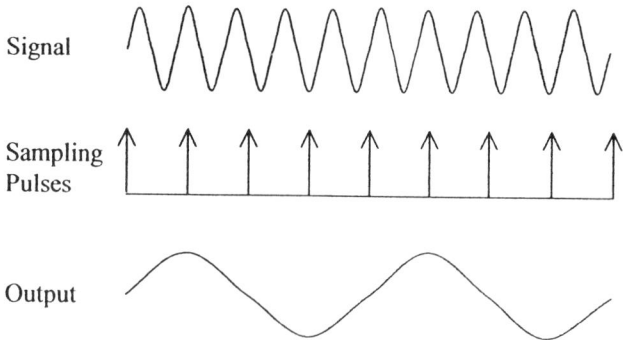

FIGURE 8.5 Sampling translates the waveform of a repetitive signal to a lower frequency.

cantly alter the bandwidth of the system. The new bandwidth corresponds to an effective pulsewidth τ_{eff} whose value is

$$\tau_{\text{eff}}^2 = \tau_{\text{pulse}}^2 + \tau_{\text{jitter}}^2. \tag{8.7}$$

As long as the timing jitter is well below the pulsewidth, it does not significantly reduce the system bandwidth.

Spatial resolution is set by the probe spot size. For frontside probing, minimum spot size is achieved when the depth of field equals the substrate thickness (focusing tighter causes the beam to diverge before it exits the substrate, and focusing less tightly decreases the beam divergence but increases the minimum spot size). For the typical 100-μm-thick substrate, the minimum spot size is about 10 μm.

When probing coplanar circuits from the back of the substrate, the obtained signal is proportional to the potential on the conductor. Testing microstrip circuits requires that the optical beam enter from the front of the circuit and pass through the fringing fields next to the conductor. The position of the laser with respect to the conductor can cause the signal level to change. However, this is a small effect and causes only a few percent variation over the spot size of the beam.

A major source of uncertainty is the reflection from the surface of the substrate. This is a large reflection, approximately 30% for GaAs. This reflected light interferes (constructively or destructively) with the probe beam that penetrates the substrate, causing the signal level to change. The interference is also extremely sensitive to substrate thickness, or more precisely to the equivalent optical path length in the substrate. The refractive index change caused by the electrical signal perturbs the optical path length in the substrate. This effect alone (known as the *etalon effect*) can generate an amplitude modulation of the probe beam independent of polarization rotation. Moreover, the etalon effect is position dependent due to variations in the thickness of the substrate. Making quantitative measurements in the presence of the etalon effect requires the calibration of the optical probe for every physical test point in the circuit. For backside probing, a single layer of antireflection coating can significantly reduce this problem.

Optical Excitation
Even though electro-optic sampling offers the flexibility of testing the signals at almost any point in the circuit, it requires that the signal be brought to the circuit by some other means. Feeding the signal by contacting probes is one possibility at lower frequencies. The other is to optically generate the signal on the wafer.

A microwave signal can be generated on a wafer by a fast photodetector that responds to the modulation of an optical signal. The excitation beam needs to have a higher than bandgap energy and be modulated at the microwave measurement frequency. The excitation beam and the probe beam

250 MEASUREMENT TECHNIQUES

are brought to the wafer at different locations. The excitation beam illuminates a photodetector, and the probe beam samples the signal in the circuit under test.

In one possible implementation, an integrated MSM photodetector in the form of a matched receive module is fabricated at the input of the circuit [5]. The fabrication of this module is fully compatible with MMICs and requires no added processing step. The receive module is connected to the circuit by airbridges and can be scribed away upon completion of the test. The output of the circuit is terminated in a 50-Ω load, also by airbridges in order to make it removable (Figure 8.6).

Instead of very high frequency modulation of the optical beam, the photodetector can be used to mix two cw optical signals. The result is the generation of an electrical output signal whose frequency is the frequency difference (beat note) between the optical signals [6]. Finally, it is possible to do the excitation of the photodetector by a short optical pulse instead of a
odulated beam. Optical pulse excitation leads to time domain testing, while

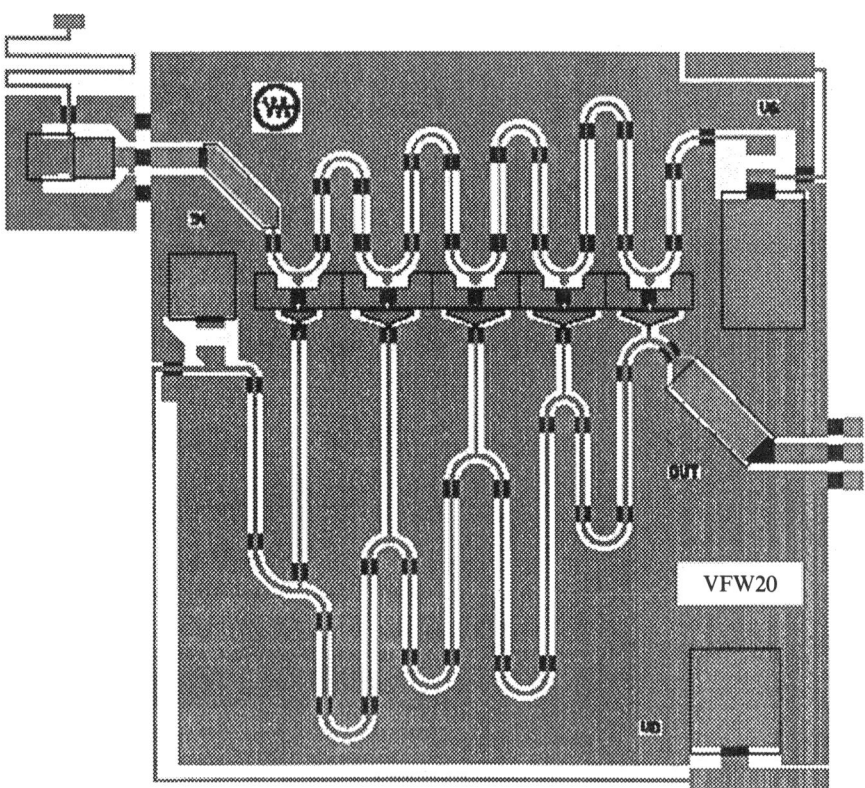

FIGURE 8.6 Integrated circuit with added photodetector and matched termination for noncontact optical probing.

modulated beam excitation allows frequency domain testing. Pulse excitation is discussed further under "photoconductive sampling."

8.2.2 Other Optical Probing Techniques

Using an external electro-optic probe is a variation of the electro-optic sampling method just discussed. A very small electro-optic crystal such as LiTaO$_3$ is brought to close proximity with the circuit where it can interact with the fringing fields. The resulting change in its refractive index of the crystal is sensed optically [7]. This approach offers the flexibility of choosing probe crystals with high electro-optic coefficients for higher sensitivity. The drawback is the small but finite capacitive loading of the test point in the circuit [8]. Another approach is *photoconductive sampling*, which is a time domain pump-probe technique for testing high speed circuits [9]. A small gap between two conductors in an integrated circuit can be made conducting for a very short period of time through excitation by a short optical pulse [10]. Gaps of this type can be added to the circuit layout to both generate an input signal for the circuit and sample the signal after propagating through the circuit (Figure 8.7). This technique is often used for time domain testing, since the signal generated by the switch approximates either an impulse or a step function. The response is mapped as a function of the time delay between the pump pulse and the probe pulse.

If the system under test responds linearly to the excitation, the Fourier transform can be applied to the resulting response to calculate the frequency-domain data. The switches are easier to calibrate than direct EO probing of

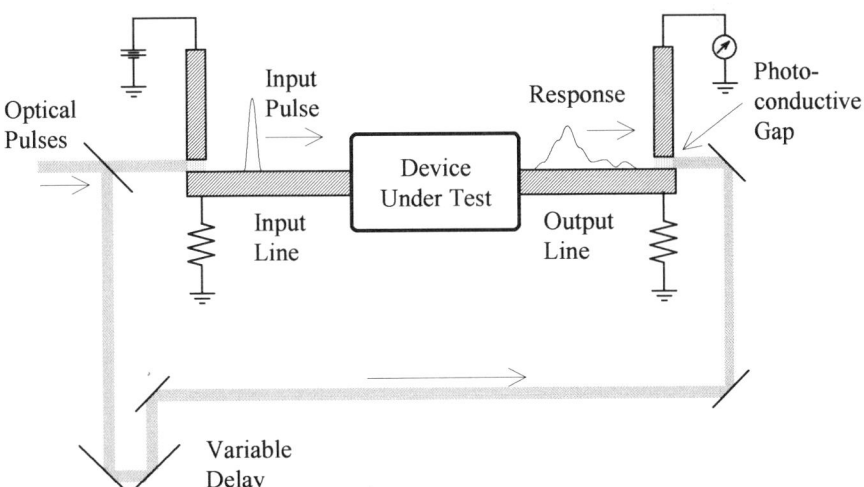

FIGURE 8.7 Photoconductive sampling uses the same laser pulse for both excitation and sensing.

the substrate. One advantage of photoconductive sampling compared with electro-optic sampling is that it uses the same optical wavelength for both excitation and probing. A disadvantage is that a specific gap is required for each test point, compared to electro-optic probing, which requires no special port for the measurement.

8.3 NETWORK ANALYZERS

Commonly used network analyzers are measurement instruments that determine the vector (magnitude and phase) S parameters of two-port devices at selected frequencies.[2] Each measurement port represents a matched termination at the reference impedance to the DUT. A source of stable signal and precise frequencies (normally a synthesized source) provides the incident signal on the DUT. The magnitude and phase of transmitted and reflected signals are measured by harmonic down conversion and heterodyne detection. The block diagrams of two of the major modules used in most network analyzers are shown in Figure 8.8. In the first module, known as the test set, the microwave signal is delivered to the DUT at the measurement frequency. Directional couplers sample the incoming signal (a_1) as well as transmitted (b_2) and reflected (b_1) signals from the test ports. Ideally, $S_{11} = b_1/a_1$ and $S_{21} = b_2/a_1$. The other two S parameters are obtained by repeating the measurement in the reverse direction. Complex ratios of the sampled signals are taken at low frequency after down conversion. The down conversion module is shown in Figure 8.8b. All measurement frequencies are down converted to the same IF frequency in order to simplify the filtering stages and the subsequent processing. A local oscillator (VTO) generates a signal at a tunable frequency f_{lo}. A harmonic generator produces a comb of harmonics of the local oscillator signal. Here, f_{lo} is picked such that one of its harmonics satisfies

$$\text{IF} = f_{\text{meas}} \pm n f_{\text{lo}}, \tag{8.8}$$

where f_{meas} is the measurement frequency and n is the harmonic number. A phase-locked loop maintains the local oscillator frequency[3] such that the IF is kept fixed (at, e.g., 20 MHz).

Preliminary S parameters are extracted from magnitudes and phases of the down-converted signals. These S parameters are not the correct parameters of the device due to (1) the undefined position of reference planes for the calculation of the phases and (2) imperfections in the hardware such as reflections from various connectors, bends, and so on. Other sources of uncertainty are finite directivity of directional couplers, finite cross talk between isolated

---
[2] Instruments known as scalar network analyzers measure only the magnitudes of S parameters.
[3] Phase locking is often done by running two synthesizers with one reference oscillator.

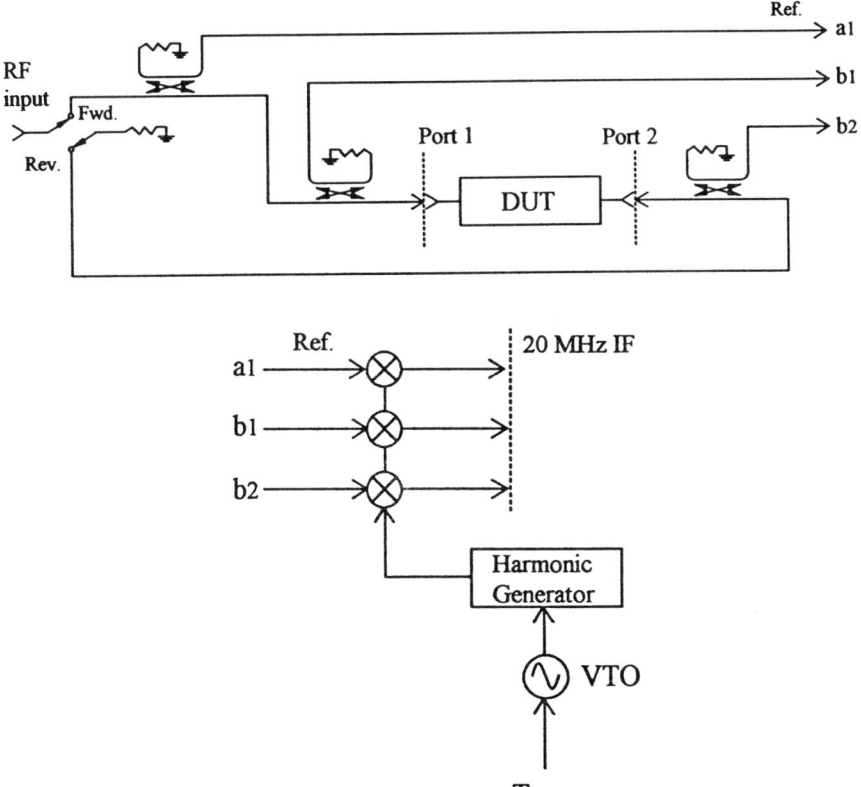

FIGURE 8.8 Two basic modules in a network analyzer; (a) test set; (b) harmonic mixer module.

channels, and imperfect matching of the signal source. Calibration steps are needed to establish the reference planes and to map the measurement space to the Smith chart plane of the device.

8.3.1 Calibration

A simple calibration for S_{11} might involve the measurement of a short circuit only. This establishes the reference plane and one point on the Smith chart. In old manual network analyzers this was often the only calibration step. Automatic analyzers go beyond this to numerically adjust the data and remove errors. It can be shown that most errors associated with imperfect directivity and finite source match can be removed with an open-short-load calibration, which gives enough information to map three points on the Smith chart ($S_{11} = 1, -1, 0$, respectively).

None of the calibration standards mentioned above are free from physical flaws. Imperfect standards add errors to the measurement that are more difficult to remove. A short-circuit standard suffers from contact resistance between the standard and the reference plane connector ($|S_{11}| < 1$). This problem of added loss is particularly severe with the so-called reference plane shorts due to the fact that the contact area is at the plane of maximum current. One approach to minimize this error is to use an offset short where the plane of contact is 90° removed from the short circuit, making it a minimum current point (Figure 8.9a).

An open-circuit standard has two problems associated with it. One is the loss due to radiation from the open end. This is normally taken care of by "shielding the open circuit," as shown in Figure 8.9b. The shield may be a waveguide below cutoff. The second problem associated with the open standard is the capacitance generated by fringing fields. This capacitance needs to be either calculated or measured and accounted for in the calibration process.

A truly matched load standard is difficult to build, especially over a broad frequency range. The solution is using either different loads for different frequencies or a "sliding load." The matched load standard is expected to allow the mapping of the center of the Smith chart. An imperfect load represents a finite magnitude and a well-defined phase. A sliding load (Figure 8.9c) is constructed such that the termination can slide inside a transmission line, thus changing its position with respect to the reference plane. The moving of the termination varies the phase of the reflected signal, keeping its magnitude the same. This process traces a circle whose center is the true center of the Smith chart.

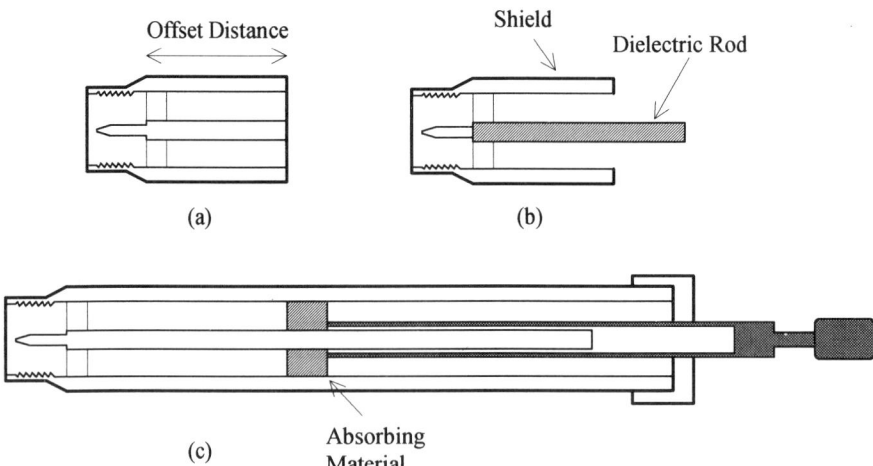

FIGURE 8.9 Compensated calibration standards: (a) offset short; (b) shielded open; (c) sliding load.

A potential source of error that cannot be removed by calibration is associated with harmonic mixers in phase-locked receivers. This is the so-called *harmonic skip*. At some measurement frequencies, it is possible for the phase lock loop to use two separate VTO frequencies multiplied by different harmonic numbers. This causes measurement errors when different harmonic numbers are used during calibration and measurement. In order to minimize this effect, synthesized sources are used as local oscillators. With synthesized sources, harmonic transition points are always fixed.

The calibration process defines a one-to-one mapping between the uncorrected space of the measured parameters to that of the Smith chart for the DUT. This process avoids addressing specific sources of error and, instead, deals with a mathematically equivalent problem: The error in the measurement is arbitrarily assigned to an unknown "black box" that we call the "error box." A two-port error box is placed before the DUT, and the measurement system is otherwise considered error free. The task of the calibration process is therefore to characterize the unknown error box and to remove it from the measurement. A two-port calibration uses two such error boxes, one before and one after the DUT (Figure 8.10).

Let us denote the measured transmission parameters of the DUT plus two error boxes by $[T_m]$, that of the device alone by $[T_d]$, and those of the two error boxes by $[T_a]$ and $[T_b]$. (Transmission parameters are also known as cascade parameters or the *ABCD* parameters.) It follows that

$$[T_m] = [T_a][T_d][T_b]. \tag{8.9}$$

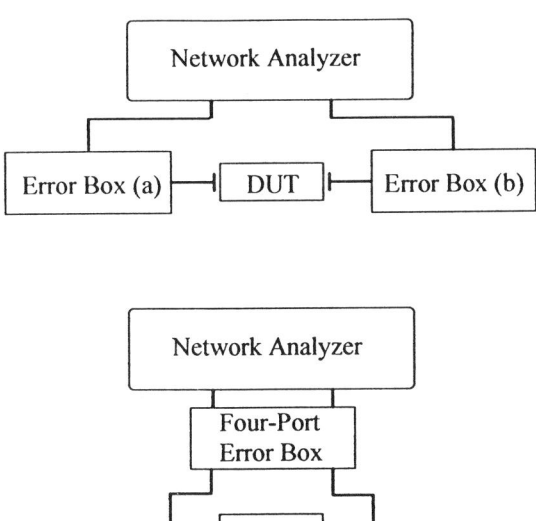

FIGURE 8.10 Calibration process assigns all system errors to two-port error boxes (a) and (b) or to a four-port error box.

In order to determine $[T_a]$ and $[T_b]$ and remove them from the measurement, it is possible to substitute the parameters of known calibration standards for $[T_d]$. The resulting three or more equations are solved by matrix inversion for $[T_a]$ and $[T_b]$ matrices. These matrices are then used to correct the measured transmission parameters:

$$[T_{\text{correct}}] = [T_a]^{-1}[T_m][T_b]^{-1}. \tag{8.10}$$

The corrected T matrix is then converted to the desired S parameters. This process is not unique, and different calibration routines can be designed. Some of the more commonly used calibration methods are the "through-short-open" calibration, the "through-short-delay" (TSD) calibration, and the "through-reflect-line" (TRL) calibration. All calibration routines are not equivalent despite the fact that mathematically they are similar. The differences lie in the accuracy of the required standards and to their susceptibility to slight errors in parameters of the standards. For example, it is generally accepted that the most accurate standard for use in calibrations is a section of transmission line. It is also one that is most often available, regardless of the nature of the measurement port. Therefore attempts have been made to make heavy use of various lengths of transmission lines in order to enhance the accuracy of the calibration. On the other hand, a standard open-circuit is one that cannot be shielded effectively in some cases, and the capacitance associated with it is prone to change by changes in its environment. Many calibration methods avoid the standard-open altogether. The TRL calibration is particularly attractive in that instead of requiring either a short or an open circuit it relies on a strong reflection.

The calibration methods discussed above do not compensate for channel cross talk in the network analyzer. Errors associated with poor isolation between the input and output channels are normally compensated separately and often by simple subtraction. More general calibration formulations are available that do compensate for leakage errors [11]. If this feature is desired, the two error boxes above should be replaced with a single four-port error box that connects the two-port network analyzer to the two-port DUT. The determination of the parameters of the four-port error box calibrates all linear error sources including cross talk. Such general solutions have not gained widespread acceptance due to their added complexity.

8.3.2 Fixture Deimbedding

In many practical measurement situations the need arises to perform the calibration at a terminal plane that is different from that of the device to be characterized. An example is when the device is placed in a test fixture, where it is connected to printed circuit transmission lines. In such cases, the input and output sections of the test fixture can be treated as "black boxes" to be characterized in the same way as in system calibration. Here, one option is combining fixture parameters with system parameters in a single calibration step at the device terminals. The other option is to calibrate the system first and then remove fixture parameters with a deimbedding procedure that, in a sense, is a second calibration. One advantage of the latter approach is that it makes it easier to pinpoint potential sources of problems, particularly errors due to lack of reproducibility in the assembly of the test fixture.

The mathematics of calibration and deimbedding are identical and fall under the category of matrix renormalization. The renormalization of a device's parameter matrix corresponds physically to changing the impedances presented to the device from the standard 50 Ω to arbitrary values depending on the inserted black boxes (Section 4.3.3). For more detailed analyses of parameter renormalization refer to the literature [12, 13].

8.4 SIX-PORT NETWORK ANALYZER

There are two basic approaches to error reduction in network analysis. One is to minimize the physical sources of error by using precision in the fabrication of the test equipment. The other is to numerically account for all sources of error and to remove them by calibration. The first approach leads to complexity in the hardware, the second approach to complexity in the software. Recent advances in network analyzers have utilized both.

The fact that it is possible to numerically compensate for the imperfections of the test equipment leads to enticing questions: Can the stringent requirements imposed on the hardware be completely removed? Is it possible to determine various magnitudes and phases of transmitted and reflected signals by heavily relying on the software? The six-port network analyzer (SPNA) was developed as a response to such questions [14]. The SPNA is a multiport interferometer that can be used to deduce both magnitude and phase of network parameters, while the hardware only needs to measure amplitude.

The basic arrangement of a six-port reflectometer consists of a microwave junction. One port of the junction is connected to the DUT. A second port brings the microwave signal in, and at the other terminals, vector sums of various proportions of input and reflected or transmitted signals are detected. With four independent amplitude detections in this manner it is possible, in principle, to determine the vector reflection coefficient. A microwave junction with four measurement ports, one port for the input signal, and one port for the DUT has a total of six ports. Thus the name *six-port network analyzer*. The advantages of SPNAs are simplicity of the hardware and high accuracy without relying on frequency precision of the signal generator. The main disadvantage is the lack of a well-developed calibration and operation software.

Consider a one-port device. The reflection coefficient ($\Gamma = S_{11}$) of the device is the ratio between the reflected and the incident amplitudes. A scalar network analyzer samples the two signals using directional couplers, thereby determining the magnitude of Γ. From a different perspective, in the complex plane, the scalar network analyzer gives enough information to confine the value of Γ to a circle centered at the origin. The following discussion will show that, in general, each scalar measurement using a multiport junction confines the reflection coefficient to a circle for which the location of the center depends on the junction. Two such measurements will give two intersecting circles yielding two solutions for Γ. A third measurement removes the ambiguity between the two solutions and identifies the correct solution, which is a unique point in the complex plane.

A schematic diagram of the six-port reflectometer is shown in Figure 8.11. The six-port junction is shown as a black box for the time being. Reflectionless power meters are attached to ports 1–4. The signal source is connected to port 5, and port 6 is the measurement port. The scattering matrix of the junction gives the relationship between ingoing and outcoming wave amplitudes, $[b] = [S][a]$. If the junction is assumed to be matched, the S matrix will be a symmetrical matrix with zero diagonal elements. Under the measurement conditions the scattering relationship is explicitly written as

$$\begin{bmatrix} b_1 \\ b_2 \\ b_3 \\ b_4 \\ b_5 \\ b_6 \end{bmatrix} = \begin{bmatrix} 0 & & & & & \\ & 0 & & S_{ij} & & \\ & & 0 & & & \\ & & & 0 & & \\ & S_{ij} & & & 0 & \\ & & & & & 0 \end{bmatrix} \begin{bmatrix} 0 \\ 0 \\ 0 \\ 0 \\ a_5 \\ a_6 \end{bmatrix}. \qquad (8.11)$$

Due to the four zero terms of the $[a]$ vector, the first four columns of the S matrix may be eliminated. Also, make the extra assumption that port 4 monitors the input signal only, such that the only nonvanishing matrix element involving port 4 is S_{45}. In that case,

$$\begin{bmatrix} b_1 \\ b_2 \\ b_3 \\ b_4 \\ b_5 \\ b_6 \end{bmatrix} = \begin{bmatrix} S_{15} & S_{16} \\ S_{25} & S_{26} \\ S_{35} & S_{36} \\ S_{45} & 0 \\ 0 & S_{56} \\ S_{56} & 0 \end{bmatrix} \begin{bmatrix} a_5 \\ a_6 \end{bmatrix}. \qquad (8.12)$$

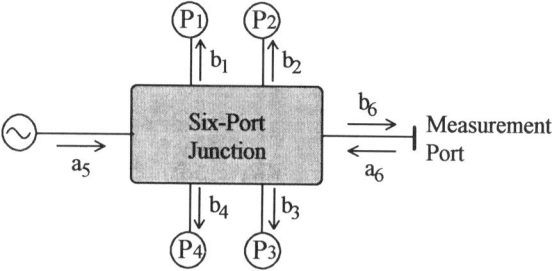

FIGURE 8.11 Schematic diagram of a six-port reflectometer.

The reflection coefficient Γ of the DUT is given by

$$a_6 = \Gamma b_6 = \Gamma S_{56} a_5 . \tag{8.13}$$

Substituting this for a_6 in Eq. (8.12), we can solve for the normalized powers:

$$\frac{|b_j|^2}{|b_4|^2} \frac{|S_{45}|^2}{|S_{56}S_{j6}|^2} = \left|\Gamma + \frac{S_{j5}}{S_{56}S_{j6}}\right|^2 . \tag{8.14}$$

The left-hand side of the equation is proportional to power measured at the jth port divided by power measured at the reference port 4. Any given value of this ratio confines Γ to a circle in the complex plane centered at $S_{j5}/(S_{56}S_{j6})$. This is a circle of constant power. If two such measurements are made, two intersecting circles are obtained, confining the reflection coefficient to two complex values. The ambiguity of having two solutions for Γ is removed by adding one more circle to the picture or making a third power measurement.

The sensitivity of the SPNA is a strong function of the location of the circles of constant power. For example, it should be apparent that, in the extreme case, it is not desirable to have two circles coincide. The locations of the centers of constant power circles are referred to as *q points* and are determined by the structure of the six-port. The circle associated with each power meter port is the locus of all reflection coefficients Γ that produce a constant power reading at that port. The radius of the circle is proportional to the measured power. The center of the circle is obtained by letting the measured power become zero. In Eq. (8.14) this corresponds to replacing the left-hand side by zero:

$$q_j = -\frac{S_{j5}}{S_{56}S_{j6}} . \tag{8.15}$$

It has been shown that for measurement of S parameters that fall inside the Smith chart (all passive components), the optimum locus of the q points is a circle of radius 2.0 and their optimum phase separation is 120° [15]. This arrangement is shown in Figure 8.12. A six-port whose q-point locations follow this recipe is called an *ideal six-port*.

Example Let us show that the combination of a matched symmetrical five-port junction and a directional coupler forms an ideal six-port (Figure 8.13).

Solution It is shown in Appendix D that a matched five-port with fivefold rotational symmetry is an equal four-way power divider with phase angles that are separated by 120°. Keeping this in mind and referring to Eq. (8.15) and Figure 8.13,

$$|S_{j5}| = |S_{j6}| = |S_{56}| = 0.5 .$$

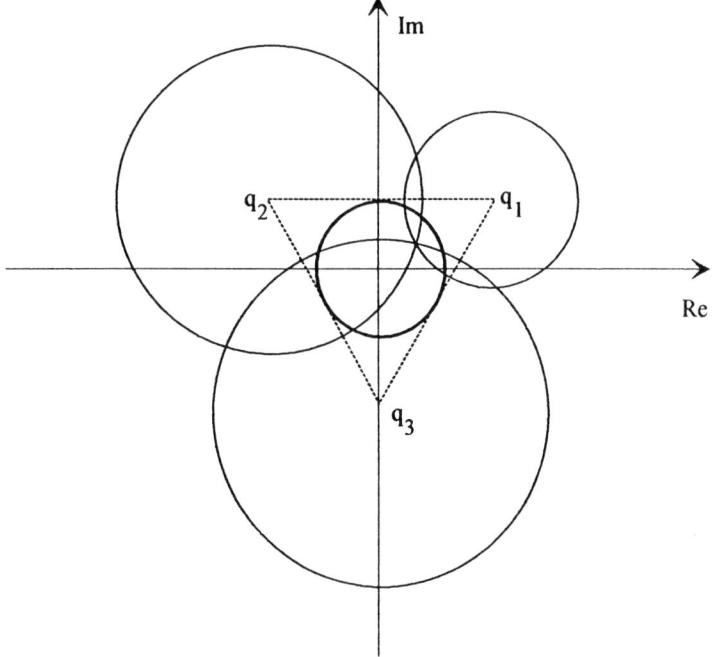

FIGURE 8.12 The q-point locations of ideal six-port and determination of complex reflection coefficient.

It immediately follows that the magnitude of q_j is equal to 2.0, as conjectured.

In order to calculate the phases of the q points, the absolute phases of the S parameters are not needed. Let us take S_{56} as the reference parameter and assign zero phase to it. The other parameters are found to be

$$S_{25} = S_{16} = S_{56} = 0.5,$$

and

$$S_{15} = S_{35} = S_{26} = S_{36} = 0.5 \angle 120.$$

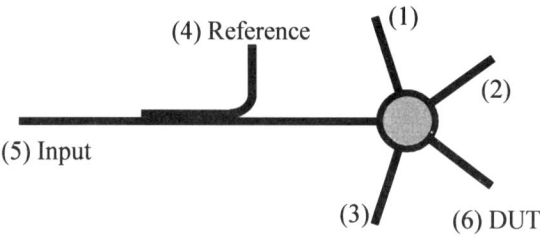

FIGURE 8.13 Ideal six-port formed by matched symmetrical five-port junction and directional coupler.

Substituting these back into Eq. (8.5), we get

$$q_1 = 2.0\angle 120, \qquad q_2 = 2.0\angle -120, \qquad q_3 = 2.0\angle 0,$$

establishing that the given combination of hardware constitutes an ideal six-port [16, 17]. ∎

8.4.1 Calibration

The complex reflection coefficient Γ of the DUT may be written in the following simple form [15]:

$$\Gamma = \frac{\sum_{i=1}^{4}(\alpha_i + j\beta_i)P_i}{\sum_{i=1}^{4}\gamma_i P_i}, \qquad (8.16)$$

where P_i is the power measured at the ith port. Once α, β, and γ are determined, the system is calibrated. Assume that the powers are normalized with respect to the power at port 4 and that $\gamma_4 P_4 = 1$. (This is equivalent to dividing through by $\gamma_4 P_4$.) The resulting equation is

$$\Gamma = \frac{(\alpha_4 + j\beta_4) + \sum_{i=1}^{3}(\alpha_i + j\beta_i)P_i}{1 + \sum_{i=1}^{3}\gamma_i P_i}. \qquad (8.17)$$

If we define P_{0i} as the power measured at port i when the DUT is a matched load, the equation above simplifies further to

$$\Gamma = \frac{\sum_{i=1}^{3}(\alpha_i + j\beta_i)(P_i - P_{0i})}{1 + \sum_{i=1}^{3}\gamma_i P_i}. \qquad (8.18)$$

The remaining nine unknowns are to be determined by solving the multiple equations that result from the use of other calibration standards. Since we have already used the matched load, the standards to be selected could be various offset shorts. It has been shown, however, that, at least with this formulation of the problem, in order to successfully extract the nine unknowns, there is a need for a standard mismatch, that is, a calibration standard whose reflection magnitude is not too close to either zero or unity [18]. The need for a standard mismatch is inconvenient at best. With some transmission lines it is difficult to fabricate an accurate standard mismatch based on geometry alone. This means that the calibration standard itself needs to be calibrated, which ties the accuracy of the six-port to that of the calibrating instrument. The dual SPNA avoids this problem.

8.4.2 Dual Six-Port Network Analyzer

Up to here we have considered only six-port reflectometers. In order to measure the full set of S parameters of a two-port device, two such reflectometers are needed. The interferometer nature of the six-port requires that to mea-

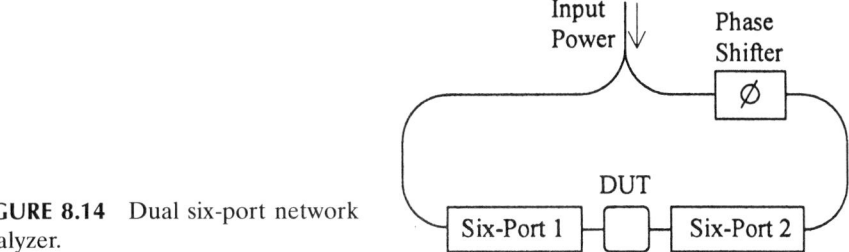

FIGURE 8.14 Dual six-port network analyzer.

sure any of the S parameters of the DUT two counterpropagating waves need to be present in each six-port. Therefore in dual six-port measurements both six-ports are simultaneously excited. The phase difference between the feeding arms is an added degree of freedom that is utilized in calibrating the dual six-port (Figure 8.14). There are two general approaches to the calibration of the dual six-port. One is to calibrate each reflectometer separately followed by a dual calibration [19]. The other is to calibrate both arms simultaneously [20]. For simultaneous calibration, the desirable TRL method may be used, which employs very reliable standards.

In conclusion, the six-port network analyzer has shown the feasibility of reducing hardware complexity in network analyzers and, instead, to rely more heavily on the software. As was seen in the above discussion, there is no step in the operation of the six-port that requires extreme frequency accuracy. The consequence is that the six-port can forego the use of synthesized signal sources. The six-port network analyzer does have a few stringent requirements that the hardware should meet. One set of requirements is the linearity and the dynamic range of the power detection. In high precision measurements, highly linear thermistor power detectors are utilized that exhibit exceptional linearity over a broad dynamic range. Diode detectors are more compact and easier to use at high frequencies but limit the dynamic range of the measurement.

8.5 OPTICAL PULSEWIDTH MEASUREMENT

The shape and width of a repetitive optical pulse can be measured by high speed sampling oscilloscopes. But even though the available bandwidth of sampling oscilloscopes is relatively high (exceeding 40 GHz at the present time) and improving, their speeds have and will continue to lag far behind what is required to measure the width of the shortest optical pulses. It has therefore been common practice to use the optical pulse itself for its width characterization. One of the simplest instruments that performs this task is the *autocorrelator*. The principle of operation of the autocorrelator is that a beam of light consisting of a train of optical pulses is split into two beams.

One beam undergoes a variable delay τ through a mechanically adjustable optical path length. The two beams are then combined in a nonlinear crystal for second-harmonic generation. The power in the second-harmonic beam is proportional to the degree of overlap between the two pulse trains. When this power is plotted against the mechanically induced delay, the result is the autocorrelation function of the optical pulse shape.

If the optical pulse shape is denoted by the intensity profile $f(t)$, the instantaneous second-harmonic signal is given by

$$A(t, \tau) \propto [f(t) + f(t - \tau)]^2 = f^2(t) + f^2(t - \tau) + 2f(t)f(t - \tau). \quad (8.19)$$

The integration time of the photodetector is much larger than either the duration of the pulse or the repetition period of the pulse train. The second-harmonic signal is therefore a function of only the variable delay τ:

$$A(\tau) \propto \int_{-\infty}^{\infty} f^2(t)\, dt + \int_{-\infty}^{\infty} f^2(t - \tau)\, dt + 2\int_{-\infty}^{\infty} f(t)f(t - \tau)\, dt. \quad (8.20)$$

The first two terms are due to the doubling of each half of the beam, and the third term is the desired autocorrelation signal. The first two terms are independent of time delay and represent a dc background. The schematic diagram of a popular type of autocorrelator known as a noncolinear beam Michelson interferometer type is shown in Figure 8.15. The autocorrelation signal and the two dc components are separated spatially, and the aperture in front of the detector blocks the dc components.

In most autocorrelators that display the signal in real time, the variable optical path length is obtained mechanically by rotating mirrors or rotating prisms. The concept of rotating mirrors is shown in Figure 8.16. It can be shown that for a small rotation angle θ, the change in the optical path length Δl is given by

$$\Delta l = 4\pi \theta D, \quad (8.21)$$

where D is the separation between mirrors. For large rotation angles, the optical delay does not remain proportional to the angle, and some nonlinearity is introduced. For a given mirror separation, there is a trade-off between this nonlinearity and the total delay range that is achieved. This trade-off is determined by the ratio of the mirror size to mirror separation [21].

Autocorrelation signal is symmetrical by nature, and its width is called the *autocorrelation width* of the pulse. There is a well-defined ratio between autocorrelation width and the actual width of the pulse that depends on pulse shape. For a Gaussian pulse this ratio is 1.41, for a single-sided exponential it is 2.0, and for a hyperbolic secant-square pulse it is 1.54 [22].

It is important to distinguish between the intensity autocorrelation that measures the pulsewidth and the amplitude autocorrelation that is the measure of the coherence length. Since the temporal coherence of light is defined

264 MEASUREMENT TECHNIQUES

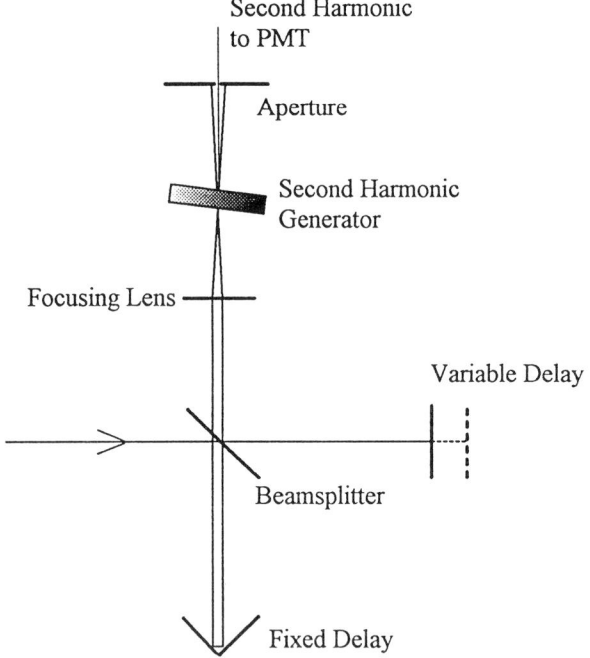

FIGURE 8.15 Noncolinear beam Michelson interferometer type autocorrelator.

as the amplitude autocorrelation width, a short coherence length CW beam will generate an autocorrelation signal with a finite width. For differential delays beyond the coherence length, the autocorrelation signal falls to the level of intensity autocorrelation, which is expected to be lower by a factor of 4, ($|2E^2|^2$ vs. $|2E|^4$). If the coherence length of an optical pulse is shorter than its pulsewidth (e.g., due to noise), the intensity autocorrelation profile

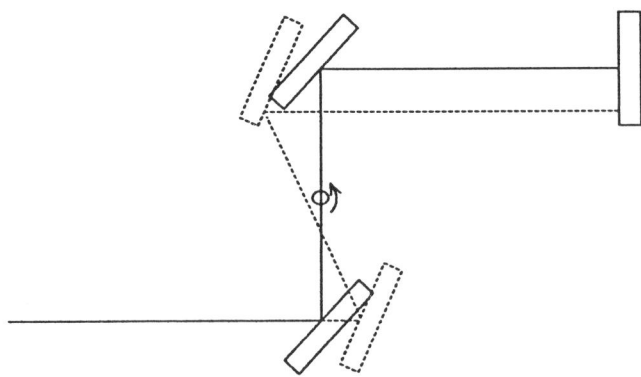

FIGURE 8.16 Rotating-mirror variable delay line utilized in some autocorrelators.

will have a central peak known as the *"coherence spike."* This is often a cause for confusion in determining the autocorrelation width of the pulse.

The autocorrelation techniques involving mechanical delay lines rely on the repetitive nature of the optical pulses to generate the autocorrelation signal. Each pulse in the pulse train gives rise to one point of the output signal. In fact, statistical variations in the pulsewidth are the cause of exponential tails in the autocorrelation signal [23]. If the variable time delay is introduced across the physical profile of the laser beam instead of sequentially from pulse to pulse, the autocorrelation width of a single pulse can be measured.

8.6 OPTICAL PULSE SHAPE CHARACTERIZATION

Correlations of higher order could, in principle, provide all the information needed to deduce the optical pulse shape. In practice, due to the required high power optical beam and the needed large bandwidth of the nonlinear process, higher order correlations are not realistic. Other proposed techniques for pulse shape characterization include convolution (involving time-reversed reflection) [24], interferometric autocorrelation [25], asymmetric correlation, and the use of the time lens.

Asymmetric correlation is identical to the intensity autocorrelation explained above, except that in one arm a dispersive medium such as glass is used to broaden the pulse. The cross correlation of the narrow and the broadened pulses gives information about the pulse shape. Some distortion of the pulse shape by the glass plate is expected.

Finally, the *time lens* method attempts to make a broadened replica of the pulse shape to be measured, such that high speed photodetectors and electronic sampling oscilloscopes can respond to its shape. The broadening of the pulse is done by a modified fiber grating pulse compressor. The compression process in a fiber grating compressor gives an output intensity whose profile is the Fourier transform of the input intensity. This is the time analog of focusing with a lens. If the compressor is modified such that there are dispersive delay lines on both sides of the optical fiber, the output intensity will

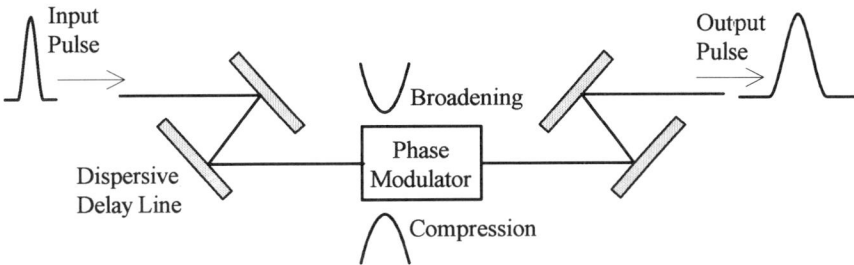

FIGURE 8.17 Configuration of time lens for distortionless compression and expansion of intensity profile.

be a compressed replica of the input pulse shape [26]. This is the time analog of imaging with a lens. Figure 8.17 shows the schematic diagram of the time lens arrangement. In this diagram the optical fiber has been replaced with an active modulator that places a linear chirp on the pulse by quadratic phase modulation. The use of the active modulator allows either compression or distortionless broadening of the pulse depending on the sign of the quadratic modulation used. Compression and broadening of picosecond pulses by factors as high as 10–20 are currently achievable by the time lens [27].

PROBLEMS

8.1 In the electro-optic sampling of a circuit on GaAs determine the optical intensity modulation obtained due to a sinusoidal signal of amplitude $V_0 = 10$ V. Assume $r_{41} = 1.4$ pm/V.

8.2 In the electro-optic sampling of a microstrip circuit, compare the magnitude of the etalon effect to the strength of the main signal.

8.3 Design a six-port reflectometer using a combination of 3- and 6-dB directional couplers. Choose the combination with q-point locations closest to an ideal SPNA.

8.4 Derive an expression as a function of the rotation angle for the optical delay in autocorrelators utilizing rotating mirrors. Determine the ratio of mirror size to mirror separation that gives a maximum nonlinearity of 10%.

8.5 Derive the ratio between the full width at half-maximum of each of the following pulse shapes to its autocorrelation width: (1) Gaussian, (2) single-sided exponential, and (3) hyperbolic secant square.

REFERENCES

[1] Cascade Microtech, Beaverton, OR.
[2] R. Majidi-Ahy, M. Shakouri, and D. Bloom, "100 GHz Active Electronic Probe for On-Wafer S-Parameter Measurements," *Electron. Lett.*, Vol. 25, No. 13, pp. 828–830, 1989.
[3] M. Rodwell, M. Riaziat, K. Weingarten, B. Auld, and D. Bloom, "Internal Microwave Propagation and Distortion Characteristics of Travelling-Wave Amplifiers Studied by Electro-Optic Sampling," *IEEE Trans. MTT*, Vol. 34, No. 12, Dec. 1986.
[4] K. Weingarten, M. Rodwell, and D. Bloom, "Picosecond Optical Sampling of GaAs Integrated Circuits," *IEEE J. Quant. Electron.*, Vol. 24, No. 2, pp. 198–220, Feb. 1988.
[5] M. Riaziat, K. Weingarten, L. Generali, D. Gerstenberger, A. Drobshoff, and L.

Ching, "All Optical Characterization of MMICs," *Proc. IEEE GaAs IC Symp.*, pp. 347–351, Oct. 1990.

[6] K. Williams et al., "6–34 GHz Offset Phase Locking of Nd:YAG 1319 nm Nonplarar Ring Lasers," *Electron. Lett.*, Vol. 25, pp. 1242–1243, 1989.

[7] J. A. Vladimis, "1 THz Bandwidth Prober for High-Speed Devices and Integrated Circuits," *Electron. Lett.*, Vol. 23, pp. 1308–1310, Nov. 1987.

[8] T. Nagatsuma et al., "Subpicosecond Sampling Using a Non-Contact Electro-Optic Probe," *J. Appl. Phys.*, Vol. 66, No. 9, pp. 4001–4009, Nov. 1989.

[9] N. Paulter, "High-Speed Optoelectronic Pulse Generation and Sampling System," *IEEE Trans. Instrum. Meas.*, Vol. 37, No. 3, pp. 449–453, Sept. 1988.

[10] D. Auston, "Impulse Response of Photoconductors in Transmission Lines," *IEEE J. Quant. Electron.*, Vol. 19, pp. 639–648, 1983.

[11] R. Speciale, "A Generalization of the TSD Network Analyzer Calibration Procedure Covering N-Port Scattering Parameter Measurements, Affected by Leakage Errors," *IEEE Trans. Microwave Theory Tech.*, Vol. 25, No. 12, pp. 1100–1115, Dec. 1977.

[12] D. Woods, "Multiport Network Analysis by Matrix Renormalization Employing Voltage Wave Parameters with Complex Renormalization," *Proc. IEE*, Vol. 124, No. 3, pp. 198–204, March 1977.

[13] R. Speciale, "Derivation of the Generalized Scattering Parameter Renormalization Transformation," *IEEE Int. Microwave Symp. Dig.*, pp. 166–169, April 1980.

[14] C. Hoer "The Six-Port Coupler: A New Approach to Measuring Voltage, Current, Power, Impedance, and Phase," *IEEE Trans. Instrum. Meas.*, Vol. 21, No. 4, pp. 466–470, Nov. 1972.

[15] G. Engen, "The Six-Port Reflectrometer: An Alternative Network Analyzer," *IEEE Trans. Microwave Theory Tech.*, Vol. 25, No. 12, pp. 1075–1079, Dec. 1977.

[16] E. R. Hansson and G. P. Riblet, "An Ideal Six-Port Network Consisting of a Matched Reciprocal Five-Port and a Perfect Directional Coupler," *IEEE Trans. Microwave Theory Tech.*, Vol. MTT-31, No. 3, pp. 284–288, March 1983.

[17] M. Riaziat and G. Zdasiuk, "Waveguide Star Junction Used in Kaband Dual Six Port Measurements," *1985 IEEE MTT-S Int. Microwave Symp. Dig.*, pp. 593–594, June 1985.

[18] D. Woods, "Analysis and Calibration Theory of the General 6-Port Reflectometer Employing Four Amplitude Detectors," *Proc. IEE*, Vol. 126, No. 2, pp. 221–228, 1979.

[19] D. Woods, "Analysis of a Dual Six-Port Network Analyzer and Derivation of Calibration Equations," *IEE Proc.*, Vol. 127, No. 8, pp. 541–548, Nov. 1980.

[20] G. Engen and C. Hoer, "Through-Reflect-Line: An Improved Technique for Calibrating the Dual Six-Port Automatic Network Analyzer," *IEEE Trans. Microwave Theory Tech.*, Vol. 27, No. 12, pp. 987–993, Dec. 1979.

[21] Z. A. Yasa and N. M. Amer, "A Rapid Scanning Autocorrelation Scheme for Continuous Monitoring of Picosecond Laser Pulses," *Opt. Commun.*, Vol. 36, No. 5, pp. 406–408, 1981.

[22] J. C. Diels et al., "Analysis of a Mode-Locked Ring Laser: Chirped Solitary Pulse Solutions," *J. Opt. Soc. Am. B,* No. 2, pp. 680–686, 1985.

[23] E. Van Stryland, "The Effect of Pulse to Pulse Variation on Ultrashort Pulsewidth Measurements," *Opt. Commun.*, Vol. 31, pp. 93–94, 1979.

[24] S. Montgomery, D. Pederson, and G. Salamo, "Intensity Profiles of Short Optical Pulses via Temporaly Reversed Pulses," *Appl. Phys. Lett.*, Vol. 49, pp. 620–621, 1986.

[25] J. Diels et al., "Control and Measurement of Ultrashort Pulse Shapes (in Amplitude and Phase) with Femtosecond Accuracy," *Appl. Opt.*, Vol. 24, No. 9, pp. 1270–1282, May 1985.

[26] B. Kolner and M. Nazarathy, "Temporal Imaging with a Time Lens," *Opt. Lett.*, Vol. 14, No. 12, pp. 630–632, June 1989.

[27] A. Godil, "Harmonic Mode Locking of Diode Pumped Lasers and Time Lens with Picosecond Resolution," Ph.D. Dissertation, Stanford University, Stanford, CA, Dec. 1992.

APPENDIX A

Waveforms and Spectra

A signal waveform $V(t)$ and its spectrum $v(f)$ are related through a Fourier transform pair that may be defined as follows:

$$v(f) = \int_{-\infty}^{\infty} V(t) e^{-j2\pi ft}\, dt,$$
$$V(t) = \int_{-\infty}^{\infty} v(f) e^{j2\pi ft}\, dt, \tag{A.1}$$

where $V(t)$ is a function of time and $v(f)$ is a function of frequency. Since the time waveform is always real, the spectrum is always hermitian (its real part is even and its imaginary part is odd).

Another way of thinking about this is as the limiting case of a Fourier series representation. Take a periodic signal $V(t)$ with a period T, where $V(t)$ is a real function of time and can be expanded in Fourier series as follows:

$$V(t) = \tfrac{1}{2} a_0 + \sum_{n=1}^{\infty} \left(a_n \cos \frac{2\pi n}{T} t + b_n \sin \frac{2\pi n}{T} t \right), \tag{A.2}$$

where

$$a_n = \frac{2}{T} \int_{-T/2}^{T/2} V(t) \cos \frac{2\pi n}{T} t\, dt,$$
$$b_n = \frac{2}{T} \int_{-T/2}^{T/2} V(t) \sin \frac{2\pi n}{T} t\, dt. \tag{A.3}$$

This is the familiar notion of a periodic signal whose spectrum consists of a fundamental frequency and its harmonics. In order to find the spectrum of a nonperiodic signal, the period T is allowed to become large. Mathematically, let $T \to \infty$, and define the new variable $f = n/T$. An increment of f is

$\delta f = 1/T$, and $a_n \to 2a(f)\delta f$, $b_n \to 2b(f)\delta f$. Summation (1.2) reduces to an integral:

$$V(t) = \int_0^\infty 2[a(f)\cos(2\pi ft) + b(f)\sin(2\pi ft)]\,df, \qquad (A.4)$$

and

$$a(f) = \int_{-\infty}^\infty V(t)\cos(2\pi ft)\,dt,$$

$$b(f) = \int_{-\infty}^\infty V(t)\sin(2\pi ft)\,dt. \qquad (A.5)$$

It should be noted that $a(f)$ is an even function of f, while $b(f)$ is odd. With this observation and by expanding the trigonometric functions of (1.4) in their exponential form, we obtain

$$V(t) = \int_{-\infty}^\infty [a(f) - jb(f)]e^{j2\pi ft}\,df. \qquad (A.6)$$

But

$$a(f) - jb(f) = \int_{-\infty}^\infty V(t)e^{-j2\pi ft}\,dt, \qquad (A.7)$$

which is the definition of $v(f)$. This results in the Fourier transform pair given in (1.1), and the hermitian nature of $v(f)$ was explicitly shown.

A.1 TIME–BANDWIDTH PRODUCT

In electrical systems one often encounters the need to know the system bandwidth required for the propagation of a pulse of a given duration. A similar question arises when one compares the response of a photodetector to an optical pulse versus the response of the same photodetector to modulated light. These quantities are of course related through a Fourier transformation. However, if pulses are assumed to have specific shapes, these questions can be answered using simple time–bandwith product numbers.

Consider the Gaussian transform pair $V(t) = \exp(-\pi t^2)$ and $v(f) = \exp(-\pi f^2)$. In the time domain if $V(t)$ is an intensity profile, the width of the pulse is defined as full width at half-maximum (FWHM), which is found by noting that half-maximum occurs at $t = \sqrt{(\ln 2/\pi)}$. Therefore,

$$(\Delta t)_{\text{FWHM}} = 2\sqrt{\frac{\ln 2}{\pi}}.$$

In the frequency domain, the bandwidth is half of the Gaussian transform width due to the fact that negative frequencies are not considered. It follows that for a pulse with a Gaussian *intensity* profile,

$$(\Delta t)(\Delta f) = 2\frac{\ln 2}{\pi} = 0.441 \, .$$

This product is independent of the scaling of the waveform. A broadening of the time signal leads to a proportional shrinking of the frequency spectrum and vice versa, keeping the product a constant. The time–bandwidth product is a function of the shape of the waveform as well as the quantity used to characterize its width. (Autocorrelation width and standard deviation are examples of other width definitions.)

In physical experiments where a transducer response needs to be taken into account, the time–bandwidth product can be modified to reflect the presence of the transducer. For example, when a light pulse is used to excite a photodetector, normally the output current is proportional to the intensity of the incident light. Since bandwidth in an electrical circuit is defined as dc to the 3-dB or half-power point but not the half-current point, an extra $\sqrt{2}$ factor appears in the time–bandwidth product of the detected optical pulse:

$$(\Delta t)(\Delta f) = \frac{0.441}{\sqrt{2}} = 0.312 \, .$$

For example, a 10-ps optical intensity pulse incident on a fast photodetector generates an electrical pulse that requires a transmission line with a bandwidth in excess of 31 GHz in order to propagate without distortion.

APPENDIX B

Noise

Problems with noise are most often encountered in receiving systems, where noise is the major impediment to the detection of weak signals. In this Appendix the definitions of useful terms and derivations of some relationships regarding noise characteristics of devices and systems are given.

B.1 TYPES OF NOISE

Noise, in general, consists of spontaneous electrical or optical fluctuations that compete with the desired signal in detection systems. There are two principal types of such fluctuations: *thermal noise* and *shot noise*. Thermal noise is caused by the random motion of conduction electrons due to their finite temperature. This kind of noise prevails in most high speed electronic components and systems. Shot noise, on the other hand, arises from the quantized nature of electrical current or light. Shot noise is important in semiconductor junctions and is often the dominant noise in optical transmission/detection systems. Other types of noise such as flicker or $1/f$ noise that are of interest at lower frequencies are not discussed here.

Assumptions that are normally valid in dealing with the characteristics of both thermal and shot noise are as follows:

1. Over the bandwidth of most detection systems, noise power is assumed to be independent of frequency. This is called *white noise*.
2. Noise amplitude varies randomly in time. The distribution of this variation is assumed to be normal or Gaussian, and thus the name *Gaussian noise*.

B.1.1 Thermal Noise

Thermal noise power N_{th} available from an electrical component in any frequency band is proportional to temperature:

$$N_{th} = kTB, \qquad (B.1)$$

where $k = 1.38 \times 10^{-23}$ J/K is Boltzmann's constant and B is the bandwidth. Here, N_{th} is the maximum available power, that is, the power dissipated in a matched load at the temperature of absolute zero (Figure B.1). In an electrical circuit, random electronic motion exists in capacitors and inductors as well as in resistors. However, the current–voltage phase relationship in reactive elements does not allow any power generation, in the same way that these elements cannot dissipate power. Therefore, it is only the resistor that can generate noise power. The noise power that appears as a voltage across a resistance R has an rms value given by

$$V = \sqrt{4kTBR}. \qquad (B.2)$$

Example The rms noise voltage at room temperature ($T = 300$) across a 50-Ω resistor when the detection bandwidth is 1.0 GHz is

$$V = \sqrt{4(1.38 \times 10^{-23})(300 \times 10^9)(50)} = 29 \, \mu V. \qquad \blacksquare$$

B.1.2 Shot Noise

Shot noise in electrical circuits arises from the quantized nature of charge transfer. It is particularly pronounced in semiconductor junctions where carriers cross the barrier independently and at random. If made audible, shot noise sounds like "a shower of pellets on a tin roof."

There is no temperature dependence with shot noise most commonly encountered. In electrical circuits shot noise is proportional to the average current I and the electronic charge e, while in optical beams it is proportional to the average optical power P and the photon energy $h\nu$. The rms shot noise current I_n is given by

$$I_n = \sqrt{2eIB}. \qquad (B.3)$$

The above equation assumes that the individual charge carriers do not interact. This assumption is violated in semiconductor devices under high injec-

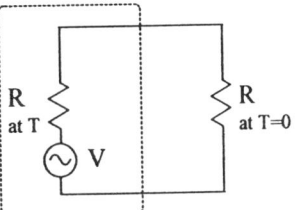

FIGURE B.1 Maximum noise power is delivered to a matched resistor.

274 NOISE

tion or when recombination becomes significant. The equation needs to be modified under such circumstances [12]. Also the expression is only valid at frequencies that satisfy $f \ll \tau$, where τ is a characteristic carrier transit time that depends on the particular device type.

The rms shot noise power in an optical beam is expressed in a similar form,

$$P_n = \sqrt{2h\nu PB}, \qquad (B.4)$$

where $B = \Delta\nu$ is the optical bandwidth.

Example Let us determine the current needed to generate the same rms shot noise power as thermal noise in a semiconductor diode with 50 Ω internal resistance. Assume room temperature operation.

Equalizing shot noise and thermal noise powers, we get

$$kTB = I_n^2 R = 2eIBR,$$

or

$$I = \frac{kT}{2eR} = \frac{(1.38 \times 10^{-23})(300)}{2(1.6 \times 10^{-19})(50)} = 0.26 \,\text{mA}. \qquad \blacksquare$$

B.2 NOISE FIGURE

Signal-to-noise ratio is a quantity that is well defined at any point in a system. It has its highest value at the signal input. Various components in the system degrade the signal-to-noise ratio depending on their internal characteristics and their placement in the system. The noise factor NF is defined for any component as the ratio of signal to noise between its input and output,

$$\text{NF} = \frac{S_i/N_i}{S_o/N_o}. \qquad (B.5)$$

The noise figure is the noise factor expressed in decibels:

$$F = 10\log(\text{NF}). \qquad (B.6)$$

Noise figure is a function of temperature and the input impedance presented to the device. Temperature is standardized to 290 K.

B.3 EFFECTIVE INPUT NOISE TEMPERATURE

The 290 K reference temperature for noise figure is convenient for most terrestrial systems. For other applications such as satellite receivers with

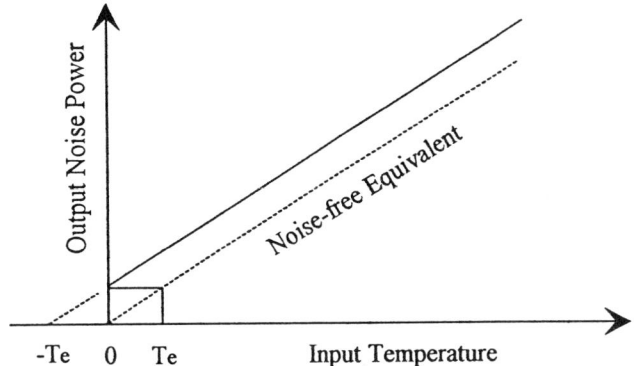

FIGURE B.2 Definition of effective input noise temperature.

antennas pointing at the cold sky or, in general, receivers with noise figures lower than 1.0 dB, another noise figure-of-merit is often used.

Consider the noise-free equivalent of the device of interest. Any noise power at the output of this device originates from the impedance at the input of the device. The *effective input noise temperature* T_e is defined as the temperature at which the input impedance has to be placed in order to generate the observed noise power at the output. This is seen more clearly as the temperature intercept of the output power versus input temperature line (Figure B.2). For a two-port transducer, T_e is related to the noise factor by the simple relationship

$$T_e = 290(\text{NF} - 1). \tag{B.7}$$

B.4 CASCADED NOISY NETWORKS

If a number of noisy networks, each with an available gain G_{an}, a noise factor NF_n, and an effective noise temperature T_{en} are cascaded, the noise factor of the combination is given by what is known as the *Friis's formula*. We simply state the formula here without proof (for more detail refer to ref. 1):

$$\text{NF} = \text{NF}_1 + \frac{\text{NF}_2 - 1}{G_{a1}} + \frac{\text{NF}_3 - 1}{G_{a1}G_{a2}} + \cdots . \tag{B.8}$$

Similarly, the effective input noise temperature of the combination is given by

$$T_e = T_{e1} + \frac{T_{e2}}{G_{a1}} + \frac{T_{e3}}{G_{a1}G_{a2}} + \cdots . \tag{B.9}$$

The above relations show that it is advantageous to start a cascade of noisy networks with a low noise amplifier. The noise contribution of the first stage is the most significant. The contribution of each consecutive network is reduced by the gain of the previous stages.

REFERENCES

[1] A. van der Ziel, *Noise in Solid State Devices and Circuits*, Wiley, New York, 1986.
[2] H. A. Haus and R. B. Adler, *Circuit Theory of Linear Noisy Networks*, Wiley, New York, 1959.

APPENDIX C

Scattering Parameters and Smith Chart

A linear network is defined as one whose output-to-input ratio is independent of the magnitude of the input. For most networks there is a limit to the validity of this constraint in terms of input power. In general, linearity is an approximation whose validity depends on the selected range of input and the accuracy of the measurement. The ratio between the output and the input is the linear network parameter. Linear network parameters are defined for a variety of inputs and outputs. Some of the most commonly used parameters are impedance, admittance, and reflection coefficient. For multiport networks, linear network parameters become elements of a matrix that relates input and output vectors. Examples of such matrices are impedance, admittance, hybrid, scattering, and cascade parameter matrices.

The electrical characteristics of the linear network is fully represented by any one of these linear matrices. For passive networks, most of the linear network matrices are convertible to other ones with simple transformation rules. For active networks, however, this is in general not the case. Particular measurement conditions are imposed on the network by the definition of the parameters. These measurement conditions affect the performance of the network and do not allow the conversion of some of the matrices to others. This point is clarified further in the following discussion by the examples of the impedance matrix that imposes open-circuit conditions on the circuit and the scattering matrix that imposes matched load conditions.

C.1 IMPEDANCE MATRIX

In a one-port network, the impedance is defined as the ratio between input voltage and current, through the relationship $V = ZI$. The impedance Z is a complex number. For multiport networks, the definition is generalized by defining the vector $[V]$ whose elements are voltages at the various ports and the vector $[I]$ representing the corresponding currents. These vectors are related through

$$[V] = [Z][I],\qquad(\text{C.1})$$

where $[Z]$ is an $n \times n$ matrix for an n-port network. The matrix element Z_{ij} relates the voltage at the ith port to the current at the jth port given that currents at all the other ports are zero. This is the so-called open-circuit condition imposed on the network when impedance parameters are measured.

Depending on the network at hand, all the elements of the impedance matrix may not be independent. The physical characteristics of the network impose symmetries and restrictions on matrix elements. One important class of networks is the *reciprocal* network, which includes all RLC combinations, and networks without any applied dc electric or magnetic field bias. For reciprocal networks the impedance matrix is symmetrical, that is, $Z_{ij} = Z_{ji}$. Another commonly encountered class is the class of lossless networks. For this class, all the elements of the impedance matrix are pure imaginary.

C.2 SCATTERING MATRIX

Elements of the scattering matrix are known as S parameters. For a one-port network, the S parameter is more commonly known as the reflection coefficient. The reflection coefficient is defined by $b = \Gamma a$, where a is the amplitude of the wave incident on the device and b is the amplitude of the reflected wave (Figure C.1). In multiport nomenclature, the same relationship is written as $b = S_{11}a$, or in general, $[b] = [S][a]$. Since incident and reflected wave amplitudes are measured by the same units, S parameters are dimensionless. The S parameters are applicable to both electrical and optical components.

The S_{ij} element of the scattering matrix is the ratio between the outgoing wave amplitude at the jth port to that of the incident wave at the ith port,

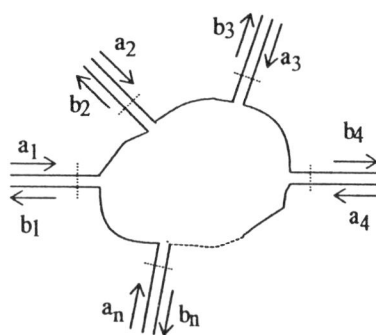

FIGURE C.1 Incident and scattered waves for a multiport network.

given that there are no other waves incident on any port. This means that all other ports should have matched terminations in order to avoid any signal from being reflected back. This condition, that during the measurement of S parameters all ports should have matched terminations, is very significant for the characterization of active devices. Some active devices such as transistors may become unstable when highly reflective loads are connected to their ports. Matched terminations minimize the possibility of unwanted oscillations. This is one of the reasons for the popularity of S parameters in high frequency measurements.

For unconditionally stable networks, S parameters can be transformed to Z parameters and vice versa. Starting with a one-port network with an impedance Z and a reflection coefficient Γ,

$$\Gamma = \frac{Z - Z_0}{Z + Z_0}, \qquad (C.2)$$

where Z_0 is the reference impedance or the impedance of the input transmission line. The same equation, when converted to matrix notation, is valid for multiport networks:

$$[S] = ([Z] - [Z_0])([Z] + [Z_0])^{-1}, \qquad (C.3)$$

where $[Z_0]$ is a diagonal matrix representing transmission line impedances at each port. Normally, a standard transmission line impedance such as 50 Ω is used at every port. The Z parameters may be normalized to this impedance, in which case $[Z_0]$ is replaced by the unity matrix $[I]$ in the above equation.

If the admittance parameters $[Y]$ are used instead of $[Z]$, the scattering matrix can be written as

$$[S] = ([Y_0] - [Y])([Y_0] + [Y])^{-1}. \qquad (C.4)$$

Symmetries of the scattering matrix are not unlike those of the impedance matrix. For instance, the scattering matrix of a reciprocal network is symmetrical, that is, $S_{ij} = S_{ji}$. For a lossless network the scattering matrix is unitary, meaning that its inverse is equal to the complex conjugate of its transpose. This can be written as

$$[S]^t[S]^* = [I]. \qquad (C.5)$$

Matrix symmetries can be used to deduce important characteristics of microwave junctions. This is illustrated in the example below as well as in Appendix D.

Example Let us show that a lossless, reciprocal three-port junction cannot be simultaneously matched at all ports.

Using matrix symmetries, the above statement is equivalent to claiming that a 3×3 symmetrical matrix with zero diagonal elements cannot be unitary. We can verify this by inspection:

$$\begin{bmatrix} 0 & S_{12} & S_{13} \\ S_{12} & 0 & S_{23} \\ S_{13} & S_{23} & 0 \end{bmatrix} \begin{bmatrix} 0 & S_{12} & S_{13} \\ S_{12} & 0 & S_{23} \\ S_{13} & S_{23} & 0 \end{bmatrix}^* = \begin{bmatrix} 1 & 0 & 0 \\ 0 & 1 & 0 \\ 0 & 0 & 1 \end{bmatrix}$$

This matrix equation has no solution.

If the reciprocity requirement is lifted, simultaneous matching can be achieved. An example is the microwave ferrite circulator that can be matched at all ports. ∎

C.3 SMITH CHART

The Smith chart gives a graphical transformation between the reflection coefficient and impedance. The complex reflection coefficient ($S = \rho e^{j\phi}$) is represented in polar coordinates. Superimposed on that are circles and lines of constant real and imaginary parts of normalized impedance ($Z_n = R_n + jX_n$). See Figure C.2.

The Smith chart provides a graphical tool for designing distributed matching networks. Consider a one-port network connected to a transmission line of impedance Z_0. At the terminal plane of the network the reflection coefficient is represented by a point on the Smith chart. At any other location along the transmission line the magnitude of the reflection coefficient remains the same but the phase angle is different. As one moves away from the load along the transmission line, the locus of S_{11} forms a circular path centered at the origin. Adding a series reactance to the transmission line is represented by moving away from the circle along a constant-resistance line. The process

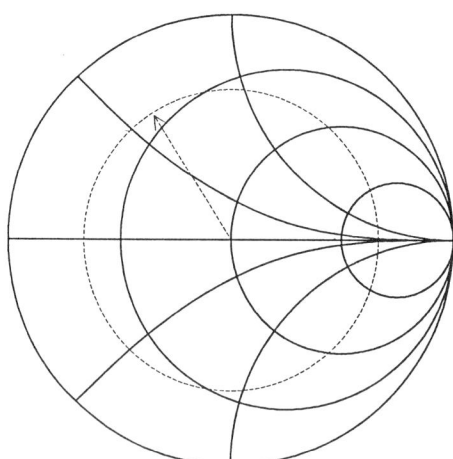

FIGURE C.2 Smith chart is a polar representation of complex S parameters. Solid circles and lines represent normalized real and imaginary parts of impedance.

of matching using the Smith chart begins by identifying the impedance of the device on the chart. Adding any section of transmission line as well as any lumped element is represented by moving along any one of the circles or lines listed above. The object of the graphical matching exercise is to end up at the origin ($S_{11} = 0$). See reference 1 for more detail.

Smith chart functions are presently automated by various circuit design software packages. In many of these packages the graphical aspect of impedance matching on the Smith chart have been preserved.

REFERENCE

[1] R. E. Collin, *Fundamentals for Microwave Engineering*, McGraw-Hill, New York, 1966.

APPENDIX D

Five-Port Symmetrical Junction

It was stated in Chapter 8 that a matched lossless reciprocal five-port can be used together with a perfect directional coupler to construct an ideal six-port network analyzer for microwave measurements. The electrical properties of the symmetrical five-port junctions that make the construction of the network analyzer possible are briefly described here.

Consider a lossless five-port junction such as the one shown in Figure D.1 with fivefold rotational symmetry. The scattering parameters of this junction form a 5×5 matrix. Assuming that the junction has no nonreciprocal elements in it, the matrix is symmetrical. Also, the fivefold rotational symmetry of the junction reduces the number of independent S parameters to 3, namely S_{11}, S_{12}, and S_{13} ($S_{15} = S_{12}, S_{14} = S_{13}$). Finally, if all of the five ports of the junction are matched, the diagonal elements of the matrix are zero, that is, $S_{11} = 0$. Since the junction is assumed lossless, the condition of being unitary can be imposed on the matrix: $[S]^t[S]^* = [I]$. Explicitly,

$$\begin{bmatrix} 0 & S_{12} & S_{13} & S_{13} & S_{12} \\ S_{12} & 0 & S_{12} & S_{13} & S_{13} \\ S_{13} & S_{12} & 0 & S_{12} & S_{13} \\ S_{13} & S_{13} & S_{12} & 0 & S_{12} \\ S_{12} & S_{13} & S_{13} & S_{12} & 0 \end{bmatrix} \begin{bmatrix} 0 & S_{12} & S_{13} & S_{13} & S_{12} \\ S_{12} & 0 & S_{12} & S_{13} & S_{13} \\ S_{13} & S_{12} & 0 & S_{12} & S_{13} \\ S_{13} & S_{13} & S_{12} & 0 & S_{12} \\ S_{12} & S_{13} & S_{13} & S_{12} & 0 \end{bmatrix}^* = \begin{bmatrix} 1 & 0 & 0 & 0 & 0 \\ 0 & 1 & 0 & 0 & 0 \\ 0 & 0 & 1 & 0 & 0 \\ 0 & 0 & 0 & 1 & 0 \\ 0 & 0 & 0 & 0 & 1 \end{bmatrix}. \quad \text{(D.1)}$$

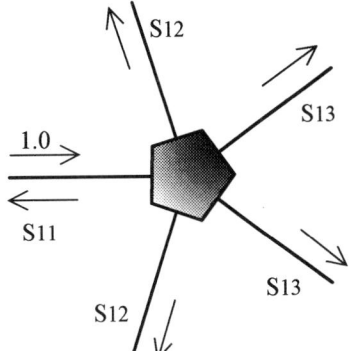

FIGURE D.1 Five-port junction with fivefold rotational symmetry.

This matrix equation gives rise to the following three independent equations,

$$\begin{aligned} 2|S_{12}|^2 + 2|S_{13}|^2 &= 1, \\ |S_{12}|^2 + S_{12}S_{13}^* + S_{12}^*S_{13} &= 0, \\ |S_{13}|^2 + S_{12}S_{13}^* + S_{12}^*S_{13} &= 0. \end{aligned} \quad \text{(D.2)}$$

It follows immediately that $|S_{12}| = |S_{13}| = 0.5$. In order to find the phase angles, the polar form is adopted: $S_{12} = 0.5e^{j\phi_2}$ and $S_{13} = 0.5e^{j\phi_3}$. Either the second or the third equation above can be used to obtain $\cos(\phi_2 - \phi_3) = -0.5$, or $(\phi_2 - \phi_3) = 120°$. In conclusion, a star junction with a fivefold rotational symmetry has three distinct S parameters. If such a lossless junction is completely matched, that is, $S_{11} = 0$, then $|S_{12}| = |S_{13}| = 0.5$, meaning that the junction is a four-way equal power divider and the phase angle between the adjacent port and the nonadjacent port transmission is 120°. These are the characteristics utilized in an ideal six-port network analyzer [4].

D.1 S-PARAMETER SENSITIVITY TO MATCHING OF STAR JUNCTION

In practice, the magnitude of S_{11} is never zero, and the properties of the five-port are less than ideal. The following analysis is aimed at quantifying this by finding the relationship between S_{12} and S_{13} when S_{11} is not zero.

Since a lossless junction has a unitary scattering matrix, the eigenvalues of the matrix all have magnitude 1. It is therefore convenient to express all the scattering parameters in terms of the eigenvalues. The complete eigenvalue problem will not be solved here, but it turns out that the symmetry of the junction reduces the number of the eigenvalues to 3. The three independent S parameters of the star junction are then written in terms of the three eigenvalues S_1, S_2, and S_3 as follows

$$S_{11} = \tfrac{1}{5}(S_1 + 2S_2 + 2S_3), \tag{D.3}$$

$$S_{12} = \tfrac{1}{5}[S_1 + 2S_2\cos(\tfrac{2}{5}\pi) - 2S_3\cos(\tfrac{1}{5}\pi)], \tag{D.4}$$

$$S_{13} = \tfrac{1}{5}[S_1 - 2S_2\cos(\tfrac{1}{5}\pi) + 2S_3\cos(\tfrac{2}{5}\pi)]. \tag{D.5}$$

For the derivation of these relationships see reference [3]. The phase angle of one of the eigenvalues can be set arbitrarily. This corresponds to the freedom in choosing the reference plane at each port. The choice we make here is

$$S_1 = -1, \qquad S_2 = e^{j\theta_2}, \qquad S_3 = e^{j\theta_3}. \tag{D.6}$$

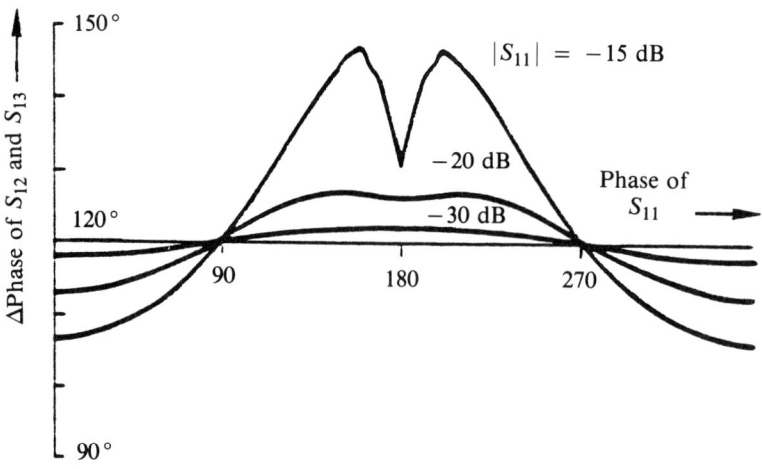

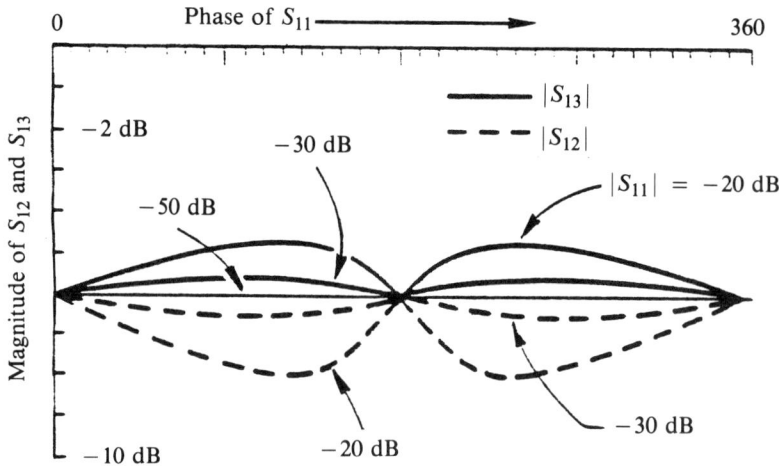

FIGURE D.2 Deviation of magnitude and phase of S_{12} from ideal as function of magnitude and phase of reflection coefficient.

Since the reference planes are now fixed, S_{11} has well-defined real and imaginary parts. Given real and imaginary parts of S_{11}, the values of θ_2 and θ_3 can be determined from the first equation. Written more explicitly, θ_2 is determined from the quadratic equation

$$\cos^2(\theta_2) - \tfrac{1}{2}[5\,\text{Re}(S_{11}) + 1]\cos(\theta_2) + \tfrac{1}{16}F(S_{11}) - \frac{25[\text{Im}(S_{11})]^2}{F(S_{11})} = 0, \quad (D.7)$$

where $F(S_{11}) = [5\,\text{Re}(S_{11}) + 1]^2 + [5\,\text{Im}(S_{11})]^2$. The other angle θ_3 is obtained by direct substitution into Eq. (D.3), and the other two S parameters are found from Eqs. (D.4) and (D.5). Exceptionally simple results are obtained if we assume that S_{11} is real:

$$\begin{aligned}\text{Re}(S_{12}) &= \text{Re}(S_{13}) = -0.25(1 + S_{11}), \\ \text{Im}(S_{12}) &= -\text{Im}(S_{13}) = 0.25[3 - 2S_{11} - 5(S_{11})^2]^{1/2}\end{aligned} \quad (D.8)$$

Equations (D.8) show that even a poorly matched junction is a good power divider under these conditions. The condition on the phase of S_{11} cannot be practically enforced, however, and it is necessary to find the behavior of the S parameters as a function of both magnitude and phase of the reflection coefficient. This is done numerically using Eqs. (D.4) and (D.5), with the results given in Figure D.2. These results show the extent of deviation from ideal behavior that can be expected from a junction with a given magnitude of S_{11}. Also, if the magnitudes of S_{11}, S_{12}, and S_{13} are known, the phase angles of S_{12} and S_{13} can be estimated from Figure D.2.

REFERENCES

[1] E. R. Hansson, and G. P. Riblet, "An Ideal Six-Port Network Consisting of a Matched Reciprocal Five-Port and a Perfect Directional Coupler," *IEEE Trans. Microwave Theory Tech.*, Vol. MTT-31, No. 3, pp. 284–288, March 1983.

[2] M. Malkomes, G. Kadisch, and H. J. Schmitt, "Optimized Microstrip Ring-Star Five-Ports for Broadband 6-Port Measurement Applications," *IEEE MTT-S Digest*, pp. 472–474, May 1984.

[3] C. G. Montgomery, R. H. Dicke, and E. M. Purcell, *Principles of Microwave Circuits*, McGraw-Hill, New York, 1948.

[4] M. Riaziat and G. Zdasiuk, "Waveguide Star Junction Used in Ka Band Dual Six-Port Network Measurements," *IEEE MTT-S Int. Microwave Symp. Dig.*, June 1985.

APPENDIX E

Effect of Feedback on FET Gain

Two important parameters needed in small-signal circuit design with an FET are the maximum available gain (MAG) and the stability factor k. Any parasitic or intentional feedback element affects both of these values. Positive feedback increases MAG and brings the value of k closer to unity. Negative feedback reduces MAG and increases the stability of the device. In this Appendix, a perturbation method is introduced for studying the effects of feedback elements on the performance of the FET. Some general conclusions follow. This is not intended as a discussion of feedback for circuit design, but rather as an analysis of various parasitic elements that can give rise to feedback and a possible method for their compensation.

In the common-source configuration of an FET, any element separating the source from ground gives rise to feedback known as series feedback. It is distinct from parallel feedback, which is achieved with external elements connecting the gate and drain terminals. Consider the very simple FET model shown in Figure E.1. Output conductance and input resistance are neglected in order to have no source of loss. A series element Z_f is introduced in the source circuit whose magnitude and phase determine the loss as seen from the input. The input impedance of this configuration is given by

$$Z_{\text{in}} = \frac{1}{j\omega C_{gs}} + Z_f + \frac{Z_f g_m}{j\omega C_{gs}}, \tag{E.1}$$

independent of the load impedance Z_l. The positive real part of Z_{in} represents input loss. As can be seen, inductive or resistive components of Z_f contribute to loss, while a capacitive Z_f gives rise to a negative resistance at the input. It is therefore possible to use a capacitive element to control the level of input loss caused by a resistive element in Z_f (or by other elements in a more general model). In the simple case of Figure E.1, a parallel RC combination with a time constant of

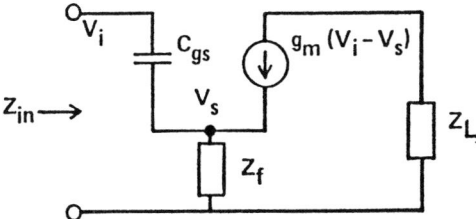

FIGURE E.1 Simple common-source FET model used for calculating the effect of feedback.

$$RC = C_{gs}/g_m \tag{E.2}$$

results in zero input loss at all frequencies. The point to be emphasized is the broadband nature of this loss compensation. Similar broadband effects can also be achieved with more complete models.

E.1 A MORE GENERAL APPROACH

Consider the unilateral FET model of Figure E.2. Series and parallel feedback elements are shown as dotted lines. In order to obtain the overall impedance matrix for the FET with series feedback, the impedance of the feedback element $Z_f = \rho_{fb} + j\xi_{fb}$ is added to every element of the impedance matrix of the unilateral FET. Similarly, for parallel feedback, the admittance of the feedback element $Y_f = \rho_{fb} + j\xi_{fb}$ is added to the admittance matrix elements of the unilateral FET.

For any given impedance or admittance matrix (imittance), the maximum available gain of the two-port is given by [1]

$$\text{MAG} = \frac{|\gamma_{21}|^2}{2\rho_{11}\rho_{22} - \text{Re}(\gamma_{12}\gamma_{21}) + \left\{[2\rho_{11}\rho_{22} - \text{Re}(\gamma_{12}\gamma_{21})]^2 - |\gamma_{12}\gamma_{21}|^2\right\}^{1/2}}, \tag{E.3}$$

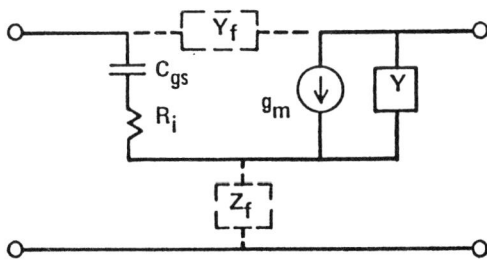

FIGURE E.2 A more general FET model with series and parallel feedback elements.

where ρ_{ij} is the real part and ξ_{ij} is the imaginary part of the (ij)th imittance matrix element γ_{ij}.

Since, for a unilateral FET, $\gamma_{12} = 0$, Eq. (E.3) reduces to

$$G_{\max} = \frac{|\gamma_{21}|^2}{4\rho_{11}\rho_{22}}. \tag{E.4}$$

When the feedback term is added to the FET, it will not stay unilateral. For studying the effects of various feedback elements on MAG, a perturbation approach is taken where the feedback effect is considered small and only the first-order correction term ΔG to Eq. (E.4) is considered:

$$\text{MAG} = G_{\max} + \Delta G + \cdots. \tag{E.5}$$

The details of this expansion will not be given here. But the results are as follows [2]: Either parallel or series resistive elements always give rise to negative feedback that reduces MAG. This is true regardless of frequency or FET parameters. For reactive feedback elements, however, a transition frequency exists above which the role of capacitive and inductive elements in giving positive or negative feedback is reversed. This is important, in particular with parasitic source inductance. At higher frequencies, source inductance can indeed reduce the stability of the device.

For series feedback, if the feedback is a parallel RC combination, the increase in gain due to capacitive feedback and the reduction in gain due to resistive feedback can have the same frequency dependence for the correct values of R_f and C_f. This leads to the possibility of broadband loss compensation usable for counteracting the effect of source resistance [2].

REFERENCES

[1] J. M. Rollett, "Stability and Power-Gain Invariants of Linear Two-Ports," *IRE Trans. Circuit Theory*, Vol. CT-9, pp. 29–32, March 1962.

[2] M. Riaziat, S. Bandy, L. Y. Ching, and G. Li, "Feedback in Distributed Amplifiers," *IEEE Trans. Microwave Theory Tech.*, Vol. 38, No. 2, pp. 212–215, Feb. 1990.

APPENDIX F

Carrier Transport in Semiconductors

In a metallic conductor the valence electrons are not bound to individual atoms and are free to contribute to electric conduction. Free-electron densities on the order of 10^{23} cm^{-3} are common. In a semiconductor, however, valence electrons are more tightly bound to individual atoms, and only thermal or optical excitation can give them enough energy to become charge carriers in electrical conduction. Also, in a semiconductor, any electron that becomes a free carrier leaves a vacancy behind, referred to as a hole. This vacancy can be filled by neighboring electrons, and therefore it can also move in the lattice and contribute to electrical conduction. In this way it acts as a positive charge carrier. It is possible for the scope of this book to make the simplifying approximation of treating these two charge carriers as classical charged particles. The following distinctions have to be made between the characteristics of a charged particle in free space and those in a crystal lattice. These distinctions arise from the quantum-mechanical nature of charge transport in the presence of the periodic potential of the semiconductor crystal.

- There are energy levels whose values are not assumable for a particle in a solid. These are *gaps* between the *energy bands*. In particular, valence electrons that are bound to atoms in a semiconductor occupy what is known as the *valence band*. Electrons that are free to conduct electricity are in a band known as the *conduction band*. Between the two is a bandgap of energy E_g, which is a characteristic of the semiconductor material. This is the amount of energy needed to excite an electron into the conduction band.
- Energy values within each band are quantized, meaning that there is a finite number of allowed energy values within a given range. No more than two particles can assume the same value of energy at the same time.

At low temperatures, this causes all energy levels to be occupied up to a level known as the *Fermi level*.
- The particle in a semiconductor has an *effective mass*, which is different from the free-particle mass. Effective mass can vary depending on the excitation state of the particle.
- In an applied electric field E, a particle in the conduction band does not accelerate as it does in vacuum. Instead, due to scattering, it moves at a constant speed v_d known as the *drift velocity*, which is proportional to the applied field:

$$v_d = \mu E. \tag{F.1}$$

Here, μ is known as the *mobility* of the particle and is given by

$$\mu = q\tau/m_e, \tag{F.2}$$

where q is the charge of the particle, m_e is its effective mass, and τ is the mean free time between collisions in the scattering process.
- The drift velocity of particles does not continue to increase linearly with the strength of the applied field. At very high fields the drift velocity reaches a value referred to as the *saturated drift velocity*, v_{sat}. In some semiconductors such as silicon, the saturated drift velocity is the maximum drift velocity. In others such as GaAs, they are not the same. This effect is described in Chapter 2.

If the free-carrier density is known in a semiconductor sample, the current that will flow when an electric field E is applied is given by

$$I_n = qAn_n\mu_n E, \tag{F.3}$$

where q is the charge of an electron, A is the cross section of the sample, and n_n is the volume free-carrier density. The subscript n refers to the negative charge carriers or electrons in the conduction band. Another contribution to the current comes from the positive charge carriers or holes. The same expression applies to this portion of the current, with subscripts n replaced with p:

$$I_p = qAn_p\mu_p E. \tag{F.4}$$

Consequently, the conductivity σ of the sample is

$$\sigma = q(n_n\mu_n + n_p\mu_p). \tag{F.5}$$

In intrinsic semiconductors (without significant impurity levels) carrier concentrations are the same for electrons and holes. Extra carriers of one kind or both can exist in the semiconductor by one of three means: (1) an increase in temperature generates more thermally excited carriers, (2) nonequilibrium processes such as optical excitation or charge injection generate extra carriers, or (3) impurities in the semiconductor can add carriers to the conduction band. A semiconductor that has extra carriers due to impurities is called *extrinsic*, and the impurity is referred to as the *dopant*. Dopants are normally either *p type* (create extra holes) or *n type* (create extra electrons). Regardless of the dopant concentration, in thermal equilibrium, the product of the electron and hole concentrations always remains constant and equal to its intrinsic value n_i^2:

$$n_n \cdot n_p = n_i^2, \qquad (F.6)$$

where n_i is a function of temperature. Room temperature values of n_i are given in Table F.1 for three types of semiconductors. Normally the impurity concentration n_d is much larger than n_i. In that case, the concentration of the majority carriers is taken to be the same as that of the ionized dopants, and the concentration of the minority carriers is then given by n_i^2/n_d. This rule is referred to as the *mass action law* and represents the equilibrium between thermal generation and the recombination rate of electrons and holes.

Example Calculate the intrinsic conductivity of silicon. If every millionth silicon atom is replaced with phosphorus, what is the new conductivity? Use values in Table F.1.

Solution The intrinsic conductivity is

$$\sigma_i = q(\mu_n + \mu_p)n_i = (4.39)10^{-6} \text{ S/cm}.$$

Phosphorus is an *n*-type dopant in silicon. If every millionth atom is replaced with phosphorus, the doping density is 5×10^{16} cm^{-3}. This value is one

TABLE F.1 Room Temperature Properties of Si, GaAs, and InP

Property	Si	GaAs	InP
Crystal structure	Diamond	Zinc Blende	Zinc Blende
Molecular weight	28.08	144.63	145.79
Density (g/cm^3)	2.33	5.32	4.79
Energy gap (eV)	1.12	1.43	1.34
Electron mobility (cm^2/V-s)	1350	8600	4600
Hole mobility (cm^2/V-s)	480	250	150
Lattice constant (Å)	5.431	5.654	5.869
Dielectric constant	11.8	12.5	12.1
Electron affinity (V)	4.01	4.07	5.34
Thermal conductivity (W/cm-°C)	1.5	0.6	0.8
Intrinsic carrier concentration n_i (cm^{-3})	1.5×10^{10}	10^7	1.1×10^7

millionth of the density of silicon atoms, which is found from Avogadro's law: One mole of silicon (28 g) contains 6.023×10^{23} silicon atoms, and the density of silicon is 2.33 g/cm^3.

Extrinsic carrier concentrations are $n_n = 5 \times 10^{16}$ and $n_p = n_i^2/n_n = 4.5 \times 10^{-3}$. The conductivity of the doped silicon is

$$\sigma = (1.6 \times 10^{-19})[(1350)(5 \times 10^{16}) + (480)(4.5 \times 10^{-3})] = 10.8 \text{ S/cm}.$$

This seemingly small atomic substitution gave rise to an increase in conductivity of greater than six orders of magnitude. ∎

Finally it should be pointed out that under an applied electric field of high enough amplitude charge carriers reach saturated velocity. Current through the semiconductor is no longer field dependent and has a maximum value given by

$$I_{\max} = qA(n_p + n_n)v_{\text{sat}}, \tag{F.7}$$

assuming that v_{sat} is the same for both carrier types.

More in-depth discussions of carrier dynamics in semiconductors can be found in references [1]–[3].

REFERENCES

[1] S. M. Sze, *Physics of Semiconductor Devices*, 2nd ed., Wiley, New York, 1981.
[2] E. S. Yang, *Fundamentals of Semiconductor Devices*, McGraw-Hill, New York, 1978.
[3] M. J. Howes and D. V. Morgan (Eds.), *Gallium Arsenide Materials, Devices, and Circuits*, Wiley, New York, 1985.

Index

Absorption, atmospheric, 2, 3
Acousto-optic modulator, 186
Active mode locking, 185, 186–187
Active waveguide, optical, 169, 170
AM mode locking, 187
Amplifier:
  distributed, 123–127
  feedback, 121
  gate periphery, 125
  reactively matched, 102
  reflection type, 20–21
Amplitude modulator, 89
Antireflection coating, 76, 82, 111
Astigmatism, 61
ATM, 230
Autocorrelation width, 263
Autocorrelator, 262
Avalanche breakdown, 12
Avalanche multiplication, 75
Avalanche transit time diode, 17
Axial ray, 163

Background light, 226
Backward coupler, 152–155
Balanced amplifier, 121
Balanced detection, 227, 228
Bandgap, 289
  direct, 48
  indirect, 49
Bandwidth-distance product, 235, 237, 238
Barker code, 89
Bend, optical waveguide, 174
Bleaching, 188
Blumlein circuit, 200, 201
Bragg diffraction, 186
Bragg reflector, 174

Bragg reflector laser, 64–67
Bragg wavelength, 66
Branching waveguide switch, 92
Branching waveguide, optical, 175
Branch line coupler, 151
Brewster's angle, 192
Built-in potential, 10–11
Buried crescent, 60
Buried heterostructure, 60
Buried waveguide, optical, 169, 170
Butterworth network, 103

Carrier transport, 289–292
Cascade parameters, 255
Channel waveguide, optical, 169
  modes, 170–174
Characteristic waveguide parameter
  (V-number), 168
Charge injection, 75
  high level, 53
Chebyshev network, 103
Chebyshev polynomials, 105
Chirped radar, 181
Chromatic dispersion, 235
Circulator, 21
Clamped gain curve, 62
Codirectional coupler, 152
Coherence length, 263
Coherence spike, 265
Coherent communications, 241
Colliding pulse mode locking, 189–190
Commensurate-length design, 109
Communications link, 221
Commutator arrangement, 212
Conditional stability, 28, 117
Conduction band, 289

293

Confinement factor, 58
Contradirectional coupler, 152
Convolution, 265
Coplanar waveguide (CPW), 88, 135–147
Coplanar waveguide (CPW) modes, 137
  finite substrate, 137
    backside metal, 140, 143
    finite ground, 142
  odd, even, CPW, slot line, 137
Corner, optical waveguide, 174
Coulomb scattering, 34
Coupling coefficient, directional coupler, 155
Coupling length, directional coupler, 155
Current gain cutoff frequency, 27

Damping coefficient, 69
Darlington method (matching network design), 105
Degenerate doping, 44
Delay line, fiber optic, 131
Depletion region, 10
Dielectric resonator oscillator (DRO), 129
Differential time delay, 163, 234
Diffusion length, 49
Direct detection, 225
Directional coupler, 150, 177–178
  switch, 91
Direct modulation, 68
Dispersion, 135
  anomalous, 235
  chromatic, 235
  compensation, 191
  coplanar waveguide (CPW), 145
  group velocity dispersion (GVD), 235
  intermodal, 163, 234
  negative, 192
  normal, 191, 235
  parameter, 235
  waveguide, 235
Dispersive delay line, 196
Distortion, 223
Distributed amplifier, 123–127
Distributed coupling, 152
Distributed feedback laser, 64–67
Dopant, 291
Double heterojunction laser, 56
Drift velocity, 290
Dual six-port network analyzer, 261

E-plane circuits, 147
Edge emitting laser diode, 53
Effective index approximation, 172, 173
Effective isotropic radiated power (EIRP), 223
Effective mass, 290

Effective noise temperature, 274
Effective temperature, 222
Electrical pulse forming, 199
Electron affinity, 13
Electron bombarded semiconductor (EBS) device, 210
Electro-optic modulator, 82, 85, 186
Electro-optic polymers, 94
Electro-optic sampling, 245–251
Embedded waveguide, optical, 169, 170
Energy bands, 289
Energy gaps (semiconductors), 37
Error box, 255
Etalon effect, 249
Extrinsic semiconductor, 291

Fabry–Perot optical resonator, 54, 57, 61
Fano's limit, 119
Feedback amplifier, 121
Feedback, electrical, 286–288
  negative, 122
  parallel, 286
  positive, 122
  series, 286
Fermat's principle, 165
Fermi level, 10, 12, 290
Fiber optic communications, 234–241
Field effect transistor, 21–32
Field overlap, 135
Field overlap integral, 145
Finline, 87, 132, 147
Five-port star junction, 282–285
Fixture deimbedding, 256
Forward biased junction, 12
Forward coupler, 152, 155–156
Fourier transform, 269
Frequency scaling, 106
Friis's formula, 275
Frozen wave generator, 202–204
Fundamental soliton, 240

GaAs laser, 56
GaAs/AlGaAs HEMT, 33
Gain-bandwidth limitation, 119
Gain-bandwidth product, 109
Gain media, laser, 182
Γ-L transition, 41
Gate capacitance, 25
Gaussian noise, 272
Geostationary orbit, 232
Graded index fiber, 164
Gradient index lens, 165
Grating, 67
Grating pair, 196–198
Group velocity dispersion (GVD), 192, 235

Guilleman network, 202
Gunn domain, 43
Gunn effect, 40

Harmonic skip, 255
HBT, *see* Heterojunction bipolar transistor
Hermitian function, 269
Heterodyne detection, 226
Heterojunction bipolar transistor (HBT), 38–40
Heterojunction LED, 51
High electron mobility transistor (HEMT), 21, 32–38
Hole burning, 63
Homojunction laser, 56
Hybrid microwave circuits, 100
Hybrid optoelectronics, 169

Ideal six-port, 259, 283
Ideality factor, 14
Image guide, 147
Imaging, time resolved, 5–8
Imittance, 287
IMPATT diode, 17
Impedance matrix, 277
Infinite substrate CPW, 137
Information capacity, 233
Injection locking, 187
Injection wave generator, 202, 210
Integrated optical passive components, 174–179
Integrated optical waveguides, 169
Integrated Services Digital Network (ISDN), 230
Integrated transmission lines, 131–147
Interdigitated couplers, 156
Interface states, 14
Interferometric autocorrelation, 265
Intermediate frequency (IF), 226
Intermodal dispersion, 163, 234
Intermodulation distortion, 223
Interstellar emission sources, 4
Intrinsic FET equivalent circuit model, 26
Inversion symmetry, 83

Junction capacitance, 11

Kerr lens mode locking, 190–191
Kuroda's transformations, 110, 112

Lange coupler, 156
Large optical cavity, 58
Laser diode, 53
Laser gain media, 182
Laser materials, semiconductor, 55

Lattice constants, semiconductor, 37
Light emitting diode (LED), 49–53
Limited space charge (LSA) mode, 43
Linear FM pulse compression, 196
Linear network parameters, 277–281
Linearly polarized modes, 167
Link budget, 222
Link margin, 223
Lithium niobate, 84, 85
Loaded waveguide, optical, 169, 170
Longitudinal modes, 61, 181
Lossless network, 278, 279
Loss tangent of substrate materials, 133
Lossy match, 120
Low earth orbit (LEO), 234

Mach–Zehnder interferometer, 90
Marginal ray, 163
Mass accelerator, 181
Mass action law, 291
Matched filtering, 196
Matching, 109
  distributed element, 109
  lumped-element, 103
Matrix renormalization, 116, 257
Maximum available gain (MAG), 28, 101
Maximum frequency of oscillation, 27
Maximum stable gain, 28, 118
Meander line, 87
Medium earth orbit (MEO), 234
Metal-semiconductor field effect transistor (MESFET), 22, 100
Metal-semiconductor junction, 12
Michelson interferometer, 263
Microstrip transmission line, 132–135
Microwave integrated circuits, 99
Microwave junctions, 148–150
Minimum noise figure, 30
Mirage, 164
Mobility, 290
Mode coupling coefficient, 135, 145
Mode locking, 185
  active, 185, 186–187
  AM, 187
  colliding pulse, 189–190
  Kerr lens, 190–191
  passive, 185, 187–189
Mode suppression, 179
Modulators, 82–91
  acousto-optic, 186
Molniya orbit, 232
Monolithic microwave integrated circuits (MMIC), 21, 99
Morse code, 220

MSM detectors, 77–82
 frequency response, 78
 transit time, limited operation, 80
Multiple access, 233
 code division, 233
 frequency division, 233
 space division, 233
 time division, 233
Multiple quantum well laser, 56
Multiplexing, 229
 frequency division, 229
 subcarrier, 229
 time division, 229
 wavelength division, 229

Natural frequency, 69
Negative differential resistivity (NDR), 40–43
Negative dispersion, 192
Network analyzers, 252
 calibration, 253
Noise, 272–276
 matching, 119
 performance of FET, 30–31
 resistance, 30
Noise figure, 274
Noncontact probing, 245
Nonlinearity, 237
Nonlinear Schrödinger equation, 238
Nonlinear transmission line, 215
Normal dispersion, 191
Normalized frequency (V-number), 167
Numerical aperture, 163

Offset short, 254
Ohmic contact, 14
One-dB compressed power, 241
Optical fiber, 162–169
Optical field confinement, 57
Optical pulse forming, 181
Oscillation condition, 128
Oscillator, FET, 127
Oscillator stabilization, 129–131
 cavity, 129
 delay line, 130
 dielectric resonator oscillator (DRO), 129
 phase locking, 130

p-n junction, 10
Parallel feedback, 286
Particle accelerator, 181
Passband ripple, 102, 103, 105
Passive mode locking, 185, 187–189
Peak-to-valley ratio, 44
Periodically loaded transmission line, 87

Photoconductive sampling, 251
Photoconductor, 76
Photodetectors, 72–82
Photodiode, 73–76
 bandwidth, 74
Photon lifetime, 70
Photophone, 234
Phototransistors, 76
Pinch-off voltage, 23
Planar waveguide modes, 170–174
 asymmetry factor, 172
Polarization separator, 178
Poling, electrical, 94
Population inversion, 53, 56
Potentially unstable device, 117
Power combiner, 175
Power divider, 175
Principal axes, 246
Principal directions, 83
Prism pair, 193–195
Profile parameter, 166
Pseudomorphic HEMT, 36
Pulse compression, 195
 linear FM, 196
Pulse forming network (PFN), 199–202
Pulse shape characterization, optical, 265
Pulsewidth measurement, optical, 262
Pumping, 53
Pump-probe, 5

Quantum counting, 225
Quantum efficiency, 73
Quantum well laser, 56
Quantum wire laser, 56
Quasi-TEM propagation, 133
Quasi-TEM transmission line, 109

Radar, 2
Radar equation, 231
Radio astronomy, 2–3
Rat-race junction, 148
Rayleigh scattering, 2
Reactively matched amplifier, 102
Receive module, 220
Reciprocal network, 278, 279
Recombination, 10
 indirect, 49
 radiative, 48
Recombination centers, 49
Rectangular waveguide, 87
Reflection-type amplifier, 20, 21
Relative permittivity of substrate materials, 133
Relaxation oscillation, 68

INDEX   **297**

Responsivity, 73
Reverse biased junction, 12
Richards' transformation, 109
Ridge waveguide, 60
  optical, 169, 170

Satellite communications, 231
Satellite orbits, 232
Saturable absorber, 187
Saturated drift velocity, 290
Saturation intensity, 188
Scattering matrix, 148, 278-280
Scattering parameters, 100
  generalized, 116
Schottky barrier, 14
Schottky contact transmission line, 215
Schottky diode, 17, 75
Schottky enhancement layer, 81
Second order system, 68
Self focusing, 190
Self phase modulation (SPM), 196, 237
Self-pulsing instability, 239
Semiconductor junctions, 10-12
Semiconductor laser, 53
Semiconductor, table of properties, 291
Separate confinement heterostructure, 58
Series feedback, 286
Shielded open, 254
Shock wave generator, 213-217
Short-gate effects, 31
Short pulse generation, 181
Shot noise, 272-274
Shot noise limited detection, 225
Sidereal day, 232
Signal-to-noise ratio, 233
Single mode fiber, 167
Six-port network analyzer (SPNA), 257
  q-points, 259
Skew rays, 167
Sliding load, 254
Slotline, 87, 132, 147
Smith Chart, 100, 280-281
Soft breakdown, 15
Solid saturable absorber, 189
Soliton, 238
  order, 240
  period, 240
Spectroscopy, 4-5
Spontaneous emission, 49
Stability circle, 117
Stability condition, oscillator, 128
Stability factor, 101, 118, 286
Standard mismatch, 261
Star junction, five-port, 282-285

Step-index fiber, 163
Stimulated emission, 53
Stopband isolation, 102, 103, 105
Stripline, 132
Strobe light, 181
Superluminescent LED, 51
Surface contacting probe, 135, 243-245
Surface emitting laser diodes, 67-68
Surface passivation, 76, 81, 82
Surface recombination, 76
Swept beam generator, 210
Synchronous Optical Network (SONET), 230
Synchronous pumping, 185

TEM transmission line, 109
Test fixtures, 243, 244
Thermal conductivity of substrate materials, 133
Thermal noise, 222, 272-273
Third order intermodulation, 224
Threshold current, 62
Through-reflect-line (TRL) calibration, 256
Through-short-delay (TSD) calibration, 256
Time lens, 265
Time-bandwidth product, 270-271
Total harmonic distortion, 223
Total internal reflection, 162
Total internal reflection (TIR) switch, 93
Transconductance, 25, 101
Transferred electron effect, 40
Transformer:
  Chebyshev, 114
  distributed, 111
  LC, 106
  multisection, 112
  quarter-wave, 111
Transit time cutoff frequency, 27
Transmission hierarchies, 229, 230
Transmission line dispersion, 135
Transmission line loss:
  conductor, 134
  dielectric, 134
  radiation, 134
Transmission line tapers, 114
Transmission parameters, 255
Transmit module, 220
Transparency, 53
Transverse modes, 61
Traveling wave modulator, 85
Tunnel diode, 44
Two-dimensional electron gas, 34

Unilateral transistor, 101, 102
Unit element (transmission line), 110, 112

Unitary matrix, 279

V-number, 168
Valence band, 289
Vertical cavity, 67
Voltage controlled oscillator (VCO), 130

Waveguide dispersion, 235

Waveguide switches, 91
Wavelength separator, 178
White noise, 272
Work function, 13

Y-matrix, 148

Zener diode, 17